T0135496

A general framework for robust analysis and control

an integral quadratic constraint based approach

Von der Fakultät Konstruktions-, Produktions- und Fahrzeugtechnik und dem
Simtech Research Centre der Universität Stuttgart
zur Erlangung der Würde eines Doktor-Ingenieurs (Dr.-Ing.) genehmigte
Abhandlung

Vorgelegt von

Joost Veenman

aus Berkel en Rodenrijs, die Niederlande

Haubtberichter: Prof. Dr. Carsten W. Scherer
Mitberichter: Prof. Dr. Olivier Sename
Mitberichter: Prof. Dr. Christian Ebenbauer

Tag der mündlichen Prüfung: 12.02.2015

Institut für Mathematische Methoden in den Ingenieurwissenschaften, Numerik und
geometrische Modellierung der Universität Stuttgart
2015

Bibliografische Information der Deutschen Nationalbibliothek

Die Deutsche Nationalbibliothek verzeichnet diese Publikation in der
Deutschen Nationalbibliografie; detaillierte bibliografische Daten sind
im Internet über http://dnb.d-nb.de abrufbar.

ISBN 978-3-8325-3963-4

Logos Verlag Berlin GmbH
Comeniushof, Gubener Str. 47,
10243 Berlin
Tel.: +49 (0)30 42 85 10 90
Fax: +49 (0)30 42 85 10 92
INTERNET: http://www.logos-verlag.de

Preface

Looking back at the last four and a half years in Germany as well as the short period before that in the Netherlands, I realize that my life has completely changed. When I moved to Germany, I was still a bachelor. Now I am a married man and a father of two daughters. On the other hand, also professionally it seem like I have gone through a metamorphosis. I learned numerous things from many people and became a (junior) researcher in the field of robust and gain-scheduling control.

The present thesis is the outcome of about five years of research; first in the Netherlands, at the Delft University of Technology, and a little later in Germany, at the University of Stuttgart[1]. I am proud of the obtained results and I would like to take a short moment to thank a few people who made all this possible.

First and foremost, I want to thank Carsten Scherer for the privilege and opportunity he has given me to be one of his students. I very much appreciate him as a supervisor and teacher. He taught me many things and despite his full agenda, he always found time to answer questions and to guide me on the way towards the completion of my Ph.D.

Further, I would like to thank my (former) colleagues Matthias Fetzer, Hakan Köroğlu, Roland Tóth, Eric Trottemant, Ilhan Polat, Julien Chaudenson, Elisabeth Scheattgen, Florian Saupe, Andres Marcos and Emre Köse, for their cooperation, help and support, as well as for the numerous discussions and all the fun.

I also would like to acknowledge the reviewers for the time and effort to read and evaluate this thesis.

Finally, special thanks goes to my wife, my family and my friends for always supporting me in everything I do.

Joost Veenman
Stuttgart, August 2014

---

[1] This project was financially supported by the German Research Foundation (DFG) within the Cluster of Excellence in Simulation Technology (EXC 310/2) at the University of Stuttgart.

Contents

Chapter 1
Introduction

1.1 Motivation

1.1.1 Robust analysis

It needs no clarification that the stability analysis of dynamical systems is of crucial importance. When engineers design a commercial airplane their number-one-goal is to fly human beings from point A to point B without falling out of the sky. In the same vein, stability is crucial in numerous other commercial and industrial applications such as e.g. cars, trains, wind-turbines, power-plants and -networks, weaponry, satellites, helicopters, Segways, petro-chemical distillation and other HVAC (heating, ventilation, and air-conditioning) processes, phones, computers, robot manipulators, electronic circuits and many many more.

Apart from stability, engineers also often need to verify whether or not some desired performance objective is achieved. For example, airplanes, trains and cars should be designed such that people experience a comfortable flight or ride. Similarly, a wind-turbine should generate a maximum amount of electrical energy; a petro-chemical distillation process should produce some high quality material; and, a satellite should provide sharp images of some exact location.

Of course dynamical systems are often rather complex, which makes the task of performing a stability and performance analysis a rather difficult one. Just think of an Airbus A380 ($l \times w \times h = 73m \times 80m \times 25m$, weight $575000kg$, operating speed $945km/h$, engines $4 \times 340kN$, etc.) that should be able to take off, land and cope with extreme velocities, large temperatures fluctuations, turbulence and continuous vibrations. An Airbus A380 is an extremely complex dynamical system and its stability and performance analysis is a very challenging problem.

In order to analyze the stability and performance properties of a dynamical system, engineers typically require a mathematical model that describes the dynamical behavior as accurately as possible. Unfortunately, high fidelity models are often difficult to obtain as well as too complex from a computational point-of-view. For these reasons engineers prefer to work with, and try to find low complexity models that still match reality as closely as possible. However, proceeding in this fashion leads to the following key question, which remains unanswered, even if the discrepancy between the model and the reality is not too large: Does stability and performance of the model also imply stability and performance of the real dynamical system?

A well-known research field that considers discrepancies between the modeled and the true dynamical behavior is called the robust approach. In robust analysis, these discrepancies are called uncertainties. The source of an uncertainty can be diverse. For example, the parameters of a model might not be known exactly, or change during operation. In this case we talk about parametric uncertainties. On the other hand, uncertainties might also originate from approximated, un-modeled and/or neglected dynamics. These are called dynamic uncertainties.

The essential idea in robust analysis is not to verify stability and performance for just one model, but rather for a whole family of models in which the true model is contained in. Indeed, although it is often rather difficult to obtain the true value of an uncertain parameter, it is usually much easier to identify some bounds. Then the uncertain parameter is confined to a so-called uncertainty set that comprises the true but unknown parameter value. Similarly, it can be very convenient to approximate some high fidelity or nonlinear model with a family of much lower fidelity or linear ones respectively. If an uncertain system is stable and achieves a desired performance objective for all uncertainties (from some set) then the system is said to be robustly stable and achieve robust performance. The big advantage of this approach is that robust stability and performance of the uncertain system also implies stability and performance of the true dynamical process.

An important and well-established set of theoretical tools for the systematic analysis of uncertain systems are called the μ-tools [122, 11]. The main idea of this approach is to consider uncertain linear time-invariant (LTI) systems that admit a so-called linear fractional representation. This means that the uncertain system can be reformulated as a feedback interconnection of a nominal LTI system and a trouble making component, which is also LTI, and which represents the uncertainties. The mathematical machinery, which essentially boils down to the computation of upper-bounds on the structured-singular value (SSV), allows us then to efficiently examine the stability and performance properties of uncertain LTI systems.

Another related but considerably more general and less-established framework within the robust analysis field is called the integral quadratic constraint (IQC) approach [116]. Also here the main idea is to consider uncertain systems that admit a linear fractional representation. However, the uncertainties are not restricted to be LTI, but can, more generally, be nonlinear, time-varying or infinite dimensional. This is achieved with IQCs, which allow us to capture the properties of a rich class of uncertainties by describing the energy distribution in the spectrum of the in- and outputs of the uncertainties. This enables us to systematically and efficiently verify the stability and performance properties for a diverse class of uncertain, possibly nonlinear, time-varying or infinite dimensional dynamical systems through the use of linear matrix inequalities (LMIs) and convex optimization techniques. It is exactly this approach that will be addresses and further developed in this thesis.

1.1.2 Robust control

Another crucial matter in systems-engineering is the design of controllers. Controllers allow us to change the stability and performance properties of a dynamical system. This can be done by measuring (a part of) the state of a system (i.e. positions, velocities, accelerations, forces, moments, voltages, currents, temperatures, etc.) and dynamically process it, in order to provide suitable commands for the actuators of the system (i.e. motors, valves, hydraulic pistons, piezoelectric actuators, etc.), such that the controller stabilizes the dynamical system and achieves some desired performance specification.

Typical control objectives are, for example, to stabilize the system (i.e. keep an airplane up in the air); to track a commanded signal (i.e. follow a specific flight path); to reject external disturbances (i.e. compensate for sudden wind gusts and turbulence); and, to avoid actuator saturation (i.e. cope with the limited deflections and finite bandwidths of the flaps of the wings). In addition, and in accordance with the previous discussion on uncertain systems, another meaningful control objective is to guarantee robust stability and to achieve robust performance (i.e. an airplane should achieve the latter objectives for all uncertainties such as e.g. temperatures, vibrations, weight variations, etc.).

The systematic design of controllers is by no means elementary. Without going into detail, let us briefly discuss a few of the possibilities. The most classical and well-known control strategy is called the proportional-integral-derivative (PID)-control approach. Although these techniques are still frequently used in practice, their application is mostly limited to single-input single-output (SISO) systems. Moreover, the tuning of PID-controllers is a rather heuristic procedure. For the design of controllers for multi-input multi-output (MIMO) systems, other more systematic techniques have been developed, such as e.g. the linear quadratic regulator (LQR) and linear quadratic gaussian (LQG) optimal control strategies [89]. However, their capability to guarantee robust stability and performance is often rather limited [39]. Of course there are many other useful control strategies. Among others, one could think of model predictive control, adaptive control, stochastic control, nonlinear control and intelligent control. It is beyond the scope of this dissertation to discuss the details here.

In this thesis we are primarily concerned with optimization based controller synthesis, which means that we translate controller synthesis scenarios into optimization problems in order to obtain stabilizing controllers that are optimal in some sense. Unfortunately, control related optimization problems are typically highly non-convex, which prevents us from efficiently finding globally optimal solutions. To overcome this trouble, a lot of research activity is currently focussed on transforming initially non-convex design problems into convex ones through the use of some dedicated LMI-techniques. For example, two well-known synthesis problems that admit a convex solution in terms of LMIs are the so-called $\mathscr{H}_\infty$- and gain-scheduled controller synthesis problems, as considered in [41, 48, 72] and [121, 174, 152] respectively. In $\mathscr{H}_\infty$-synthesis the goal is to find an LTI controller that stabilizes a weighted LTI system, while the $\mathscr{H}_\infty$-norm from the disturbance input to the performance output is minimized; and in gain-scheduling synthesis, the goal is to find a more general linear time-varying (LTV) controller that stabilizes a weighted LTV system, while the induced $\mathscr{L}_2$-gain from the disturbance input to the performance output is minimized. The applied techniques are essential for the synthesis results that will be presented in this thesis.

It is important to note that the IQC-tools are not only useful for the robustness analysis of uncertain systems, but also for the systematic synthesis of robust controllers. In this thesis we consider a rather general IQC-based synthesis configuration, which is called the robust gain-scheduled controller synthesis problem. Unfortunately, this problem is, in its full generality, very hard to convexify. However, for a large class of specialized design scenarios, it is possible to exploit additional structure in order to transform the initially non-convex optimization problem into a convex one. This is, for example, the case for the $\mathscr{H}_\infty$- and gain-scheduled controller synthesis problems that we discussed earlier. Another well-known specialized design scenario that also admits a convex solution is the so-called robust estimator synthesis problem [51, 180, 165]. In robust estimation the goal is to design robust filters that optimally estimate non-measured variables in uncertain dynamical systems. Similarly, it is possible to solve the so-called 'dual' robust feedforward controller synthesis problem [54, 103]. Here the goal is to design robust filters that improve reference tracking performance specifications. It is one of the main goals of this thesis to reveal how other, more general scenarios can be handled as well.

1.2 Aim and main goals of the thesis

Despite the general and flexible nature of the IQC-analysis approach, and although it has already been developed in the late nineties, it never really evolved into a well-established and widely applied method, such as e.g. the μ-tools did. Out of the many possible causes, the author believes that the following two are essential in this matter:

- There is no tutorial analysis overview that is easily accessible for engineers.
- There are only a few synthesis results for specialized robust design scenarios.

To this end, we endeavor to establish a general framework for the systematic analysis of uncertain systems and the synthesis of robust controllers based on integral quadratic constraints. This leads us to the following main goal of this thesis:

The main goal of this thesis is to further develop the existing IQC-theory on both the analysis as well as the synthesis side.

To achieve this objective and in view of the latter discussion, we will first develop a solid foundation for our results. This includes an easy-to-access tutorial introduction on IQC-analysis as well as an extensive survey on the formulation and parameterization of IQC-multipliers. Subsequently, we will further reveal the potential of the IQC-framework

- by presenting new insights on the analysis side;
- by showing how a number of interesting robust gain-scheduled design problems can be systematically and efficiently handled with convex optimization techniques and dedicated LMI-tools.

We emphasize that the underlying technical subgoals are given in the introduction of each chapter separately. They require a thorough development of the theoretical background first.

1.3 Outline and contributions

The core of this thesis consists of five main chapters. The content and the most important contributions of these chapters are summarized below.

In Chapter 2 we introduce the general setup and develop the most relevant tools that are needed for a proper IQC-analysis. In addition to that we present an extensive survey on the formulation and parametrization of IQC-multipliers as well as some new insights in this matter. Also we provide a novel proof of the IQC-theorem, which opens the way for new and interesting applications. The contributions of this chapter can be summarized as follows:

- an easy-to-access tutorial introduction on IQC-analysis (Section 2.2-2.5)
- an extensive survey on the formulation and parametrization of IQC-multipliers, along with some new insights and generalizations (Section 2.6-2.7):
  - a new IQC for time-invariant parametric uncertainties (Section 2.6.3.2)
  - a new IQC for multi-variable sector conditions (Section 2.6.8.2)

- a proof of asymptotic exactness of the Zames-Falb multiplier parametrization for odd-monotone nonlinearities (Section 2.6.8.3)
- a new IQC for dynamic generalized quadratic-performance (Section 2.7.5)

- a concise numerical investigation on the choice of different basis functions for the IQC-multipliers (Section 2.9.1)
- a novel reformulation of the IQC-analysis theorem (Section 2.10)
- a dissipation based proof of the IQC-analysis theorem (Section 2.10)

In Chapter 3 we make a transition from robust analysis towards robust synthesis. In this preparatory chapter we formulate the most general synthesis setup that will be considered in this work: the robust gain-scheduled controller synthesis problem. This is convenient since all synthesis scenarios that will be considered in the subsequent chapters can be obtained from this problem by specialization. In addition, we present the LMI solutions for the previously discussed $\mathscr{H}_\infty$- and gain-scheduled controller synthesis problems, along with the applied techniques. The latter play an essential role for the synthesis results that follow later on. There are no technical contributions in this chapter.

In Chapter 4 we present our first synthesis result. It is shown how the existing solutions on robust estimation and gain-scheduled controller synthesis can be unified into one general configuration in order to design robust gain-scheduled estimators. The main contributions of this chapter are

- an LMI-solution for the existence of such robust gain-scheduled estimators;
- an LMI-solution for the 'dual' robust gain-scheduled feedforward controller synthesis problem.

In Chapter 5 we continue with a significantly more general and generic synthesis configuration, which encompasses a rich class of concrete design scenarios. Among others, this includes robust estimation, robust feedforward control, generalized l_2-synthesis, open-loop controller synthesis and gain-scheduled output-feedback control with uncertain performance weights. The contributions of this chapter are listed below:

- a convex solution in terms of LMIs for a generic feasibility problem that encompasses all the latter design scenarios
- an illustrative overview of further design scenarios that are covered and that have not been considered in the literature before, such as e.g.:

  - robust control synthesis without control channel uncertainties
  - robust gain-scheduled observer design
  - robust control synthesis with unstable weights

In Chapter 6 we present the last synthesis tools of this thesis. Analogously to the existing μ-tools we present an alternative algorithm for the systematic synthesis of robust controllers based on IQCs. The suggested algorithm enables us to perform robust controller synthesis for a significantly larger class of uncertainties if compared to the existing methods. The key to the results are new insights in the factorization of the IQC-multipliers, which allows us to speed up the synthesis process through the use of warm-start techniques. The contributions of this chapter are the following:

- a streamlined and more insightful proof of an existing IQC-multiplier factorization
- a novel state-space construction of a one-sided Wiener-Hopf factorization
- the application of these two factorizations in a robust controller synthesis algorithm with warm-start options

In Chapter 7 we conclude the thesis with some general conclusions and an overview of recommendations for future research. We emphasize that many of the results are supported by illustrative numerical examples. Also we have differed some useful information to the appendix. This includes the explanation of symbols and abbreviations in Appendix A and B respectively, as well as some new and existing auxiliary results in Appendix C-G. An English and German summary of the thesis and a short biography of the author are found at the very end of the Thesis.

Finally, this dissertation is the result of various papers that have already been published in several conference proceedings, journals and books. In particular, this concerns the following papers: [170, 190, 191, 192, 193, 194, 195, 198, 196, 197, 199, 200]. Throughout this thesis we will emphasize in which of these articles the corresponding results have appeared. Further, although the majority of the material has been unified, streamlined and rewritten, we stress that some of the passages overlap. Also this will be explicitly indicated to a reasonable extent.

Chapter 2
Stability analysis with integral quadratic constraints

2.1 Introduction

In this chapter we are concerned with the analysis of uncertain dynamical systems. We build upon a powerful framework, also called the integral quadratic constraint approach [116], which enables us to efficiently perform robust stability and performance analysis via finite dimensional linear matrix inequalities (LMIs) and convex optimization techniques. Integral quadratic constraints (IQCs) are very useful in capturing the properties of a rich class of uncertainties, such as e.g. time-invariant and (rate-bounded) time-varying parametric uncertainties, linear time-invariant (LTI) dynamic uncertainties, delay uncertainties, norm-bounded uncertainties and passive uncertainties as well as sector-bounded and slope-restricted nonlinearities. Moreover, in addition to that, the IQC-framework also allows us to consider a diverse class of performance specifications as well as to incorporate spectral information of signals. Altogether, this enables us to examine a rather diverse class of complex dynamical systems.

The goal of this chapter is fourfold. First of all, this is a preparatory chapter in which we provide a brief tutorial on the standard IQC-analysis tools as they are known from the literature [116]. This will form the foundation on which we build our synthesis results in the subsequent chapters. The second goal is to provide an extensive survey on the formulation and parameterization of a rich class of IQC-multipliers. Although, apart from a couple of small contributions, there are no essential novelties found in this part of the chapter, it is stressed that no such survey is available in the literature. Moreover, in view of the recent developments on the synthesis side of the IQC-framework, such an overview will strongly support our results from a theoretical and a practical point-of-view. Thirdly, being aware of the rather mathematical machinery, another goal of this chapter is to present the IQC-analysis framework,

whenever possible, in an as nontechnical fashion as possible, in order to make the
tools more accessible for engineers. The fourth and final goal of this chapter is to
establish a connection between the IQC-framework and the so-called dissipativity
approach [205, 206]. This leads to the largest contribution of this chapter, which is a
novel proof of the IQC-theorem, with possibly new and interesting applications.

This chapter is organized as follows. First, in Section 2.2, we will formally introduce a standard feedback interconnection, which will form the basis for the IQC-
framework in Section 2.3. Subsequently, in Section 2.4, we will employ the IQC-
framework in order to obtain an infinite dimensional feasibility test for robust stability and performance analysis. In Section 2.5 it is then shown how to turn this
test into a genuine semi-definite program. We will continue by providing an extensive overview of IQC-multipliers for uncertainties, nonlinearities and performance
specifications in Sections 2.6, 2.7 and 2.8 respectively, followed by some illustrative
examples in Section 2.9. Subsequently, in Section 2.10, we give an alternative proof
of the IQC-theorem, and we conclude the chapter with a summary in Section 2.11.

2.2 A specific feedback interconnection

One of the key ingredients of the IQC-approach is to consider uncertain dynamical
systems $\Sigma(\Delta, G)$ that are composed of a nominal part G and a trouble making component Δ that represents the uncertainties. The generic setup is depicted in Figure 2.1.
Loosely speaking, this allows us to view the uncertainties Δ as perturbations that act
on the nominal system G through feedback. Stability of $\Sigma(\Delta, G)$ is then guaranteed
if for all uncertainties Δ (from some set) the feedback interconnection of Figure 2.1
remains stable.

In this thesis we restrict ourselves to nominal systems G that are finite dimensional, linear, causal and stable, while the trouble making component Δ can, more
generally, be infinite dimensional and nonlinear as well. To be mathematically precise,
the uncertain dynamical system $\Sigma(\Delta, G)$ is defined through the equations

$$q = Gp + \mu,$$
$$p = \Delta(q) + \eta,$$
(2.1)

where

- the nominal plant $G \in \mathscr{RH}_\infty^{n_q \times n_p}$ is a real rational and proper transfer matrix
  without poles in the closed right-half complex plane,
- the trouble making component $\Delta : \mathscr{L}_{2e}^{n_q} \to \mathscr{L}_{2e}^{n_p}$ is a bounded and causal operator
  which is allowed to vary in a certain class Δ,
- the exogenous disturbance inputs μ and η are signals in $\mathscr{L}_{2e}^{n_q}$ and $\mathscr{L}_{2e}^{n_p}$ respectively.

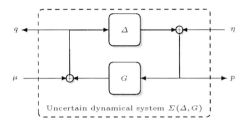

Fig. 2.1 Standard feedback interconnection for robust stability analysis

We refer the reader to Appendix A for further details about the symbols that are used.

Given G and $\boldsymbol{\Delta}$, the central task in robust stability analysis is to characterize whether the feedback interconnection (2.1) remains stable for all $\Delta \in \boldsymbol{\Delta}$. To be precise, let us rewrite the feedback interconnection (2.1) as

$$\begin{pmatrix} I & -G \\ -\Delta & I \end{pmatrix} \begin{pmatrix} q \\ p \end{pmatrix} := \begin{pmatrix} q \\ -\Delta(q) \end{pmatrix} + \begin{pmatrix} -Gp \\ p \end{pmatrix} = \begin{pmatrix} \mu \\ \eta \end{pmatrix} \tag{2.2}$$

and introduce the following three definitions.

Definition 2.1 (Well-posedness). The feedback interconnection (2.1) is well-posed if for each $\mathrm{col}(\mu,\eta) \in \mathscr{L}_{2e}$ there exists a unique response $\mathrm{col}(q,p) \in \mathscr{L}_{2e}$ satisfying (2.2) such that the map $\mathrm{col}(\mu,\eta) \to \mathrm{col}(q,p)$ is causal.

Definition 2.2 (Stability). The feedback interconnection (2.1) is stable if it is well-posed and if the $\mathscr{L}_2$-gain of the map $\mathrm{col}(\mu,\eta) \to \mathrm{col}(q,p)$ is bounded.

Definition 2.3 (Robust stability). The feedback interconnection (2.1) is robustly stable if it is stable for all $\Delta \in \boldsymbol{\Delta}$.

Let us emphasize that in this thesis we are neither concerned with the modeling of uncertain dynamical systems nor with the reformulation of these systems into the standard form of Figure 2.1. This is a whole research area by itself [210, 30, 21, 62, 61, 185]. We assume that the uncertain systems are given in accordance with the descriptions from above. Further note that, although the setup in Figure 2.1 might seem restrictive, both G as well as Δ might be highly structured. Moreover, Δ might be time-varying, non-linear or even infinite dimensional. It will become clearer in the sequel that this allows us to consider a rather diverse class of uncertain dynamical system.

2.3 Stability analysis with integral quadratic constraints

With the latter framework in mind, let us now discuss the central stability result from [116], which is based on integral quadratic constraints (IQCs). Two signals $p \in \mathscr{L}_2^{n_p}$ and $q \in \mathscr{L}_2^{n_q}$ are said to satisfy the IQC defined by the IQC-multiplier $\Pi = \Pi^* \in \mathscr{RL}_\infty^{(n_q+n_p)\times(n_q+n_p)}$ if

$$IQC(\Pi, q, p) := \left\langle \begin{pmatrix} q \\ p \end{pmatrix}, \begin{pmatrix} \Pi_{11} & \Pi_{12} \\ \Pi_{12}^* & \Pi_{22} \end{pmatrix} \begin{pmatrix} q \\ p \end{pmatrix} \right\rangle \geq 0. \tag{2.3}$$

With $p = \Delta(q)$ and by imposing suitable constraints on the IQC-multiplier Π such that

$$IQC(\Pi, q, \Delta(q)) \geq 0 \quad \forall q \in \mathscr{L}_2^{n_q}, \tag{2.4}$$

it becomes possible to describe the energy distribution in the spectrum of $\mathrm{col}(q, \Delta(q))$, and, hence, to capture properties of uncertainties, nonlinearities and signals. In order to illustrate the abstract looking concept, let us give an elementary example.

Example 2.1. Let us choose $\Pi_{11} := I$, $\Pi_{12} := 0$ and $\Pi_{22} := -I$ with compatible dimensions. Then the IQC (2.4) holds for all bounded and causal Δ with $\|\Delta\|_{\mathrm{i2}} \leq 1$. This is clear since

$$\int_0^\infty \begin{pmatrix} q(t) \\ \Delta(q)(t) \end{pmatrix}^* \begin{pmatrix} I & 0 \\ 0 & -I \end{pmatrix} \begin{pmatrix} q(t) \\ \Delta(q)(t) \end{pmatrix} dt = \int_0^\infty \|q(t)\|^2 - \|\Delta(q)(t)\|^2 dt \geq 0$$

holds for all $q \in \mathscr{L}_2^{n_q}$ if and only if $\|\Delta\|_{\mathrm{i2}} \leq 1$.

In addition to norm-bounds, it is possible to capture many other properties of uncertainties with IQCs. We refer the reader to Section 2.6 for an extensive overview of examples. Let us now state the well-known result of [116].

Theorem 2.1. *Assume that*

1. *for all $\tau \in [0,1]$ the feedback interconnection of G and $\tau\Delta$ is well-posed;*
2. *for all $\tau \in [0,1]$ and some $\Pi = \Pi^* \in \mathscr{RL}_\infty^{(n_q+n_p)\times(n_q+n_p)}$ the IQC (2.4) is satisfied for Δ replaced by $\tau\Delta$;*
3. *the following frequency domain inequality (FDI) is satisfied:*

$$\begin{pmatrix} G(i\omega) \\ I \end{pmatrix}^* \Pi(i\omega) \begin{pmatrix} G(i\omega) \\ I \end{pmatrix} \prec 0 \quad \forall \omega \in \mathbb{R} \cup \{\infty\}. \tag{2.5}$$

Then the feedback interconnection (2.1) is stable.

Instead of giving the original proof of [116] here, we will provide a novel alternative proof based on dissipativity arguments at the end of this chapter in Section 2.10. As discussed in the introduction, this leads to another interpretation of the IQC-theorem with possibly new and interesting applications.

Remark 2.1. In general, $\Pi : i\mathbb{R} \to \mathbb{C}^{(n_q+n_p)\times(n_q+n_p)}$ can be any measurable essentially bounded Hermitian-valued function. However, throughout this thesis we will restrict our attention to IQC-multipliers of the form $\Pi = \Pi^* \in \mathcal{RL}_\infty^{(n_q+n_p)\times(n_q+n_p)}$.

Remark 2.2. If $\tau\Delta$, $\tau \in [0,1]$, satisfies the IQC (2.4) for $\Pi_1,\ldots,\Pi_k$, then $\tau\Delta$ also satisfies the IQC (2.4) for the conic combination $\sum_{i=1}^{k} p_i\Pi_i$, $p_i \geq 0$. Hence, since each IQC-multiplier Π_i, $i = 1,\ldots,k$ yields a stability test, the conic combination IQC-multipliers leads to a (possibly) better stability test, which can be easily applied.

Remark 2.3. Suppose that the IQC (2.4) holds with $\Pi_{11} \succcurlyeq 0$ on $\mathbb{C}^0$ and $\Pi_{22} \preccurlyeq 0$ on $\mathbb{C}^0$. Then the mapping $p \to IQC(\Pi,q,p)$ is concave and (2.4) automatically implies that $IQC(\Pi,q,\tau\Delta(q)) \geq 0$ for all $q \in \mathcal{L}_2^{n_q}$ and for all $\tau \in [0,1]$.

Remark 2.4. Sometimes it is not possible to verify the first two conditions in Theorem 2.1 if $\tilde{\Delta}(\tau) := \tau\Delta$ depends linearly on τ. For bounded and causal operators $\Delta : \mathcal{L}_{2e}^{n_q} \to \mathcal{L}_{2e}^{n_p}$, we can replace the conditions with weaker ones, by allowing $\tilde{\Delta}(\tau)$ to be a continuous function that satisfies $\tilde{\Delta}(0) = 0$ and $\tilde{\Delta}(1) = \Delta$. For further readings we refer the reader to [73, 85] (see also [24, 25, 92]).

Remark 2.5. Sometimes the last two conditions in Theorem 2.1 imply that the first one (i.e. the well-posedness assumption) is satisfied. For example, this is the case if Δ is LTI [171]. Although well-posed is important and should not be ignored, we stress that it is not the main theme of this chapter; it will not be further addressed in the sequel. For further details we refer the reader to e.g. [171].

2.4 Robust stability and performance analysis

2.4.1 Robust stability analysis

If the conditions in Theorem 2.1 are satisfied, then stability of the feedback interconnection (2.1) is guaranteed for all $\tau\Delta$ with $\tau \in [0,1]$. On the other hand, in applications one typically considers larger classes of uncertainties, which we denote by $\boldsymbol{\Delta}$. If not specified otherwise, we tacitly assume throughout this thesis that $\boldsymbol{\Delta}$ satisfies the following property.

Assumption 2.1. $\Delta \in \boldsymbol{\Delta}$ implies that $\tau\Delta \in \boldsymbol{\Delta}$ for all $\tau \in [0,1]$.

If a given uncertainty set $\boldsymbol{\Delta}$ does not satisfy Assumption 2.1, it is often possible to normalize it through loop-transformation techniques such that it does satisfy Assumption 2.1.

In order to handle larger classes of uncertainties, we typically construct a whole family of IQC-multipliers $\boldsymbol{\Pi} \subset \mathscr{RL}_{\infty}^{(n_q+n_p) \times (n_q+n_p)}$ such that the IQC (2.4) holds for all $\Pi \in \boldsymbol{\Pi}$ and for all $\Delta \in \boldsymbol{\Delta}$. Then we can employ Theorem 2.1 for the stability analysis of a whole class of Δ's as follows.

Corollary 2.1. *Assume that*

1. *for all $\Delta \in \boldsymbol{\Delta}$ the feedback interconnection of G and Δ is well-posed;*
2. *for all $\Delta \in \boldsymbol{\Delta}$ and for all $\Pi \in \boldsymbol{\Pi}$ the IQC (2.4) is satisfied;*
3. *there exists some $\Pi \in \boldsymbol{\Pi}$ for which the following FDI is satisfied:*

$$
\begin{pmatrix} G \\ I \end{pmatrix}^* \Pi \begin{pmatrix} G \\ I \end{pmatrix} \prec 0 \text{ on } \mathbb{C}^0. \tag{2.6}
$$

Then the feedback interconnection (2.1) is robustly stable.

Since the first two conditions in Corollary 2.1 are satisfied for all $\Delta \in \boldsymbol{\Delta}$ with Assumption 2.1 we do not have to bother about τ. This simplifies the notation as well as the readability of the material in the sequel. Further note that we dropped the notational dependency of (2.5) on ω and used the alternative and more compact notation of (2.6), where $\mathbb{C}^0 := i\mathbb{R} \cup \{\infty\}$ denotes the extended imaginary axis.

2.4.2 Robust performance analysis

Often, we are not only interested in robust stability, but also in robust performance. For this purpose, we augment the feedback interconnection in Figure 2.1 with an additional performance channel as depicted in Figure 2.2. The plant is now defined through the linear fractional representation

$$
\begin{pmatrix} q \\ z \end{pmatrix} = \underbrace{\begin{pmatrix} G_{qp} & G_{qw} \\ G_{zp} & G_{zw} \end{pmatrix}}_{G} \begin{pmatrix} p \\ w \end{pmatrix}, \quad p = \Delta(q), \tag{2.7}
$$

which we assume to be well-posed for all $\Delta \in \boldsymbol{\Delta}$ (i.e. $I - G_{qp}\Delta$ has a causal inverse for all $\Delta \in \boldsymbol{\Delta}$). Again $G \in \mathscr{RH}_{\infty}^{(n_q+n_z) \times (n_p+n_w)}$ denotes the nominal plant, while $p \to q$ represents the uncertainty channel and $w \to z$ the performance channel on which we would like to impose performance criteria through an IQC

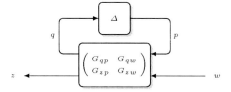

Fig. 2.2 Standard feedback interconnection for robust stability and performance analysis

$$IQC(\Pi_{\mathrm{p}}, z, w) = \left\langle \begin{pmatrix} z \\ w \end{pmatrix}, \begin{pmatrix} \Pi_{\mathrm{p}11} & \Pi_{\mathrm{p}12} \\ \Pi_{\mathrm{p}12}^* & \Pi_{\mathrm{p}22} \end{pmatrix} \begin{pmatrix} z \\ w \end{pmatrix} \right\rangle \le -\epsilon \|w\|_2^2. \tag{2.8}$$

Here $\epsilon > 0$ and we assume that Π_{p} is confined to some given set

$$\boldsymbol{\Pi}_{\mathrm{p}} \subset \left\{ \Pi_{\mathrm{p}} \in \mathscr{RL}_{\infty}^{(n_z+n_w)\times(n_z+n_w)} : \Pi_{\mathrm{p}11} \succcurlyeq 0 \text{ on } \mathbb{C}^0 \right\}.$$

For an overview of concrete performance IQC-multipliers we refer the reader to Section 2.7. Robust performance is now achieved if (2.7) is robustly stable (i.e. if $I - G_{qp}\Delta$ has a bounded and causal inverse for all $\Delta \in \boldsymbol{\Delta}$) and if for all trajectories of the uncertain system the latter inequality holds true. This yields the following extension of Corollary 2.1.

Lemma 2.1. *Robust stability and performance is achieved if*

1. *for all $\Delta \in \boldsymbol{\Delta}$ the linear fractional representation (2.7) is well-posed;*
2. *for all $\Delta \in \boldsymbol{\Delta}$ and for all $\Pi \in \boldsymbol{\Pi}$ the IQC (2.4) is satisfied;*
3. *there exist some $\Pi \in \boldsymbol{\Pi}$ and $\Pi_{\mathrm{p}} \in \boldsymbol{\Pi}_{\mathrm{p}}$ for which the following FDI holds true:*

$$\begin{pmatrix} G_{qp} & G_{qw} \\ I & 0 \\ \hdashline G_{zp} & G_{zw} \\ 0 & I \end{pmatrix}^* \begin{pmatrix} \Pi & 0 \\ 0 & \Pi_{\mathrm{p}} \end{pmatrix} \begin{pmatrix} G_{qp} & G_{qw} \\ I & 0 \\ \hdashline G_{zp} & G_{zw} \\ 0 & I \end{pmatrix} \prec 0 \text{ on } \mathbb{C}^0. \tag{2.9}$$

Proof. Since $\Pi_{\mathrm{p}11} \succcurlyeq 0$ we infer that (2.9) implies

$$\begin{pmatrix} G_{qp} \\ I \end{pmatrix}^* \Pi \begin{pmatrix} G_{qp} \\ I \end{pmatrix} \prec 0 \text{ on } \mathbb{C}^0.$$

By Corollary 2.1 we conclude that $I - G_{qp}\Delta$ has a bounded and causal inverse for all $\Delta \in \boldsymbol{\Delta}$, which in turn implies robust stability. Hence, $w \in \mathscr{L}_2^{n_w}$ implies that p, q and z are all in $\mathscr{L}_2^\bullet$. The Fourier transforms $\hat{p}$, $\hat{q}$, $\hat{w}$ and $\hat{z}$ of p, q, w and z respectively, are thus well-defined. For robust performance, observe that since (2.9) is strict we can replace the right-hand size of (2.9) with $-\epsilon I$ for some small $\epsilon > 0$ such that

$$
\begin{pmatrix} G_{qp} & G_{qw} \\ I & 0 \\ \hline G_{zp} & G_{zw} \\ 0 & I \end{pmatrix}^* \begin{pmatrix} \Pi & 0 \\ 0 & \Pi_{\mathrm{p}} \end{pmatrix} \begin{pmatrix} G_{qp} & G_{qw} \\ I & 0 \\ \hline G_{zp} & G_{zw} \\ 0 & I \end{pmatrix} \preccurlyeq -\epsilon I \text{ on } \mathbb{C}^0
$$

persists to hold. Right- and left-multiplying this inequality with $\mathrm{col}(\hat{p}(i\omega),\hat{w}(i\omega))$ and its conjugate transpose and exploiting the system description (2.7) then leads to

$$
\begin{pmatrix} \hat{q}(i\omega) \\ \hat{p}(i\omega) \end{pmatrix}^* \Pi(i\omega) \begin{pmatrix} \hat{q}(i\omega) \\ \hat{p}(i\omega) \end{pmatrix} + \begin{pmatrix} \hat{z}(i\omega) \\ \hat{w}(i\omega) \end{pmatrix}^* \Pi_{\mathrm{p}}(i\omega) \begin{pmatrix} \hat{z}(i\omega) \\ \hat{w}(i\omega) \end{pmatrix} \leq -\epsilon(\|\hat{p}(i\omega)\|^2 + \|\hat{w}(i\omega)\|^2)
$$

for all $\omega \in \mathbb{R} \cup \{\infty\}$. Integration and exploiting $p = \Delta(q)$ finally yields

$$
IQC(\Pi, q, \Delta(q)) + IQC(\Pi_{\mathrm{p}}, z, w) \leq -\epsilon(\|p\|_2^2 + \|w\|_2^2),
$$

which clearly implies (2.8) since $IQC(\Pi, q, \Delta(q)) \geq 0$ for all $q \in \mathscr{L}_2^{n_q}$. ∎

2.4.3 Mixed uncertainties

As one of the main advantages of the IQC-framework, it is possible to systematically check stability for plants that are affected by multiple uncertainties. For this purpose, let us consider plants with a linear fractional dependency on $\Delta_1, \ldots, \Delta_k$, where Δ_i, $i = 1, \ldots, k$ take their values in $\boldsymbol{\Delta}_i$. In particular we are interested in those plants that admit a linear fractional representation of the form (2.7) with G and Δ being structured as

$$
\begin{pmatrix} G_{qp} & G_{qw} \\ G_{zp} & G_{zw} \end{pmatrix} := \begin{pmatrix} G_{q_1 p_1} & \cdots & G_{q_1 p_k} & G_{q_1 w} \\ \vdots & \ddots & \vdots & \vdots \\ G_{q_k p_1} & \cdots & G_{q_k p_k} & G_{q_k w} \\ \hline G_{zp_1} & \cdots & G_{zp_k} & G_{zw} \end{pmatrix} \quad \text{and} \quad \Delta(q) := \begin{pmatrix} \Delta_1(q_1) \\ \vdots \\ \Delta_k(q_k) \end{pmatrix}
$$

respectively. Then (2.7) needs to be well-posed for all $\Delta_i \in \boldsymbol{\Delta}_i$, $i = 1, \ldots, k$. With the individual Δ_i-blocks satisfying the IQCs

$$
IQC(\Pi_i, q_i, \Delta_i(q_i)) \geq 0 \quad \forall q_i \in \mathscr{L}_2^{n_{q_i}}, \quad i = 1, \ldots, k,
$$

we can then simply define one composite IQC

$$IQC(\Pi, q, \Delta(q)) \geq 0 \quad \forall q \in \mathscr{L}_2^{n_q}$$

with $q := \text{col}(q_1, \ldots, q_k)$ and

$$\Pi := \text{daug}(\Pi_1, \ldots, \Pi_k) = \begin{pmatrix} \Pi_{1,11} & & 0 & \vdots & \Pi_{1,12} & & 0 \\ & \ddots & & \vdots & & \ddots & \\ 0 & & \Pi_{k,11} & \vdots & 0 & & \Pi_{k,12} \\ \text{-} & \text{-} & \text{-} & \text{-} & \text{-} & \text{-} & \text{-} \\ \Pi_{1,12}^* & & 0 & \vdots & \Pi_{1,22} & & 0 \\ & \ddots & & \vdots & & \ddots & \\ 0 & & \Pi_{k,12}^* & \vdots & 0 & & \Pi_{k,22} \end{pmatrix}$$

and apply Lemma 2.1 without any additional proof or complications. This makes the IQC-framework a very generic and flexible approach for the analysis of uncertain systems. As an example, we refer the reader to [135] for an extensive illustration of how easily the IQC-framework comprises and substantially generalizes a number of stand-alone stability results for bilateral teleoperation.

2.5 From infinite to finite dimensional feasibility tests

Although we have presented some of the most important ideas and results of the IQC-approach, we still did not establish a simple test that is computationally tractable. In fact, verifying robust stability and performance with Lemma 2.1 boils down to finding unknown rational matrix variables $\Pi \in \boldsymbol{\Pi}$ and $\Pi_p \in \boldsymbol{\Pi}_p$ for which the FDI (2.9) is satisfied for all $\omega \in \mathbb{R} \cup \{\infty\}$. In order to overcome this issue, we need to find suitable parametrizations of Π and Π_p such that (2.9) yields a linear constraint on some set of unknown parameters. Subsequently, we need to transform the resulting inequality, which still will be semi-infinite in ω, into a computationally tractable test.

2.5.1 Parameterizing IQCs

In this thesis we consider families of IQC-multipliers that are parameterized as

$$\boldsymbol{\Pi} = \{\Psi^* P \Psi : P \in \boldsymbol{P}\}$$

with an LMIable set $\boldsymbol{P}$ of real symmetric matrices $P \in \mathbb{S}^{\bullet}$ (i.e. it can be represented as the projection of the feasible set of an LMI constraint) and with some fixed and typically tall transfer matrix $\Psi \in \mathscr{RH}_{\infty}^{\bullet \times (n_q + n_p)}$. As will become clearer in the sequel, we choose this structure because it is particularly well-suited for obtaining LMIs. For the time being, we assume no additional knowledge about the structure of Ψ and $P \in \boldsymbol{P}$. The IQC (2.4) now reads as

$$IQC(\Psi^* P\Psi, q, \Delta(q)) \geq 0 \quad \forall q \in \mathscr{L}_2^{n_q}.$$

Again, to illustrate the concept, let us give an example.

Example 2.2. In Section 2.6.1 we will encounter IQC-multipliers of the form

$$\Pi := \begin{pmatrix} \pi_{11} I_{n_q} & 0 \\ 0 & -\pi_{11} I_{n_p} \end{pmatrix}, \quad \mathscr{RL}_{\infty}^{1 \times 1} \ni \pi_{11} \geq 0 \text{ on } \mathbb{C}^0.$$

In accordance with the previous discussion, we can parameterize this IQC-multiplier as

$$\Psi^* P\Psi := \begin{pmatrix} \star \\ \star \end{pmatrix}^* \begin{pmatrix} P_{11} \otimes I_{n_q} & 0 \\ 0 & -P_{11} \otimes I_{n_p} \end{pmatrix} \begin{pmatrix} \psi_\nu \otimes I_{n_q} & 0 \\ 0 & \psi_\nu \otimes I_{n_p} \end{pmatrix}. \quad (2.10)$$

Here $P_{11} \in \mathbb{S}^{\nu+1}$ is a free matrix variable and $\psi_\nu \in \mathscr{RH}_{\infty}^{(\nu+1) \times 1}$ is a fixed basis-function with a McMillan degree of ν, which can, for example, be chosen as

$$\psi_\nu(i\omega) = \begin{pmatrix} 1 \\ \frac{1}{(i\omega - \rho)} \\ \frac{1}{(i\omega - \rho)^2} \\ \vdots \\ \frac{1}{(i\omega - \rho)^\nu} \end{pmatrix} \quad \text{or} \quad \psi_\nu(i\omega) = \begin{pmatrix} 1 \\ \frac{(i\omega + \rho)}{(i\omega - \rho)} \\ \frac{(i\omega + \rho)^2}{(i\omega - \rho)^2} \\ \vdots \\ \frac{(i\omega + \rho)^\nu}{(i\omega - \rho)^\nu} \end{pmatrix}, \quad (2.11)$$

with $\rho < 0$ and $\nu \in \mathbb{N}_0$. This leads to an affine parametrization of the IQC-multiplier in the matrix variable P_{11}. Moreover, as we will see in the sequel, the IQC-multiplier (2.10) as well as the inertia constraint

$$\pi_{11} = \psi_\nu^* P_{11} \psi_\nu \geq 0 \text{ on } \mathbb{C}^0 \quad (2.12)$$

have the right structure for obtaining LMIs.

In complete analogy, let us also parameterize the family of performance multipliers $\boldsymbol{\Pi}_{\mathrm{p}}$ as

$$\boldsymbol{\Pi}_{\mathrm{p}} = \left\{ \Psi_{\mathrm{p}}^* P_{\mathrm{p}} \Psi_{\mathrm{p}} : P_{\mathrm{p}} \in \boldsymbol{P}_{\mathrm{p}} \right\}$$

with, again, an LMIable set $\boldsymbol{P}_{\mathrm{p}}$ of real symmetric matrices $P_{\mathrm{p}} \in \mathbb{S}^{\bullet}$ and with some fixed transfer matrix $\Psi_{\mathrm{p}} \in \mathscr{RH}_{\infty}^{\bullet \times (n_z + n_w)}$. The performance IQC (2.8) now reads as

$$IQC(\Psi_{\mathrm{p}}^* P_{\mathrm{p}} \Psi_{\mathrm{p}}, z, w) \leq -\epsilon \|w\|_2^2$$

and robust stability and performance can be characterized as follows.

Corollary 2.2. *Robust stability and performance is achieved if*

1. for all $\Delta \in \boldsymbol{\Delta}$ the linear fractional representation (2.7) is well-posed;
2. for all $\Delta \in \boldsymbol{\Delta}$ and for all $\Pi = \Psi^* P \Psi$ with $P \in \boldsymbol{P}$ the IQC (2.4) is satisfied;
3. there exist some $P \in \boldsymbol{P}$ and $P_{\mathrm{p}} \in \boldsymbol{P}_{\mathrm{p}}$ for which the following FDI holds:

$$\begin{pmatrix} G_{qp} & G_{qw} \\ I & 0 \\ \hline G_{zp} & G_{zw} \\ 0 & I \end{pmatrix}^* \begin{pmatrix} \Psi^* P \Psi & 0 \\ 0 & \Psi_{\mathrm{p}}^* P_{\mathrm{p}} \Psi_{\mathrm{p}} \end{pmatrix} \begin{pmatrix} G_{qp} & G_{qw} \\ I & 0 \\ \hline G_{zp} & G_{zw} \\ 0 & I \end{pmatrix} \prec 0 \text{ on } \mathbb{C}^0. \qquad (2.13)$$

Observe that the FDI (2.13) is affine in the matrix variables P and P_{p}. Hence, if both $\boldsymbol{P}$ and $\boldsymbol{P}_{\mathrm{p}}$ are LMIable sets, we obtain an semi-infinite convex feasibility test for robust stability and performance of the feedback system (2.1).

Remark 2.6. Note that the length of the basis-function (2.11) plays an essential role for the proposed algorithms in this dissertation. By choosing $\nu = 0$, we restrict the IQC-multipliers to be non-dynamic, which might yield conservative results. Taking $\nu \in \mathbb{N}$ allows for dynamics in the IQC-multipliers which often leads to a considerable reduction of conservatism, at the cost of a higher computational load. The particular basis-functions in (2.11) have the beneficial property that any $\psi \in \mathscr{RH}_{\infty}$ can be approximated arbitrarily closely in the $\mathscr{H}_{\infty}$-norm by $\lambda_{\nu}^{\top} \psi_{\nu}$ for a suitable $\lambda_{\nu} \in \mathbb{R}^{\nu+1}$ if ν is chosen sufficiently large [134, 148]. This implies that e.g. the parameterization (2.12) is sufficiently rich to approximate general $\pi_{11} > 0$ on $\mathbb{C}^0$ up to an arbitrary small error without the need for frequency gridding. On the other hand, for a fixed length $\nu > 0$, the achievable approximation error does certainly depend on the location of the pole ρ. Since no general recipe is available for an appropriate choice, we recommend a line-search over ρ for lengths $\nu \in \{1, 2, 3, 4\}$. Let us finally emphasize, though, that our algorithms do not rely on the particular selection of ψ_{ν} as in (2.12) but that it can be applied with any ψ_{ν} as long as it is stable. We refer the reader to Section 2.9.1 for some further numerical evaluations.

2.5.2 The KYP Lemma

In order to verify whether or not (2.13) is satisfied, we need to find some $P \in \boldsymbol{P}$ and $P_{\mathrm{p}} \in \boldsymbol{P}_{\mathrm{p}}$ such that the FDI (2.13) holds true for all, and hence an infinite number of frequencies $\omega \in \mathbb{R} \cup \{\infty\}$. One way to do this is to fix a finite grid of properly distributed frequencies and simply check whether the resulting set of LMIs are all satisfied. However, this is rather ad-hoc since critical frequencies might be missed. On the other hand, it is also possible to follow an alternative procedure which is computationally more expensive, but which gives us guarantees for all frequencies. $\omega \in \mathbb{R} \cup \{\infty\}$. Indeed, in this section we exploit another fundamental result, which generalizes the well-known positive-real and bounded-real lemma (see e.g. [2, 48]) and which is called the Kalman-Yakubovich-Popov or simply KYP-lemma. In particular, we consider the following version, a slight generalization of [139].

Lemma 2.2. *Let* $P \in \mathbb{S}^{\bullet}$ *and let* $G \in \mathscr{R}\mathscr{L}_{\infty}^{\bullet \times \bullet}$ *admit the realization* (A, B, C, D) *with* $A \in \mathbb{R}^{n_G \times n_G}$ *and* $\mathrm{eig}(A) \cap \mathbb{C}^0 = \emptyset$. *The following two statements are equivalent:*

1. $G^* P G \prec 0$ *on* $\mathbb{C}^0$.
2. *There exists a matrix* $X \in \mathbb{S}^{n_G}$ *such that*

$$\begin{pmatrix} I & 0 \\ A & B \\ C & D \end{pmatrix}^{\top} \begin{pmatrix} 0 & X & 0 \\ X & 0 & 0 \\ 0 & 0 & P \end{pmatrix} \begin{pmatrix} I & 0 \\ A & B \\ C & D \end{pmatrix} \prec 0.$$

The corresponding equivalence for

- *non-strict inequalities holds, if, in addition, the pair* (A, B) *is controllable.*
- *equalities holds, if, in addition,* A *is Hurwitz and the pair* (A, B) *is controllable.*

The part of the proof which involves inequalities has been given by [139]. The remainder is found in Appendix D. For further details on the KYP-lemma we refer the reader to [2, 204, 139, 8, 70] and references therein. Since condition (2.13) can now be checked numerically, we have introduced all the necessary ingredients in order to formulate a finite dimensional convex feasibility test for robust stability and performance. For this purpose, we introduce the realization

$$\begin{pmatrix} \Psi & 0 \\ 0 & \Psi_{\mathrm{p}} \end{pmatrix} \begin{pmatrix} G_{qp} & G_{qw} \\ I & 0 \\ \hline G_{zp} & G_{zw} \\ 0 & I \end{pmatrix} = \left[\begin{array}{c|cc} A & B_1 & B_2 \\ \hline C_1 & D_{11} & D_{12} \\ C_2 & D_{21} & D_{22} \end{array} \right], \qquad (2.14)$$

where $A \in \mathbb{R}^{\bullet \times \bullet}$, $\mathrm{eig}(A) \subset \mathbb{C}^-$, and obtain the following result.

Corollary 2.3. *Robust stability and performance is achieved if*

1. for all $\Delta \in \mathbf{\Delta}$ the linear fractional representation (2.7) is well-posed;

2. for all $\Delta \in \mathbf{\Delta}$ and for all $\Pi = \Psi^* P \Psi$ with $P \in \mathbf{P}$ the IQC (2.4) is satisfied;

3. there exists some $X \in \mathbb{S}^\bullet$, $P \in \mathbf{P}$ and $P_{\mathrm{p}} \in \mathbf{P}_{\mathrm{p}}$ for which the following LMI holds true:

$$
\begin{pmatrix} I & 0 & 0 \\ \hline A & B_1 & B_2 \\ \hline C_1 & D_{11} & D_{12} \\ C_2 & D_{21} & D_{22} \end{pmatrix}^\top
\begin{pmatrix} 0 & X & 0 & 0 \\ X & 0 & 0 & 0 \\ \hline 0 & 0 & P & 0 \\ 0 & 0 & 0 & P_{\mathrm{p}} \end{pmatrix}
\begin{pmatrix} I & 0 & 0 \\ \hline A & B_1 & B_2 \\ \hline C_1 & D_{11} & D_{12} \\ C_2 & D_{21} & D_{22} \end{pmatrix} \prec 0.
\qquad (2.15)
$$

Hence, if $\mathbf{P}$ and $\mathbf{P}_{\mathrm{p}}$ are LMIable, we obtain a finite dimensional convex feasibility test for robust stability and performance. Given $\mathbf{P}$ and $\mathbf{P}_{\mathrm{p}}$, feasibility of the LMI (2.15) can be easily checked numerically with tools such as e.g. Matlab, LMIlab, CVX, Sedumi, SDPT3, Yalmip and many more [50, 10, 179, 184, 128, 60, 187, 109, 8, 58, 59].

2.6 IQC-multipliers for uncertainties and nonlinearities

The possibility of IQCs are rather comprehensive. The main purpose of this section is to illustrate this by giving a thorough overview of some new and existing IQC-multipliers with an explicit emphasis on their parametrization. Once more we emphasize that no such survey is accessible in the literature.

2.6.1 Uncertain LTI dynamics

The IQC-framework allows us to formulate various IQC-multipliers for uncertain LTI dynamics. As will be discussed in Section 2.8.1, these are closely related to the well-known D-scalings from μ-theory. In the subsequent subsections we provide a brief overview of possibilities.

2.6.1.1 Full-block uncertain LTI dynamics

Let us start by considering the class of LTI dynamic full-block operators $p = \Delta q$ defined by $\hat{p}(i\omega) := \Delta(i\omega)\hat{q}(i\omega)$ with $\|\Delta\|_\infty \leq \alpha$. Here α is some given non-negative constant, while $\hat{p}$ and $\hat{q}$ denote the Fourier transforms of p and q respectively (as will be tacitly assumed in the sequel). In short, this means that Δ is confined to the set of LTI dynamic full-block uncertainties

$$\boldsymbol{\Delta}_{\mathrm{lti,dyn,fb}} := \left\{ \Delta \in \mathscr{H}_\infty^{n_p \times n_q} : \|\Delta\|_\infty \leq \alpha \right\}.$$

LTI uncertainties of this form, and in particular the norm constraint $\|\Delta\|_\infty \leq \alpha$ can be captured with the following IQC-multiplier.

IQC-Multiplier 2.1. *The IQC* (2.4) *holds with*

$$\Psi^* P \Psi := \begin{pmatrix} \star \\ \star \end{pmatrix}^* \begin{pmatrix} P_{11} \otimes I_{n_q} & 0 \\ 0 & -P_{11} \otimes I_{n_p} \end{pmatrix} \begin{pmatrix} \alpha \psi_\nu \otimes I_{n_q} & 0 \\ 0 & \psi_\nu \otimes I_{n_p} \end{pmatrix} \qquad (2.16)$$

for all $\Delta \in \boldsymbol{\Delta}_{\mathrm{lti,dyn,fb}}$ *if*

$$\psi_\nu^* P_{11} \psi_\nu \geq 0 \text{ on } \mathbb{C}^0. \qquad (2.17)$$

Here $P_{11} \in \mathbb{S}^{\nu+1}$ *is a free matrix variable and* $\psi_\nu \in \mathscr{RH}_\infty^{(\nu+1)\times 1}$ *is a fixed basis-function with a McMillan degree of* ν *which can, for example, be chosen as in* (2.11).

Let us observe that the FDI (2.17) can be equivalently turned into an LMI. By choosing the minimal realization

$$\psi_\nu = \left[\begin{array}{c|c} A_\nu & B_\nu \\ \hline C_\nu & D_\nu \end{array} \right] = \left[\begin{array}{ccccc|c} \rho & 0 & \cdots & \cdots & 0 & 1 \\ 1 & \ddots & \ddots & \ddots & \vdots & 0 \\ 0 & \ddots & \ddots & \ddots & \vdots & \vdots \\ \vdots & \ddots & \ddots & \ddots & 0 & \vdots \\ 0 & \cdots & 0 & 1 & \rho & 0 \\ \hline & & 0 & & & 1 \\ & & I_\nu & & & 0 \end{array} \right] \qquad (2.18)$$

for the first basis-function in (2.11), we can infer, by the KYP-lemma that (2.17) is equivalent to the existence of the symmetric matrix $X_\nu \in \mathbb{S}^\nu$ such that the following LMI holds true:

$$\begin{pmatrix} I & 0 \\ A_\nu & B_\nu \\ C_\nu & D_\nu \end{pmatrix}^\top \begin{pmatrix} 0 & X_\nu & 0 \\ X_\nu & 0 & 0 \\ 0 & 0 & P_{11} \end{pmatrix} \begin{pmatrix} I & 0 \\ A_\nu & B_\nu \\ C_\nu & D_\nu \end{pmatrix} \succcurlyeq 0. \qquad (2.19)$$

As discussed in Remark 2.6, our algorithms do not rely on the particular selection of ψ_ν as in (2.18), but can be applied with any $\psi_\nu = (A_\nu, B_\nu, C_\nu, D_\nu)$ as long as A_ν is Hurwitz and (A_ν, B_ν) is controllable. Alternatively, we can e.g. choose the minimal realization

$$
\psi_\nu = \left[\begin{array}{c|c} A_\nu & B_\nu \\ \hline C_\nu & D_\nu \end{array} \right] =
\left[\begin{array}{cccc|c}
\rho & 2\rho & \cdots & 2\rho & -\sqrt{2|\rho|} \\
0 & \ddots & \ddots & \vdots & \vdots \\
\vdots & \ddots & \ddots & 2\rho & \vdots \\
0 & \cdots & 0 & \rho & -\sqrt{2|\rho|} \\
\hline
0 & \cdots & \cdots & 0 & 1 \\
0 & \cdots & 0 & \sqrt{2|\rho|} & \vdots \\
\vdots & \iddots & \iddots & \vdots & \vdots \\
0 & \sqrt{2|\rho|} & \cdots & \sqrt{2|\rho|} & \vdots \\
\sqrt{2|\rho|} & \cdots & \cdots & \sqrt{2|\rho|} & 1
\end{array} \right]
$$

for the second basis-function in (2.11). We explicitly provided the latter two realizations because they are used in our numerical examples in Section 2.9.

Proof. Combining (2.4) and (2.16) yields the inequality

$$
\int_{-\infty}^{\infty} \hat{q}(i\omega)^*(\alpha\psi_\nu(i\omega) \otimes I_{n_q})^*(P_{11} \otimes I_{n_q})(\alpha\psi_\nu(i\omega) \otimes I_{n_q})\hat{q}(i\omega)-
$$
$$
\hat{q}(i\omega)^* \Delta(i\omega)^*(\star)^*(P_{11} \otimes I_{n_p})(\psi_\nu(i\omega) \otimes I_{n_p})\Delta(i\omega)\hat{q}(i\omega)\ d\omega \geq 0 \quad \forall q \in \mathscr{L}_2^{n_q}.
$$

Through simple manipulations we can rewrite this as

$$
\int_{-\infty}^{\infty} \hat{q}(i\omega)^* \left(\alpha^2 \psi_\nu(i\omega)^* P_{11} \psi_\nu(i\omega) I_{n_q} \right) \hat{q}(i\omega)-
$$
$$
\hat{q}(i\omega)^* \Delta(i\omega)^* \left(\psi_\nu(i\omega)^* P_{11} \psi_\nu(i\omega) I_{n_p} \right) \Delta(i\omega)\hat{q}(i\omega)\ d\omega \geq 0 \quad \forall q \in \mathscr{L}_2^{n_q}.
$$

Since (2.17) is a SISO transfer function, the latter constraint is easily seen to be implied by

$$
\psi_\nu^* P_{11} \psi_\nu \left(\alpha^2 I_{n_q} - \Delta^* \Delta \right) \succcurlyeq 0 \text{ on } \mathbb{C}^0,
$$

which holds true if (2.17) is satisfied and if $\Delta \in \mathbf{\Delta}_{\text{lti,dyn,fb}}$.

■

2.6.1.2 Diagonally repeated LTI uncertainties

It often happens that the operator Δ has the specialized structure $\Delta := \delta I_{n_q}$ defined by $\hat{p}(i\omega) = \delta(i\omega)\hat{q}(i\omega)$ with $\|\delta\|_\infty \leq \alpha$. This yields the specialized set of LTI dynamic diagonally repeated uncertainties

$$
\mathbf{\Delta}_{\text{lti,dyn,dr}} := \left\{ \delta I_{n_q} : \delta \in \mathscr{H}_\infty,\ \|\delta\|_\infty \leq \alpha \right\}.
$$

The extra structure allows us to introduce more freedom in the IQC-multiplier 2.1 in the following sense.

IQC-Multiplier 2.2. *The IQC* (2.4) *holds with*

$$
\Psi^* P \Psi := \begin{pmatrix} \alpha \psi_\nu \otimes I_{n_q} & 0 \\ 0 & \psi_\nu \otimes I_{n_p} \end{pmatrix}^* \begin{pmatrix} P_{11} & 0 \\ 0 & -P_{11} \end{pmatrix} \begin{pmatrix} \alpha \psi_\nu \otimes I_{n_q} & 0 \\ 0 & \psi_\nu \otimes I_{n_p} \end{pmatrix}
\tag{2.20}
$$

for all $\Delta \in \Delta_{\mathrm{lti,dyn,dr}}$ *if*

$$
(\psi_\nu \otimes I_{n_q})^* P_{11} (\psi_\nu \otimes I_{n_q}) \succcurlyeq 0 \text{ on } \mathbb{C}^0.
\tag{2.21}
$$

Here $P_{11} \in \mathbb{S}^{n_q(\nu+1)}$ *is again a free matrix variable and* $\psi_\nu \in \mathcal{RH}_\infty^{(\nu+1) \times 1}$ *a fixed basis-function of McMillan degree* ν.

Observe that also the FDI (2.21) can be easily turned into an LMI. By choosing the minimal realization

$$
\psi_\nu \otimes I_n = \left[\begin{array}{c|c} A_\nu \otimes I_n & B_\nu \otimes I_n \\ \hline C_\nu \otimes I_n & D_\nu \otimes I_n \end{array} \right]
\tag{2.22}
$$

and $n = n_q$, we can infer by the KYP-lemma that (2.21) is equivalent to the existence of some symmetric matrix $X_\nu \in \mathbb{S}^{n_q \nu}$ such that the following LMI holds true:

$$
\begin{pmatrix} I & 0 \\ A_\nu \otimes I_{n_q} & B_\nu \otimes I_{n_q} \\ C_\nu \otimes I_{n_q} & D_\nu \otimes I_{n_q} \end{pmatrix}^\top \begin{pmatrix} 0 & X_\nu & 0 \\ X_\nu & 0 & 0 \\ 0 & 0 & P_{11} \end{pmatrix} \begin{pmatrix} I & 0 \\ A_\nu \otimes I_{n_q} & B_\nu \otimes I_{n_q} \\ C_\nu \otimes I_{n_q} & D_\nu \otimes I_{n_q} \end{pmatrix} \succcurlyeq 0.
\tag{2.23}
$$

Proof. Combining (2.4) and (2.20) yields the inequality

$$
\int_{-\infty}^{\infty} \hat{q}(i\omega)^* (\alpha \psi_\nu(i\omega) \otimes I_{n_q})^* P_{11} (\alpha \psi_\nu(i\omega) \otimes I_{n_q}) \hat{q}(i\omega) -
$$

$$
\hat{q}(i\omega)^* \delta(i\omega)^* (\psi_\nu(i\omega) \otimes I_{n_p})^* P_{11} (\psi_\nu(i\omega) \otimes I_{n_p}) \delta(i\omega) \hat{q}(i\omega) \, d\omega \geq 0 \quad \forall q \in \mathcal{L}_2^{n_q}.
$$

Clearly this is implied by

$$
(\alpha^2 - \delta^* \delta)(\psi_\nu \otimes I_{n_q})^* P_{11} (\psi_\nu \otimes I_{n_q}) \succcurlyeq 0 \text{ on } \mathbb{C}^0
$$

if $\|\delta\|_\infty \leq \alpha$ and if (2.21) holds true. Hence, the IQC (2.4) holds with (2.20) for all $\Delta \in \Delta_{\mathrm{lti,dyn,dr}}$.

∎

2.6.2 Arbitrarily fast time-varying parametric uncertainties

The IQC-framework also allows us to capture the properties of various classes of arbitrarily fast time-varying parametric uncertainties. In the subsequent subsections we provide a selected overview of possibilities.

2.6.2.1 Real repeated parameters

Let us first consider the class of linear time-varying (LTV) real diagonally repeated operators $p = \Delta q$ defined by $p(t) := \delta(t)q(t)$. Here δ is assumed to be a real time-varying parameter $\delta : [0, \infty) \to \Lambda$ with

$$\Lambda := \{\delta \in \mathbb{R} : |\delta| \leq \alpha\},$$

such that Δ is contained in the set

$$\boldsymbol{\Delta}_{\mathrm{ltv,re,dr}} := \left\{ \delta I_{n_q} : \delta \in PC([0, \infty), \Lambda) \right\}.$$

Here $PC([0, \infty), \Lambda)$ denotes the space of piecewise continuous functions $[0, \infty) \to \Lambda$. Operators from the set $\boldsymbol{\Delta}_{\mathrm{ltv,re,dr}}$ can be considered with the following well-known static DG-multiplier.

IQC-Multiplier 2.3. *The IQC* (2.4) *holds with*

$$\Psi^* P \Psi := \begin{pmatrix} \alpha I_{n_q} & 0 \\ 0 & I_{n_q} \end{pmatrix}^* \begin{pmatrix} P_{11} & P_{12} \\ P_{12}^\top & -P_{11} \end{pmatrix} \begin{pmatrix} \alpha I_{n_q} & 0 \\ 0 & I_{n_q} \end{pmatrix} \tag{2.24}$$

for all $\Delta \in \boldsymbol{\Delta}_{\mathrm{ltv,re,dr}}$, *if* P *is confined to*

$$\boldsymbol{P}_{\mathrm{dg}} := \left\{ P = \begin{pmatrix} P_{11} & P_{12} \\ P_{12}^\top & -P_{11} \end{pmatrix} \in \mathbb{S}^{2n_q} : P_{11} \succcurlyeq 0, \ P_{12} = -P_{12}^\top \right\}.$$

Proof. Observe that (2.4) combined with (2.24) yields the inequality

$$\int_0^\infty \left(\alpha^2 - \delta(t)^2\right) q(t)^\top P_{11} q(t) + \alpha \delta q(t)^\top (P_{12} + P_{12}^\top) q(t) dt \geq 0 \quad \forall q \in \mathscr{L}_2^{n_q},$$

which clearly holds true if $|\delta| \leq \alpha$ and $P \in \boldsymbol{P}_{\mathrm{dg}}$. We conclude that the IQC (2.4) holds with (2.24) for all $\Delta \in \boldsymbol{\Delta}_{\mathrm{ltv,re,dr}}$. ∎

2.6.2.2 Coefficients from a polytope

Let us proceed with a much more generic class of LTV real full-block operators that take their values from a polytope. For this purpose, consider the LTV operator $\Delta := \tilde{\Delta} \circ \delta$ which is assumed to be a linear function of the time-varying parameter vector $\delta : [0, \infty) \to \Lambda$. Here the map $\tilde{\Delta} : \mathbb{R}^k \to \mathbb{R}^{n_p \times n_q}$ is defined by

$$\tilde{\Delta}(\delta) := \sum_{i=1}^{k} \delta_i \Omega_i$$

for some fixed matrices $\Omega_i \in \mathbb{R}^{n_p \times n_q}$, $\delta \in \mathbb{R}^k$, and we assume that δ takes its values in the compact polytope

$$\Lambda := \mathrm{co}\left\{\delta^1, \ldots, \delta^m\right\}$$

with $\delta^j = (\delta_1^j, \ldots, \delta_k^j)$, $j \in \{1, \ldots, m\}$ as generator points. Without loss of generality Λ contains the origin. Then Δ is contained in the set

$$\boldsymbol{\Delta}_{\mathrm{ltv,re,fb}} := \left\{\tilde{\Delta} \circ \delta : \delta \in PC([0, \infty), \Lambda)\right\}$$

and defines the operator $p(t) = \tilde{\Delta}(\delta(t))q(t)$. For operators of this particular form we can consider the following so-called static full-block multipliers.

IQC-Multiplier 2.4. *The IQC* (2.4) *holds with*

$$\Psi^* P \Psi := \begin{pmatrix} I_{n_q} & 0 \\ 0 & I_{n_p} \end{pmatrix}^* \begin{pmatrix} P_{11} & P_{12} \\ P_{12}^\top & P_{22} \end{pmatrix} \begin{pmatrix} I_{n_q} & 0 \\ 0 & I_{n_p} \end{pmatrix}. \qquad (2.25)$$

for all $\Delta \in \boldsymbol{\Delta}_{\mathrm{ltv,re,fb}}$, *if* P *is confined to*

$$\boldsymbol{P}_{\mathrm{fb}} := \left\{P \in \mathbb{S}^{n_q + n_p} : \Phi(P, \delta) \succcurlyeq 0 \quad \forall \delta \in \Lambda\right\}.$$

Here the map $\Phi(P, \delta)$ *is defined by*

$$\Phi(P, \delta) := \begin{pmatrix} I_{n_q} \\ \tilde{\Delta}(\delta) \end{pmatrix}^\top \begin{pmatrix} P_{11} & P_{12} \\ P_{12}^\top & P_{22} \end{pmatrix} \begin{pmatrix} I_{n_q} \\ \tilde{\Delta}(\delta) \end{pmatrix}. \qquad (2.26)$$

Proof. First pre- and post multiply (2.26) with $q(t)^\top$ and $q(t)$ respectively. Subsequently integrating on $[0, \infty)$ then directly yields the desired result.

∎

Let us observe that $\boldsymbol{P}_{\mathrm{fb}}$ consists of an infinite family of LMIs. For the actual implementations we need to find relaxations (i.e. close approximations) described by finitely many LMIs. The need for such relaxations is one of the fundamental reasons for conservatism in robustness analysis. In the sequel we discuss a selected overview of relaxation schemes that have been suggested in the literature.

Convex-hull relaxation

One of the most elementary, but possibly conservative, available relaxation schemes is the so called convex-hull relaxation (see e.g. [116, 148]). This amounts to imposing the additional constraint $P_{22} \preccurlyeq 0$ on P and leads to the following finite family of LMIs:

$$\boldsymbol{P}_{\mathrm{ch}} := \left\{ P \in \mathbb{S}^{n_q + n_p} : \Phi(P, \delta^j) \succcurlyeq 0, \quad P_{22} \preccurlyeq 0 \quad \forall j = 1, \ldots, m \right\}.$$

Due to the additional constraint $P_{22} \preccurlyeq 0$ the mapping $\delta \to \Phi(P, \delta)$ is concave. Positivity of its values at all the generators δ^j, $j = 1, \ldots, m$ hence implies positivity for all $\delta \in \Lambda$ and hence positivity of $\Phi(P, \delta)$ for all $\Delta \in \boldsymbol{\Delta}_{\mathrm{ltv,re,fb}}$. We conclude that $\boldsymbol{P}_{\mathrm{ch}} \subseteq \boldsymbol{P}_{\mathrm{fb}}$.

Partial convexity

It is often the case that the mapping $\tilde{\Delta} : \mathbb{R}^k \to \mathbb{R}^{n_p \times n_q}$ is defined by

$$\tilde{\Delta}(\delta) := \mathrm{diag}(\delta_1 I_{n_{q1}}, \ldots, \delta_k I_{n_{qk}}) = \sum_{i=1}^{k} \delta_i J_{n_{qi}} J_{n_{qi}}^\top, \quad |\delta_i| \leq 1, \qquad (2.27)$$

such that Λ is now a hyper rectangular box with center zero, and $J_{n_{qi}}$ is taken in accordance with the column partition $I_{n_q} = (J_{n_{q1}}, \ldots, J_{n_{qk}})$. For this special case, we can exploit the DG-multipliers from Section 2.6.2.1 and confine P to

$$\boldsymbol{P}_{\mathrm{dg,e}} := \left\{ P = \begin{pmatrix} \mathrm{diag}(P_{11,1}, \ldots, P_{11,k}) & \mathrm{diag}(P_{12,1}, \ldots, P_{12,k}) \\ \mathrm{diag}(P_{12,1}, \ldots, P_{12,k})^\top & -\mathrm{diag}(P_{11,1}, \ldots, P_{11,k}) \end{pmatrix} \in \mathbb{S}^{2n_q} : \right.$$
$$\left. P_{11,i} \succcurlyeq 0, \quad P_{12,i} = -P_{12,i}^\top \quad \forall i = 1, \ldots, k \right\}.$$

Here we stress that $\boldsymbol{P}_{\mathrm{dg,e}}$ is a potentially worse approximation of $\boldsymbol{P}_{\mathrm{fb}}$ if compared to $\boldsymbol{P}_{\mathrm{ch}}$ [172]. On the other hand, we can exploit the extra structure and refine the previous relaxation scheme by replacing the inertia constraint $P_{22} \preccurlyeq 0$ with

$$J_{n_{qi}}^\top P_{22} J_{n_{qi}} \preccurlyeq 0, \quad i = 1, \ldots, k, \qquad (2.28)$$

such that P is now confined to

$$\boldsymbol{P}_{\mathrm{pc}} := \left\{ P \in \mathbb{S}^{n_q + n_p} : \Phi(P, \delta^j) \succcurlyeq 0 \;\; \forall j = 1, \ldots, m, \;\; J_{n_{qi}}^\top P_{22} J_{n_{qi}} \preccurlyeq 0 \;\; \forall i = 1, \ldots, k \right\}.$$

Due to (2.28), we conclude that for any $P \in \boldsymbol{P}_{\mathrm{pc}}$ the mapping $\delta \to \Phi(P, \delta)$ is concave in each δ_i for $i = 1, \ldots k$ [49]. Let us emphasize that $\boldsymbol{P}_{\mathrm{dg,e}} \subseteq \boldsymbol{P}_{\mathrm{ch}} \subseteq \boldsymbol{P}_{\mathrm{pc}} \subseteq \boldsymbol{P}_{\mathrm{fb}}$, such that $\boldsymbol{P}_{\mathrm{pc}}$ is a potentially better approximation of $\boldsymbol{P}_{\mathrm{fb}}$, and hence might lead to less conservative numerical results.

Pólya relaxation

As an other alternative to the convex-hull relaxation, one could also employ the so called Pólya relaxation [137, 35, 138, 153, 154, 99]. For this purpose, let us recall that δ takes its values in the compact polytope

$$\Lambda = \mathrm{co}\left\{\delta^1, \ldots, \delta^m\right\} := \left\{\sum_{j=1}^m \alpha_j \delta^j : \alpha_j \geq 0, \ \sum_{j=1}^m \alpha_j = 1\right\}.$$

One can show that the mapping $\alpha \to \Phi(P, \sum_{j=1}^m \alpha_j \delta^j)$ is a homogeneous polynomial of degree two and can be written as

$$\sum_{j=1}^m \sum_{n=1}^m \alpha_j \alpha_n \begin{pmatrix} I_{n_q} \\ \sum_{i=1}^k \delta_i^j \Omega_i \end{pmatrix}^\top P \begin{pmatrix} I_{n_q} \\ \sum_{i=1}^k \delta_i^n \Omega_i \end{pmatrix} = \sum_{j=1}^m \sum_{n=1}^m \alpha_j \alpha_n \Phi(P, \delta^j, \delta^n),$$

(2.29)

where

$$\Phi(P, \delta^j, \delta^n) := \begin{pmatrix} I_{n_q} \\ \tilde{\Delta}(\delta^j) \end{pmatrix}^\top P \begin{pmatrix} I_{n_q} \\ \tilde{\Delta}(\delta^n) \end{pmatrix}.$$

Moreover, this in turn can be reformulated as

$$\Phi(P, \delta^j, \delta^n) = \sum_{j=1}^m \alpha_j^2 \Phi(P, \delta^j, \delta^j) + \sum_{n=j+1}^m \sum_{j=1}^m \alpha_j \alpha_n \left(\Phi(P, \delta^j, \delta^n) + \Phi(P, \delta^n, \delta^j)\right).$$

Hence, since all coefficients α_j are non-negative, we can infer that $\Phi(P, \delta) \succcurlyeq 0$ for all $\delta \in \Lambda$ if

$$\begin{aligned} \Phi(P, \delta^j, \delta^j) &\succcurlyeq 0 && \forall j = 1, \ldots, m, \\ \Phi(P, \delta^j, \delta^n) + \Phi(P, \delta^n, \delta^j) &\succcurlyeq 0 && \forall j = 1, \ldots, m, \ n = j+1, \ldots, m. \end{aligned}$$

(2.30)

We arrive at the following finite family of LMIs:

$$\boldsymbol{P}_{\mathrm{pr}} := \left\{P \in \mathbb{S}^{n_q + n_p} : (2.30) \ \text{holds}\right\}.$$

It is not difficult to show that this so-called zero'th order Pólya relaxation is as least as good as (and often better than) the convex-hull relaxation. We conclude that $\boldsymbol{P}_{\mathrm{ch}} \subseteq \boldsymbol{P}_{\mathrm{pr}} \subseteq \boldsymbol{P}_{\mathrm{fb}}$. Moreover, the relaxation quality can be straightforwardly improved with guaranteed asymptotic exactness. For further details we refer the reader to [35, 138, 153, 154].

Sum-of-Squares Relaxation

Another technique that has a large variety of applications in control is the so-called matrix sum-of-squares (SOS) approach [27, 128, 133, 67, 66, 153, 154, 164]. A polynomial matrix $\Phi_\mathrm{p}(\delta)$ of dimension $n_q \times n_q$ is said to be SOS if there exists a (possibly tall and non-square) polynomial matrix $T_\mathrm{p}(\delta)$ such that

$$\Phi_\mathrm{p}(\delta) = T_\mathrm{p}(\delta)^\top T_\mathrm{p}(\delta).$$

The terminology is motivated by the 1-dimensional case (i.e. $n_q = 1$): With $T_\mathrm{p}(\delta) = \mathrm{col}(t_1(\delta),\ldots,t_m(\delta)) \in \mathbb{R}^m$ we obtain $\Phi_\mathrm{p}(\delta) = \sum_{j=1}^m t_j(\delta)^2$, which obviously is a sum-of-squares of polynomials.

To verify whether a polynomial matrix $\Phi_\mathrm{p}(\delta)$ is SOS one can choose a basis of (pairwise different) monomials $u_\mathrm{p}(\delta) = \mathrm{col}(u_{\mathrm{p}1}(\delta),\ldots,u_{\mathrm{p}m}(\delta))$ and represent $T_\mathrm{p}(\delta)$ with coefficients $X_{\mathrm{p}1},\ldots,X_{\mathrm{p}m}$ as

$$T_\mathrm{p}(\delta) = X_\mathrm{p}U_\mathrm{p}(\delta), \quad X_\mathrm{p} = \begin{pmatrix} X_{\mathrm{p}1} & \cdots & X_{\mathrm{p}m} \end{pmatrix}, \quad U_\mathrm{p}(\delta) = u_\mathrm{p}(\delta) \otimes I_{n_q}.$$

It is then said that $\Phi_\mathrm{p}(\delta)$ is SOS with respect to $U_\mathrm{p}(\delta)$ if there exists a coefficient matrix X_p with $\Phi_\mathrm{p}(\delta) = U_\mathrm{p}(\delta)^\top X_\mathrm{p}^\top X_\mathrm{p} U_\mathrm{p}(\delta)$. Note that this is convexified by the elementary change of variables $Y_\mathrm{p} = X_\mathrm{p}^\top X_\mathrm{p}$, which leads to the following result (see e.g. [153, 164]).

Lemma 2.3. $\Phi_\mathrm{p}(\delta)$ *is SOS with respect to* $U_\mathrm{p}(\delta)$ *if and only if there exists some* $Y_\mathrm{p} \succcurlyeq 0$ *with* $\Phi_\mathrm{p}(\delta) = U_\mathrm{p}(\delta)^\top Y_\mathrm{p} U_\mathrm{p}(\delta)$.

Let us now go back to our problem of finding a finite family of LMIs such that $\Phi(P,\delta) \succcurlyeq 0$ for all $\delta \in \Lambda$. For this purpose, we equivalently redescribe Λ as

$$\Lambda = \{\delta : g_1(\delta) \leq 0,\ldots,g_m(\delta) \leq 0\},$$

where $g_j(\delta)$, $j = 1,\ldots,m$ are real-valued polynomial mappings. By weak Lagrange duality (see e.g. [110, 153]) it is then possible to infer that $\Phi(P,\delta) \succcurlyeq 0$ persists to hold for all $\delta \in \Lambda$ if there exist SOS matrices $S_1(\delta),\ldots,S_m(\delta)$ such that

$$\Phi(P,\delta) + S_1(\delta)g_1(\delta) + \cdots + S_m(\delta)g_m(\delta) \quad \text{is SOS.} \tag{2.31}$$

Indeed, since $S_j(\delta) \succcurlyeq 0$ and $g_j(\delta) \leq 0$ for all $\delta \in \Lambda$, we have that

$$\Phi(P,\delta) \succcurlyeq -S_1(\delta)g_1(\delta) - \cdots - S_m(\delta)g_m(\delta) \succcurlyeq 0 \quad \forall \delta \in \Lambda.$$

Hence $\Phi(P,\delta) \succcurlyeq 0$ for all $\delta \in \Lambda$. Note that this condition turns out to be exact if Λ is a compact polytope and if $\Phi(P,\delta)$ is strictly positive for all $\delta \in \Lambda$ [163].

In order to render (2.31) computational, one can now simply choose basis-functions $U_j(\delta)$ with monomials (as above) and symmetric matrix variables $Y_j \succcurlyeq 0$, such that

$$\Phi(P,\delta) + \sum_{j=1}^{m} U_j(\delta)^\top Y_j U_j(\delta) g_j(\delta) = U_0(\delta)^\top Y_0 U_0(\delta),$$

which clearly is a linear constraint in the variables P, Y_0 and Y_j, $j = 1, \dots m$.

Example 2.3. Let us briefly address a concrete example from [99] for rectangular and ellipsoidal regions. For this purpose, suppose that $\tilde{\Delta} : \mathbb{R}^k \to \mathbb{R}^{n_p \times n_q}$ is defined by (2.27), but where δ is now confined to

$$\Lambda := \left\{ U(\delta) : U(\delta)^\top R_j U(\delta) \geq 0 \ \ \forall j = 1, \dots, m \right\}. \tag{2.32}$$

Here $U(\delta) := \mathrm{col}(1, \delta_1, \dots, \delta_k) \otimes I_v$ is again a basis-function and R a symmetric matrix that identifies the region. In this fashion we can construct all sorts of regions as follows:

- **Rectangular regions.** With $R_i = \mathrm{diag}\left(\alpha_i^2, 0, \dots, 0, -1, 0, \dots, 0\right)$, $i = 1, \dots, k$, and -1 being located at the ith diagonal entry of R_i we ensure that δ takes its values in the rectangular region $\Lambda := \{\delta_i \in \mathbb{R} : |\delta_i| \leq \alpha_i, \ i = 1, \dots, k\}$.
- **Ellipsoidal regions.** With $R = \mathrm{diag}\left(1, -\alpha_1^{-2}, \dots, -\alpha_k^{-2}\right)$ we ensure that δ takes its values in the ellipsoidal region $\Lambda := \{\delta_i \in \mathbb{R} : \sum_{i=1}^{k} \delta_i^2 \alpha_i^{-2} \leq 1\}$. Note that regions of this form are in particular interesting for rate-bounded parametric uncertainties as discussed in Section 2.6.3.

The infinite set of LMI constraints $\boldsymbol{P}_{\mathrm{fb}}$ can now be approximated by the finite set of LMIs

$$\boldsymbol{P}_{\mathrm{sos}} := \left\{ P \in \mathbb{S}^{2n_q} : J^\top P J \otimes I_v - \sum_{j=1}^{m} R_j \otimes S_j - Q \succcurlyeq 0, \quad \mathbb{S}^{n_q} \ni S_j \succcurlyeq 0, \right.$$

$$\left. Q = [Q_{ij}]_{i=1,\dots,n_q, \ j=1,\dots,n_q}, \quad Q_{ij} \in \mathbb{R}^{(n_q v) \times (n_q v)}, \quad Q_{ii} = 0, \quad Q_{ij} = -Q_{ji} \right\}.$$

Here $J = \mathrm{diag}\left(I_{n_q}, (J_{n_{q1}} J_{n_{q1}}^\top, \dots, J_{n_{qk}} J_{n_{qk}}^\top)\right)$. Again we conclude that $\boldsymbol{P}_{\mathrm{sos}} \subseteq \boldsymbol{P}_{\mathrm{fb}}$. For a proof and further details we refer the reader to [99].

2.6.3 Time-invariant parametric uncertainties

Another class of uncertainties that can be captured by IQCs is the class of time-invariant parametric uncertainties. This class is closely related to the one described in the previous section. However, the time-invariant nature of the uncertainties in this section allows us to include dynamics in the IQC-multipliers and systematically reduce conservatism, if compared to the class or arbitrarily fast varying parameters.

2.6.3.1 Real repeated parameters

Analogously to Section 2.6.2, let us first consider the class of LTI real diagonally repeated parametric operators $p = \Delta q$ defined by $p(t) := \delta q(t)$ with $\delta \in \mathbb{R}$, $|\delta| \le \alpha$ and $\alpha \ge 0$ being some given non-negative constant. In short, this confines Δ to the set

$$\Delta_{\mathrm{lti,re,dr}} := \left\{ \delta I_{n_q} : |\delta| \le \alpha \right\}.$$

Operators from the set $\Delta_{\mathrm{lti,re,dr}}$ can be considered with the so-called dynamic DG-multiplier.

IQC-Multiplier 2.5. *The IQC (2.4) holds with*

$$\Psi^* P \Psi := \begin{pmatrix} \alpha \psi_\nu \otimes I_{n_q} & 0 \\ 0 & \psi_\nu \otimes I_{n_q} \end{pmatrix}^* \begin{pmatrix} P_{11} & P_{12} \\ P_{12}^\top & -P_{11} \end{pmatrix} \begin{pmatrix} \alpha \psi_\nu \otimes I_{n_q} & 0 \\ 0 & \psi_\nu \otimes I_{n_q} \end{pmatrix} \tag{2.33}$$

for all $\Delta \in \Delta_{\mathrm{lti,re,dr}}$ if

$$(\psi_\nu \otimes I_{n_q})^* P_{11}(\psi_\nu \otimes I_{n_q}) \succcurlyeq 0 \text{ on } \mathbb{C}^0 \tag{2.34}$$

and

$$\left(\psi_\nu \otimes I_{n_q} \right)^* \left(P_{12} + P_{12}^\top \right) \left(\psi_\nu \otimes I_{n_q} \right) = 0 \text{ on } \mathbb{C}^0. \tag{2.35}$$

Here $P_{11} \in \mathbb{S}^{n_q(\nu+1)}$ and $P_{12} \in \mathbb{R}^{(n_q(\nu+1)) \times (n_q(\nu+1))}$ are free matrix variables and $\psi_\nu \in \mathscr{RH}_\infty^{(\nu+1) \times 1}$ a fixed basis-function of McMillan degree ν.

As show in Section 2.6.1, the inequality (2.34) can be equivalently turned into the LMI (2.23). On the other hand, the equality constraint (2.35) is satisfied if and only if there exists some $Y_\nu \in \mathbb{S}^{n_q \nu}$ for which the following linear matrix equation is satisfied:

$$\begin{pmatrix} I & 0 \\ A_\nu \otimes I_{n_q} & B_\nu \otimes I_{n_q} \\ C_\nu \otimes I_{n_q} & D_\nu \otimes I_{n_q} \end{pmatrix}^\top \begin{pmatrix} 0 & Y_\nu & 0 \\ Y_\nu & 0 & 0 \\ 0 & 0 & P_{12} + P_{12}^\top \end{pmatrix} \begin{pmatrix} I & 0 \\ A_\nu \otimes I_{n_q} & B_\nu \otimes I_{n_q} \\ C_\nu \otimes I_{n_q} & D_\nu \otimes I_{n_q} \end{pmatrix} = 0. \tag{2.36}$$

Proof. Combining (2.4) and (2.33) yields the inequality

$$\int_{-\infty}^{\infty} (\alpha^2 - \delta^2) \hat{q}(i\omega)^* (\psi_\nu(i\omega) \otimes I_{n_q})^* P_{11}(\psi_\nu(i\omega) \otimes I_{n_q}) \hat{q}(i\omega) +$$

$$\delta \alpha \hat{q}(i\omega)^* (\psi_\nu(i\omega) \otimes I_{n_p})^* (P_{12} + P_{12}^\top)(\psi_\nu(i\omega) \otimes I_{n_p}) \hat{q}(i\omega) \, d\omega \ge 0 \quad \forall q \in \mathscr{L}_2^{n_q},$$

which obviously holds true for all $|\delta| \le \alpha$ if (2.34) and (2.35) are both satisfied. Hence, the IQC (2.4) holds with (2.33) for all $\Delta \in \boldsymbol{\Delta}_{\mathrm{lti,re,dr}}$.

∎

Remark 2.7. Note that the IQC-multiplier 2.5 trivially simplifies to the one discussed in Section 2.6.2.1 for LTV real diagonally repeated parametric uncertainties, just by restricting the McMillan degree of ψ_ν to zero (i.e. $\nu = 0$).

2.6.3.2 Coefficients from a polytope

In case of more than one LTI real diagonally repeated parametric uncertainty, the DG-multipliers from the previous section might be conservative. As an alternative, one can also apply the less-conservative but more computationally demanding dynamic full-block multipliers. For this purpose, consider the LTI real full-block operator $p = \Delta q$ defined by $p(t) := \tilde{\Delta}(\delta)q(t)$. Here the map $\tilde{\Delta} : \mathbb{R}^k \to \mathbb{R}^{n_p \times n_q}$ is defined by

$$\tilde{\Delta}(\delta) := \sum_{i=1}^{k} \delta_i \Omega_i$$

for some parameter vector $\delta \in \mathbb{R}^k$ and fixed matrices $\Omega_i \in \mathbb{R}^{n_p \times n_q}$. We assume that δ takes its values in the polytope

$$\Lambda := \mathrm{co}\left\{\delta^1, \dots, \delta^m\right\} \subseteq \mathbb{R}^k,$$

with

$$\delta^j = (\delta_1^j, \dots, \delta_k^j), \quad j \in \{1, \dots, m\}$$

as generator points. Without loss of generality we assume again that Λ contains the origin. Then Δ is contained in the set

$$\boldsymbol{\Delta}_{\mathrm{lti,re,fb}} := \left\{\tilde{\Delta}(\delta) : \delta \in \Lambda\right\}$$

For operators of this form we can consider the following dynamic full-block multiplier.

IQC-Multiplier 2.6. *The IQC (2.4) holds with*

$$\Psi^* P \Psi := \begin{pmatrix} \psi_\nu \otimes I_{n_q} & 0 \\ 0 & \psi_\nu \otimes I_{n_p} \end{pmatrix}^* \begin{pmatrix} P_{11} & P_{12} \\ P_{12}^\top & P_{22} \end{pmatrix} \begin{pmatrix} \psi_\nu \otimes I_{n_q} & 0 \\ 0 & \psi_\nu \otimes I_{n_p} \end{pmatrix} \quad (2.37)$$

for all $\Delta \in \boldsymbol{\Delta}_{\mathrm{lti,re,fb}}$ if

$$\begin{pmatrix} \star \\ \star \end{pmatrix}^* \begin{pmatrix} P_{11} & P_{12} \\ P_{12}^\top & P_{22} \end{pmatrix} \begin{pmatrix} \psi_\nu \otimes I_{n_q} \\ (\psi_\nu \otimes I_{n_p})\tilde{\Delta}(\delta^j) \end{pmatrix} \succ 0 \text{ on } \mathbb{C}^0 \quad \forall j = 1, \dots, m \quad (2.38)$$

and

$$(\psi_\nu \otimes I_{n_p})^* P_{22} (\psi_\nu \otimes I_{n_p}) \preccurlyeq 0 \text{ on } \mathbb{C}^0 \tag{2.39}$$

Here $P \in \mathbb{S}^{(n_q + n_p)(\nu+1)}$ is again a free matrix variable and $\psi_\nu \in \mathscr{RH}_\infty^{(\nu+1)\times 1}$ a fixed basis-function of McMillan degree ν.

Proof. Since (2.39) is satisfied, the mapping

$$\delta \to \begin{pmatrix} \psi_\nu \otimes I_{n_q} \\ (\psi_\nu \otimes I_{n_p})\tilde{\Delta}(\delta) \end{pmatrix}^* \begin{pmatrix} P_{11} & P_{12} \\ P_{12}^\top & P_{22} \end{pmatrix} \begin{pmatrix} \psi_\nu \otimes I_{n_q} \\ (\psi_\nu \otimes I_{n_p})\tilde{\Delta}(\delta) \end{pmatrix}$$

is concave. Hence, positivity of its values at all the generators δ^j, $j = 1, \ldots, m$ implies that (2.38) is satisfied for all $\delta \in \Lambda$. Pre- and post multiplying (2.38) with $\hat{q}(i\omega)^*$ and $\hat{q}(i\omega)$ respectively, and integrating on $(-\infty, \infty)$ finally implies that the IQC (2.4) is satisfied with (2.37) for all $\Delta \in \mathbf{\Delta}_{\mathrm{lti,re,fb}}$.   ∎

Remark 2.8. Again we emphasize that also the latter IQC-multiplier simplifies to the one discussed in Section 2.6.2.2, just by choosing $\nu = 0$. Further note that, in view of Section 2.6.2.2, we have applied the convex-hull relaxation in order to render (2.38) computational. Other relaxation schemes are applicable in an obvious fashion.

Let us finally observe that both (2.38) and (2.39) can be equivalently turned into LMIs. With the minimal realization (2.22) and $n = n_p$, we infer that the FDI (2.39) is equivalent to the existence of some $X_\nu \in \mathbb{S}^{n_p \nu}$ for which the following LMI holds true:

$$\begin{pmatrix} I & 0 \\ A_\nu \otimes I_{n_p} & B_\nu \otimes I_{n_p} \\ C_\nu \otimes I_{n_p} & D_\nu \otimes I_{n_p} \end{pmatrix}^\top \begin{pmatrix} 0 & X_\nu & 0 \\ X_\nu & 0 & 0 \\ 0 & 0 & P_{22} \end{pmatrix} \begin{pmatrix} I & 0 \\ A_\nu \otimes I_{n_p} & B_\nu \otimes I_{n_p} \\ C_\nu \otimes I_{n_p} & D_\nu \otimes I_{n_p} \end{pmatrix} \preccurlyeq 0.$$

On the other hand, observe that the outer-factor of (2.38) can be written as

$$\left(\psi_\nu \otimes I_{n_p + n_q}\right) \check{\Delta}(\delta^j) := \left(\psi_\nu \otimes I_{n_p + n_q}\right) \begin{pmatrix} I_{n_q} \\ \tilde{\Delta}(\delta^j) \end{pmatrix}$$

Hence, we can choose the realization (2.22), now with $n = n_p + n_q$, and infer that the jth FDI in (2.38) is equivalent to the existence of some $Y_{\nu,j} \in \mathbb{S}^{\nu(n_p + n_q)}$ for which

$$\begin{pmatrix} \star \\ \star \\ \star \end{pmatrix}^\top \begin{pmatrix} 0 & Y_{\nu,j} & 0 \\ Y_{\nu,j} & 0 & 0 \\ 0 & 0 & P \end{pmatrix} \begin{pmatrix} I & 0 \\ A_\nu \otimes I_{n_p + n_q} & \left(B_\nu \otimes I_{n_p + n_q}\right)\check{\Delta}(\delta^j) \\ C_\nu \otimes I_{n_p + n_q} & \left(D_\nu \otimes I_{n_p + n_q}\right)\check{\Delta}(\delta^j) \end{pmatrix} \succ 0$$

is satisfied.

2.6.4 Rate-bounded time-varying parametric uncertainties

In this section we consider a more sophisticated class of IQC-multipliers that captures
the properties of LTV parametric uncertainties that have a bounded rate-of-variation.
For this purpose, we slightly modify the class of LTV real diagonally repeated operators from Section 2.6.2.1 and introduce another class of operators $p = \Delta_{n_q} q$ defined
by $p(t) := \delta(t)q(t)$. Here the subscript n_q has been added to Δ in order to explicitly
indicate the number of repetitions of δ. We assume that δ is an LTV rate-bounded
parameter $\delta : [0, \infty) \to \Lambda \subset \mathbb{R}^2$, where Λ is convex and contains the origin. If compared
to the definitions in Section 2.6.2.1, δ is not a piecewise continuous but a continuously
differentiable function

$$\delta \in \boldsymbol{\delta} := \left\{ C^1([0,\infty), \mathbb{R}) : (\delta(t), \dot{\delta}(t)) \in \Lambda \ \forall t \geq 0 \right\},$$

such that Δ_{n_q} is now confined to

$$\boldsymbol{\Delta}_{\mathrm{ltv,rb}} := \left\{ \delta I_{n_q} : \delta \in \boldsymbol{\delta} \right\}.$$

Since δ is continuously differentiable, we can as well define the operator

$$p_{\mathrm{e}} := \mathscr{V}_{\Delta_{n_q}} q = (s\Delta_{n_q} - \Delta_{n_q} s)q$$

defined by

$$p_{\mathrm{e}}(t) := \frac{d\delta}{dt}(t)q(t) = \frac{dp}{dt}(t) - \delta(t)\frac{dq}{dt}(t).$$

We will exploit this so-called variation operator in order to derive a stability test for
LTV rate-bounded uncertainties. IQC-tests developed for this class of operators have
been addressed in [183, 84, 64, 99] and are based on the so-called Swapping lemma.

Lemma 2.4. *Suppose that $\psi_\nu \otimes I_{n_q}$ admits the realization (2.22) with $n = n_q$ and
let us define the realizations*

$$\psi_{\nu,\mathrm{l}} \otimes I_{n_q} := \left[\begin{array}{c|c} A_\nu \otimes I_{n_q} & I_\nu \otimes I_{n_q} \\ \hline C_\nu \otimes I_{n_q} & 0 \end{array} \right], \quad \psi_{\nu,\mathrm{r}} := \left[\begin{array}{c|c} A_\nu \otimes I_{n_q} & B_\nu \otimes I_{n_q} \\ \hline I_\nu \otimes I_{n_q} & 0 \end{array} \right].$$

For the given constants $k_1 = n_q(\nu + 1)$ and $k_2 = n_q\nu$ we have

$$\underbrace{\begin{pmatrix} \psi_\nu \otimes I_{n_q} & \psi_{\nu,\mathrm{l}} \otimes I_{n_q} \\ 0 & I \end{pmatrix}}_{\Psi_{\mathrm{le}}} \underbrace{\begin{pmatrix} \Delta_{n_q} \\ \mathscr{V}_{\Delta_{k_2}}(\psi_{\nu,\mathrm{r}} \otimes I_{n_q}) \end{pmatrix}}_{\Delta_{\mathrm{re}}} = \underbrace{\begin{pmatrix} \Delta_{k_1} & 0 \\ 0 & \mathscr{V}_{\Delta_{k_2}} \end{pmatrix}}_{\Delta_{\mathrm{le}}} \underbrace{\begin{pmatrix} \psi_\nu \otimes I_{n_q} \\ \psi_{\nu,\mathrm{r}} \otimes I_{n_q} \end{pmatrix}}_{\Psi_{\mathrm{re}}}$$

Here the extended operators Δ_{le} and Δ_{re} respectively take their values in the sets

$$\boldsymbol{\Delta}_{\mathrm{le}} := \left\{ \Delta_{\mathrm{le}} = \mathrm{diag}\left(\Delta_{k_1}, \mathscr{V}_{\Delta_{k_2}}\right) : \delta \in \varLambda \right\},$$
$$\boldsymbol{\Delta}_{\mathrm{re}} := \left\{ \Delta_{\mathrm{re}} = \mathrm{col}\left(\Delta_{n_q}, \mathscr{V}_{\Delta_{k_2}}\left(\psi_{\nu,\mathrm{r}} \otimes I_{n_q}\right)\right) : \delta \in \varLambda \right\}.$$

Proof. For notational simplicity, assume that $n_q = 1$. Then we have

$$
\begin{aligned}
\Delta_{\nu+1}\psi_\nu - \psi_\nu\Delta_1 &= C_\nu\Delta_\nu\left(sI - A_\nu\right)^{-1}B_\nu - C_\nu\left(sI - A_\nu\right)^{-1}\Delta_\nu B_\nu \\
&= C_\nu\left(sI - A_\nu\right)^{-1}\left[\left(sI - A_\nu\right)\Delta_\nu - \Delta_\nu\left(sI - A_\nu\right)\right]\left(sI - A_\nu\right)^{-1}B_\nu \\
&= C_\nu\left(sI - A_\nu\right)^{-1}\left[s\Delta_\nu - \Delta_\nu s\right]\left(sI - A_\nu\right)^{-1}B_\nu \\
&= \psi_{\nu,\mathrm{l}}\,\mathscr{V}_{\Delta_\nu}\,\psi_{\nu,\mathrm{r}}
\end{aligned}
$$

which proves the claim.

■

It is now possible formulate the following IQC-multiplier for the class of LTV real diagonally repeated and rate-bounded parametric uncertainties.

IQC-Multiplier 2.7. *Consider the following two realizations:*

$$
\Psi_{\mathrm{le}} = \left[\begin{array}{c|cc}
A_\nu \otimes I_{n_q} & B_\nu \otimes I_{n_q} & I_\nu \otimes I_{n_q} \\
\hline
C_\nu \otimes I_{n_q} & D_\nu \otimes I_{n_q} & 0 \\
0 & 0 & I_\nu \otimes I_{n_q}
\end{array}\right], \quad
\Psi_{\mathrm{re}} = \left[\begin{array}{c|c}
A_\nu \otimes I_{n_q} & B_\nu \otimes I_{n_q} \\
\hline
C_\nu \otimes I_{n_q} & D_\nu \otimes I_{n_q} \\
I_\nu \otimes I_{n_q} & 0
\end{array}\right].
$$

The IQC (2.4) holds with

$$
\Psi^* P \Psi := \begin{pmatrix} \Psi_{\mathrm{re}} & 0 \\ 0 & \Psi_{\mathrm{le}} \end{pmatrix}^* \begin{pmatrix} P_{11} & P_{12} \\ P_{12}^\top & P_{22} \end{pmatrix} \begin{pmatrix} \Psi_{\mathrm{re}} & 0 \\ 0 & \Psi_{\mathrm{le}} \end{pmatrix} \tag{2.40}
$$

for all $\Delta_{\mathrm{re}} \in \boldsymbol{\Delta}_{\mathrm{re}}$ and Δ replaced by Δ_{re} if P is confined to

$$
\boldsymbol{P}_{\mathrm{rb}} := \left\{ P \in \mathbb{S}^{2\nu(2n_q+1)} : \begin{pmatrix} I \\ \tilde{\Delta}_{\mathrm{le}}(\delta,\dot\delta) \end{pmatrix}^\top \begin{pmatrix} P_{11} & P_{12} \\ P_{12}^\top & P_{22} \end{pmatrix} \begin{pmatrix} I \\ \tilde{\Delta}_{\mathrm{le}}(\delta,\dot\delta) \end{pmatrix} \succcurlyeq 0 \quad \forall \delta \in \varLambda \right\}.
$$

Here $\tilde{\Delta}_{\mathrm{le}}(\delta,\dot\delta) := \mathrm{diag}(\delta I_{k_1}, \dot\delta I_{k_2})$.

In order to use the IQC-multiplier 2.7 it is required to augment the nominal plant G with an extra zero block as

$$
G_{\mathrm{e}} = \left(G \ \ 0_{n_q \times n_q \nu} \right).
$$

Hence, by employing the IQC-multiplier 2.7 we actually check whether or not $I - G_{\mathrm{e}}\Delta_{\mathrm{re}}$ has a bounded and causal inverse for all $\Delta_{\mathrm{re}} \in \boldsymbol{\Delta}_{\mathrm{re}}$. It is a matter of direct verification that this implies that $I - G\Delta_{n_q}$ has a bounded and causal inverse for all $\Delta_{n_q} \in \Delta_{\mathrm{ltv,rb}}$. Further observe that the set $\boldsymbol{P}_{\mathrm{rb}}$ is an infinite family of LMIs that has the very same features as the set $\boldsymbol{P}_{\mathrm{fb}}$. Hence, $\boldsymbol{P}_{\mathrm{rb}}$ can be easily relaxed to a finite family of LMIs by applying one of the relaxation schemes discussed in Section 2.6.2.2. In particular we refer to the SOS relaxation scheme for elliptic regions with smooth boundaries as discussed in Example 2.3, since $\delta(t)$ and $\dot{\delta}(t)$ might both be smooth functions of t.

Remark 2.9. We emphasize that the results straightforwardly generalize to the case of more than one repeated parameter. It is even possible to consider correlations between different parameters and variations [100]. Further, we refer to [100, 101] for an extensive research on the estimation of the quality of the latter IQC-multiplier based on lower bound computations.

Proof. Observe that the inequality

$$\begin{pmatrix} I \\ \tilde{\Delta}_{\mathrm{le}}(\delta,\dot{\delta}) \end{pmatrix}^{\top} \begin{pmatrix} P_{11} & P_{12} \\ P_{12}^{\top} & P_{22} \end{pmatrix} \begin{pmatrix} I \\ \tilde{\Delta}_{\mathrm{le}}(\delta,\dot{\delta}) \end{pmatrix} \succcurlyeq 0 \quad \forall \delta \in \Lambda$$

implies that the IQC

$$\int_0^{\infty} \begin{pmatrix} q_f(t) \\ p_f(t) \end{pmatrix}^{\top} \begin{pmatrix} P_{11} & P_{12} \\ P_{12}^{\top} & P_{22} \end{pmatrix} \begin{pmatrix} q_f(t) \\ p_f(t) \end{pmatrix} dt \geq 0$$

is satisfied for $p_f = \Delta_{\mathrm{le}} q_f$ for all $q_f \in \mathscr{L}_2$ and for all $\Delta_{\mathrm{le}} \in \boldsymbol{\Delta}_{\mathrm{le}}$. By Lemma 2.4 we have that

$$p_f = \Psi_{\mathrm{le}} p_e = \Delta_{\mathrm{le}} q_f \qquad (2.41)$$

for $p_e = \Delta_{\mathrm{re}} q_e$ and $q_f = \Psi_{\mathrm{re}} q_e$. If we denote by $\hat{p}_e$, $\hat{q}_e$, $\hat{p}_f$ and $\hat{q}_f$ the Fourier transforms of p_e, q_e, p_f and q_f respectively, then the latter IQC is equivalent to (Parseval)

$$\int_{-\infty}^{\infty} \begin{pmatrix} \hat{q}_f(i\omega) \\ \hat{p}_f(i\omega) \end{pmatrix}^{*} \begin{pmatrix} P_{11} & P_{12} \\ P_{12}^{\top} & P_{22} \end{pmatrix} \begin{pmatrix} \hat{q}_f(i\omega) \\ \hat{p}_f(i\omega) \end{pmatrix} d\omega \geq 0$$

for all $q_f \in \mathscr{L}_2$ and for all $\Delta_{\mathrm{le}} \in \boldsymbol{\Delta}_{\mathrm{le}}$. By now exploiting (2.41) we finally infer that

$$\int_{-\infty}^{\infty} \begin{pmatrix} \hat{q}_e(i\omega) \\ \hat{p}_e(i\omega) \end{pmatrix}^{*} \begin{pmatrix} \Psi_{\mathrm{re}} & 0 \\ 0 & \Psi_{\mathrm{le}} \end{pmatrix}^{*} \begin{pmatrix} P_{11} & P_{12} \\ P_{12}^{\top} & P_{22} \end{pmatrix} \begin{pmatrix} \Psi_{\mathrm{re}} & 0 \\ 0 & \Psi_{\mathrm{le}} \end{pmatrix} \begin{pmatrix} \hat{q}_e(i\omega) \\ \hat{p}_e(i\omega) \end{pmatrix} d\omega \geq 0$$

holds for all $q_e \in \mathscr{L}_2$ and for all $\Delta_{\mathrm{re}} \in \boldsymbol{\Delta}_{\mathrm{re}}$, which proves that the IQC (2.4) holds with (2.40) for all $\Delta_{\mathrm{re}} \in \boldsymbol{\Delta}_{\mathrm{re}}$.

∎

2.6.5 Time-delay uncertainties

In this section we present two IQC-multipliers for diagonally repeated LTI time-delay operators $\tilde{\Delta}$ defined by multiplication with $e^{-i\omega\alpha}I_{n_q}$ in the frequency domain for $0 \leq \alpha \leq \beta$. In order to apply the IQC-results for operators of this form we follow the standard procedure [43, 210, 178] and rearrange the feedback interconnection of Figure 2.1 as depicted in Figure 2.3.

Here W_d is a stable, but possibly non-proper transfer matrix, while $G := (I - \tilde{G})^{-1}\tilde{G}W_d$ is the new nominal plant, which is assumed to be proper and stable, and $\Delta := W_d^{-1}(\tilde{\Delta} - I)$ is a weighted and shifted version of of the original delay operator $\tilde{\Delta}$. We emphasize that we require $\tilde{G}$ to be stable as well in order to guarantee stability of the original uncertain interconnection. This is clear since

$$(I - G\Delta)^{-1} = \left(I - (I - \tilde{G})^{-1}\tilde{G}W_d W_d^{-1}(\tilde{\Delta} - I)\right)^{-1} =$$
$$= \left(I - (I - \tilde{G})^{-1}\tilde{G}(\tilde{\Delta} - I)\right)^{-1} = (I - \tilde{G}\tilde{\Delta})^{-1}(I - \tilde{G}).$$

2.6.5.1 Delay-multiplier 1

The first IQC-multiplier that we consider has been suggested by [86]. Here it is assumed that $W_d(s) = sI$ is the differentiation operator, such that Δ is defined by multiplication with $\frac{1}{i\omega}\left(e^{-i\omega\alpha} - 1\right)I_{n_q}$ in the frequency domain and confined to the set

$$\Delta_{\text{td1}} := \left\{ \tfrac{1}{i\omega}\left(e^{-i\omega\alpha} - 1\right)I_{n_q} : \alpha \in [0, \beta] \right\}.$$

For operators of this form we can employ the IQC-multiplier 2.2 as presented in Section 2.6.1. Then the IQC (2.4) holds with (2.20) for all $\alpha \in [0, \beta]$, if the FDI (2.21) or the corresponding LMI (2.23) holds true. To see this, we need to prove that $\Delta_{\text{td1}} \subset \Delta_{\text{lti,dyn,dr}}$ and hence that $\|\Delta\|_\infty \leq \alpha$ for all $\alpha \in [0, \beta]$ as is shown next.

Proof. Let us first observe that

$$\left(\tfrac{1}{i\omega}\left(e^{-i\omega\alpha} - 1\right)\right)^* \left(\tfrac{1}{i\omega}\left(e^{-i\omega\alpha} - 1\right)\right) =$$
$$= \tfrac{2}{\omega^2}\left(1 - \tfrac{e^{i\omega\alpha} + e^{-i\omega\alpha}}{2}\right) =$$
$$= \tfrac{2}{\omega^2}(1 - \cos(\alpha\omega)) = \alpha^2 \left(\tfrac{2}{\alpha\omega}\right)^2 \sin^2\left(\tfrac{\alpha\omega}{2}\right).$$

Now let $y = \frac{\alpha\omega}{2}$ and note that $f(y) := \left(\frac{\sin y}{y}\right)^2$ satisfies $f(y) \leq 1$ since $|\sin(y)| \leq |y|$ for all $y \in \mathbb{R}$ and $\lim_{y\to 0} f(y) = 1$. We conclude that $\|\Delta\|_\infty \leq \alpha$ for all $\alpha \in [0, \beta]$. ∎

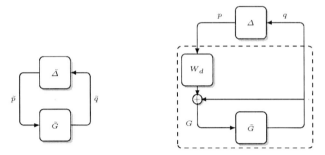

Fig. 2.3 Reformulated feedback interconnection for delay uncertainties

Remark 2.10. Note that the latter IQC-multiplier can as well be applied for arbitrarily fast time-varying time-delays by simply restricting the McMillan degree of ψ_ν to be zero. For more work on time-varying time-delays we refer the reader to [95, 97].

2.6.5.2 Delay-multiplier 2

The second IQC-multiplier has been reported in [173, 135] and is a refinement of the standard approach that is used in μ-analysis [122, 43, 210, 178]. Also here we reformulate the feedback interconnection of Figure 2.3, but now with $W_d = I$. Consequently, the new delay operator Δ is defined by multiplication with $\left(e^{-i\omega\alpha} - 1\right)I_{n_q}$ in the frequency domain and confined to the set

$$\Delta_{\mathrm{td2}} := \left\{ \left(e^{-i\omega\alpha} - 1\right)I_{n_q} : \alpha \in [0,\beta] \right\}.$$

Delay uncertainties of this particular form can be approximated by a generic frequency dependent stable weight

$$\psi_{\mathrm{td}}(s) = 2\frac{\left(s + \frac{2}{\beta\theta}\right)\left(s + \frac{\epsilon}{\beta}\right)}{\left(s^2 - 2\frac{\theta\cos(\theta^2)}{\beta}s + \frac{\theta^2}{\beta^2}\right)} \tag{2.42}$$

with $\theta = \frac{\pi}{2}$ and some small $\epsilon > 0$, in order to ensure that

$$|\Delta| \leq |\psi_{\mathrm{td}}| \ \text{ on } \ \mathbb{C}^0 \tag{2.43}$$

for all $\Delta \in \Delta_{\mathrm{td2}}$. This enables us to formulate the following IQC-multiplier.

IQC-Multiplier 2.8. *The IQC (2.4) holds with*

$$\Psi^* P \Psi := \begin{pmatrix} \star \\ \star \\ \star \end{pmatrix}^* \begin{pmatrix} P_{11} & 0 & 0 \\ 0 & 0 & P_{22} \\ 0 & P_{22} & P_{22} - P_{11} \end{pmatrix} \begin{pmatrix} \psi_\nu \psi_{\mathrm{td}} \otimes I_{n_q} & 0 \\ \psi_\nu \otimes I_{n_q} & 0 \\ 0 & \psi_\nu \otimes I_{n_q} \end{pmatrix} \quad (2.44)$$

for all $\Delta \in \boldsymbol{\Delta}_{\mathrm{td}2}$, if (2.21) and the following FDI are satisfied:

$$(\psi_\nu \otimes I_{n_q})^*(P_{22} - P_{11})(\psi_\nu \otimes I_{n_q}) \preccurlyeq 0 \text{ on } \mathbb{C}^0. \quad (2.45)$$

As before, the FDI (2.21) can be equivalently turned into the LMI (2.23). Similarly, the FDI (2.45) is satisfied if and only if there exists some $Y_\nu \in \mathbb{S}^{n_q \nu}$ for which the following LMI is satisfied:

$$\begin{pmatrix} I & 0 \\ A_\nu \otimes I_{n_q} & B_\nu \otimes I_{n_q} \\ C_\nu \otimes I_{n_q} & D_\nu \otimes I_{n_q} \end{pmatrix}^\top \begin{pmatrix} 0 & Y_\nu & 0 \\ Y_\nu & 0 & 0 \\ 0 & 0 & P_{22} - P_{11} \end{pmatrix} \begin{pmatrix} I & 0 \\ A_\nu \otimes I_{n_q} & B_\nu \otimes I_{n_q} \\ C_\nu \otimes I_{n_q} & D_\nu \otimes I_{n_q} \end{pmatrix} \preccurlyeq 0.$$
$$(2.46)$$

Proof. Observe that the IQC (2.4) for the multiplier (2.44) is satisfied if

$$(\psi_\nu \psi_{\mathrm{td}} \otimes I_{n_q})^* P_{11} (\psi_\nu \psi_{\mathrm{td}} \otimes I_{n_q}) - \Delta^* (\psi_\nu \otimes I_{n_q})^* P_{11} (\psi_\nu \otimes I_{n_q}) \Delta +$$
$$+ \Delta^* (\psi_\nu \otimes I_{n_q})^* P_{22} (\psi_\nu \otimes I_{n_q}) + (\psi_\nu \otimes I_{n_q})^* P_{22} (\psi_\nu \otimes I_{n_q}) \Delta +$$
$$+ \Delta^* (\psi_\nu \otimes I_{n_q}) P_{22} (\psi_\nu \otimes I_{n_q}) \Delta \succcurlyeq 0 \text{ on } \mathbb{C}^0$$

for all $\Delta \in \boldsymbol{\Delta}_{\mathrm{td}2}$. By exploiting the structure of Δ and $\psi_{\mathrm{td}} \otimes I_{n_q}$ we can rewrite this as

$$\left(\psi_{\mathrm{td}}^* \psi_{\mathrm{td}} \otimes I_{n_q} - \Delta^* \Delta \right) (\psi_\nu \otimes I_{n_q})^* P_{11} (\psi_\nu \otimes I_{n_q}) +$$
$$+ (\Delta^* + \Delta + \Delta^* \Delta)(\psi_\nu \otimes I_{n_q})^* P_{22} (\psi_\nu \otimes I_{n_q}) \succcurlyeq 0 \text{ on } \mathbb{C}^0 \quad (2.47)$$

for all $\Delta \in \boldsymbol{\Delta}_{\mathrm{td}2}$. Now observe that (2.43) translates into

$$\psi_{\mathrm{td}}^* \psi_{\mathrm{td}} \otimes I_{n_q} - \Delta^* \Delta \succcurlyeq 0 \text{ on } \mathbb{C}^0 \quad \forall \Delta \in \boldsymbol{\Delta}_{\mathrm{td}2}$$

Moreover, it is a matter of direct verification that

$$\Delta^* + \Delta + \Delta^* \Delta = 0 \text{ on } \mathbb{C}^0 \quad \forall \Delta \in \boldsymbol{\Delta}_{\mathrm{td}2}. \quad (2.48)$$

Hence, inequality (2.47) is satisfied for all $\Delta \in \boldsymbol{\Delta}_{\mathrm{td}2}$ if the FDI (2.21) or the corresponding LMI (2.23) holds true.

Finally, since $\boldsymbol{\Delta}_{\mathrm{td}2}$ does not satisfy Assumption 2.1, the IQC (2.4) must hold for all $\tau \boldsymbol{\Delta}_{\mathrm{td}2}$, $\tau \in [0,1]$. In accordance with Remark 2.3, this is simply achieved by enforcing $\Pi_{22} \preccurlyeq 0$ on $\mathbb{C}^0$ and, hence, by imposing the additional constraint (2.45). $\blacksquare$

2.6.6 Passive uncertainties/nonlinearities

Let us also consider the class of nonlinear bounded and causal operators $\Delta : \mathscr{L}_{2e}^{n_q} \to \mathscr{L}_{2e}^{n_q}$ with

$$\Delta \in \pmb{\Delta}_{\mathrm{ps}} := \left\{ \Delta : \langle q, \Delta_{\mathrm{ps}}(q) \rangle \geq 0 \;\; \forall q \in \mathscr{L}_2^{n_q} \right\}.$$

For operators from the set $\pmb{\Delta}_{\mathrm{ps}}$ we can use the following passivity-multiplier.

IQC-Multiplier 2.9. *For all* $\Delta \in \pmb{\Delta}_{\mathrm{ps}}$ *the IQC* (2.4) *is satisfied with*

$$\Psi^* P \Psi := \begin{pmatrix} I_{n_q} & 0 \\ 0 & I_{n_q} \end{pmatrix}^* \begin{pmatrix} 0 & I_{n_q} \\ I_{n_q} & 0 \end{pmatrix} \begin{pmatrix} I_{n_q} & 0 \\ 0 & I_{n_q} \end{pmatrix}. \tag{2.49}$$

The IQC-theorem hence covers the well-known passivity-theorem without requiring any additional proof (see e.g. [37]).

Proof. Evaluating (2.4) for the IQC-multiplier (2.49) yields the inequality

$$\int_{-\infty}^{\infty} \hat{q}(i\omega)^* \widehat{\Delta(q)}(i\omega) + \widehat{\Delta(q)}(i\omega)^* \hat{q}(i\omega) d\omega \geq 0 \;\; \forall q \in \mathscr{L}_2^{n_q},$$

which clearly holds if and only if $\Delta \in \pmb{\Delta}_{\mathrm{ps}}$.

∎

Although the passivity theorem allows us to analyze the robustness properties of complex system interconnections, the results are often conservative. This can be improved by means of the so-called Multiplier theorem (see e.g. [209, 37, 5, 7]). For this purpose, consider the more general nonlinear bounded and causal operator $\Delta : \mathscr{L}_{2e}^{n_q} \to \mathscr{L}_{2e}^{n_p}$ which we allow to vary in some set $\pmb{\Delta}_M$ with Assumption 2.1 as well as the following class of multipliers

$$\pmb{M} := \left\{ M \in \mathscr{RL}_{\infty}^{n_q \times n_p} : \langle q, M\Delta(q) \rangle \geq 0 \;\; \forall q \in \mathscr{L}_2^{n_q}, \;\; \forall \Delta \in \pmb{\Delta}_M \right\}.$$

Here we assume M to be parameterized as $M = (\psi_\nu \otimes I_{n_q}) P_{12} (\psi_\nu \otimes I_{n_p})$ with P_{12} being any matrix in $\mathbb{R}^{(n_q \nu) \times (n_p \nu)}$. Then we can consider the following IQC-multiplier.

IQC-Multiplier 2.10. *For all* $M \in \pmb{M}$ *with* $M = (\psi_\nu \otimes I_{n_q}) P_{12} (\psi_\nu \otimes I_{n_p})$ *and hence for all* $\Delta \in \pmb{\Delta}_M$ *the IQC* (2.4) *is satisfied with*

$$\Psi^* P \Psi := \begin{pmatrix} \psi_\nu \otimes I_{n_q} & 0 \\ 0 & \psi_\nu \otimes I_{n_p} \end{pmatrix}^* \begin{pmatrix} 0 & P_{12} \\ P_{12}^\top & 0 \end{pmatrix} \begin{pmatrix} \psi_\nu \otimes I_{n_q} & 0 \\ 0 & \psi_\nu \otimes I_{n_p} \end{pmatrix}. \tag{2.50}$$

The proof is identical to the one from above. We conclude that the IQC-theorem seamlessly generalizes the Multiplier theorem. As discussed in [116], the conditions on M in the IQC-framework are much weaker than the ones required in e.g. [209, 37, 5, 7] (i.e. the factorizability of M as $M_1^* M_2$ with M_1, M_1^{-1}, M_2 and M_2^{-1} all being stable). The price to be paid is the mild restriction that Δ_M satisfies Assumption 2.1.

Remark 2.11. One can show that, as long as Π_{22} in (2.3) satisfies the additional constraint $\Pi_{22} \preccurlyeq 0$ on $\mathbb{C}^0$, the IQC-theorem is actually equivalent to the Multiplier theorem. We refer the reader to [46, 29, 104] for a proof and further details.

2.6.7 Norm-bounded uncertainties/nonlinearities

Let us recall the IQC-multiplier for norm-bounded uncertainties from Example 2.1 and discuss it in a little more detail. For this purpose, consider the (possibly) nonlinear operator $\Delta : \mathscr{L}_{2e}^{n_q} \to \mathscr{L}_{2e}^{n_p}$, which we assume to be confined to the set

$$\Delta_{\mathrm{sg}} := \{\Delta : \|\Delta\|_{\mathrm{i2}} \le \alpha\}.$$

Here α is some fixed non-negative constant. Then we can consider the following IQC-multiplier.

IQC-Multiplier 2.11. *For all* $\Delta \in \Delta_{\mathrm{sg}}$ *the IQC* (2.4) *is satisfied with*

$$\Psi^* P \Psi := \begin{pmatrix} \alpha I_{n_q} & 0 \\ 0 & I_{n_p} \end{pmatrix}^* \begin{pmatrix} I_{n_q} & 0 \\ 0 & -I_{n_p} \end{pmatrix} \begin{pmatrix} \alpha I_{n_q} & 0 \\ 0 & I_{n_p} \end{pmatrix}. \tag{2.51}$$

The proof is identical to the one shown in Example 2.1. The extra message of this section is that we recover the well-known small-gain theorem without requiring any additional proof (see e.g. [37]). Further we would like to stress that, although the small-gain theorem allows us to analyze the robustness properties of highly complex system interconnections, it might not be of to much use in practice. In fact, only imposing a norm-bound on the operator Δ might lead to overly conservative results, especially in case that Δ has structure. On the other hand, conservatism can be reduced, even in the unstructured case, by applying the IQC-multiplier 2.1 for norm-bounded LTI uncertainties. Indeed, the IQC (2.4) also holds with (2.16) for all $\Delta \in \Delta_{\mathrm{sg}}$ if the FDI (2.17) or the corresponding LMI (2.19) holds true for $\nu = 0$ (i.e. static multipliers). Moreover, if the integrand of the resulting IQC is non-negative for all $\omega \in \mathbb{R} \cup \{\infty\}$, it is even possible to allow for $\nu > 0$ (i.e. dynamic multipliers). The proof is identical to the one in Section 2.6.1.

2.6.8 Sector bounded and slope-restricted nonlinearities

In many applications it is essential to analyze systems that are affected by nonlinearities in the control channel. For example, consider the feedback interconnection in Figure 2.4. Here

$$\tilde{p}(t) := \mathrm{sat}(\tilde{q}(t)) = \min\{|\tilde{q}(t)|, \alpha\} \mathrm{sign}(\tilde{q}(t)), \quad \tilde{q} \in \mathscr{L}_2, \quad \alpha > 0,$$

is a saturation nonlinearity that expresses the limitations of the actuator in the control system. In order to analyze such a feedback interconnection within the IQC-framework, we need to reformulate the feedback interconnection as shown in Figure 2.5. Here we 'pulled out' the saturation nonlinearity in order to be able to consider a nominal plant $G := (I - \tilde{K}\tilde{G})^{-1}\tilde{K}\tilde{G}$, which we, as usual, require to be stable. This is done by transforming the saturation nonlinearity into a dead-zone nonlinearity

$$p(t) := \mathrm{dzn}(q(t)) = q(t) - \mathrm{sat}(q(t)), \quad q \in \mathscr{L}_2.$$

Nonlinearities such as saturation and dead-zones can be identified as so-called sector bounded and slope restricted nonlinearities, whose properties can be captured with sector conditions and Zames-Falb multipliers as discussed next. We emphasize that most parts in this subsection have already appeared in our recent paper [198].

2.6.8.1 Circle criteria

Consider the nonlinear operator $p = \Delta(q) = \mathrm{col}(\Delta_1(q_1), \ldots, \Delta_k(q_k))$ defined by $p(t) = \Delta(q)(t) := \tilde{\Delta}(q(t))$ for $t \geq 0$ and let the nonlinearities $\tilde{\Delta}_i : \mathbb{R} \to \mathbb{R}$ satisfy $\tilde{\Delta}_i(0) = 0$ as well as the sector constraints

$$\left(\tilde{\Delta}_i(\tilde{q}) - \alpha_i \tilde{q}\right)\left(\beta_i \tilde{q} - \tilde{\Delta}_i(\tilde{q})\right) \geq 0 \tag{2.52}$$

for all $\tilde{q} \in \mathbb{R}$ with fixed non-negative constants $\alpha_i \leq 0 \leq \beta_i$. For the described nonlinearities we can consider the following IQC-multiplier.

IQC-Multiplier 2.12. *The IQC* (2.4) *holds with*

$$\Psi^* P \Psi := \begin{pmatrix} \star \\ \star \end{pmatrix}^* \begin{pmatrix} 0 & \mathrm{diag}(p_1, \ldots, p_k) \\ \star & 0 \end{pmatrix} \begin{pmatrix} -\mathrm{diag}(\alpha_1, \ldots, \alpha_k) & I_{n_q} \\ \mathrm{diag}(\beta_1, \ldots, \beta_k) & -I_{n_q} \end{pmatrix} \tag{2.53}$$

for $\Delta(q)$ *with* $k = n_q$ *if* $p_i \geq 0$, $i = 1, \ldots, k$.

Proof. Observe that

$$2p_i \int_0^\infty \left(\tilde{\Delta}_i(q_i(t)) - \alpha_i q_i(t)\right)\left(\beta_i q_i(t) - \tilde{\Delta}_i(q_i(t))\right) dt \geq 0, \quad i = 1, \ldots, k,$$

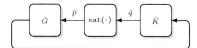

Fig. 2.4 Feedback interconnection with a saturation nonlinearity

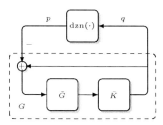

Fig. 2.5 Reformulated feedback interconnection for saturation nonlinearities

holds for all $q_i \in \mathscr{L}_2$, for all $p_i \geq 0$ and for the given constants $\alpha_i \leq 0 \leq \beta_i$, $i = 1, \ldots, k$. Direct computations reveal that the latter inequality can be identically written as

$$\int_0^\infty \begin{pmatrix} q_i(t) \\ \tilde{\Delta}_i(q_i(t)) \end{pmatrix}^\top \begin{pmatrix} -2\alpha_i\beta_i p_i & (\alpha_i + \beta_i)p_i \\ (\alpha_i + \beta_i)p_i & -2p_i \end{pmatrix} \begin{pmatrix} q_i(t) \\ \tilde{\Delta}_i(q_i(t)) \end{pmatrix} dt \geq 0, \quad i = 1, \ldots, k.$$

We conclude that the IQC (2.4) is satisfied with (2.53) for $\Delta(q)$ with $k = n_q$.

2.6.8.2 Full-block multiplier circle criteria

Observe that in the latter section we have constructed diagonal multipliers which share their structure with the nonlinearity $p = \Delta(q) = \mathrm{col}(\Delta_1(q_1), \ldots, \Delta_k(q_k))$. It is interesting to note that indirectly described full-block multipliers can as well be employed, with the benefit of reducing conservatism. In fact, with $\Theta(\delta) := \mathrm{diag}(\delta_1, \ldots, \delta_k)$ and the set of LMI descriptions

$$\boldsymbol{P}_{\mathrm{sc}} = \left\{ P \in \mathbb{S}^{2k} : \begin{pmatrix} I \\ \Theta(\delta) \end{pmatrix}^\top P \begin{pmatrix} I \\ \Theta(\delta) \end{pmatrix} \succcurlyeq 0, \quad \delta_i \in [\alpha_i, \beta_i] \ \forall i = 1, \ldots, k \right\},$$

we arrive at the following result.

IQC-Multiplier 2.13. *If $P \in \boldsymbol{P}_{\mathrm{sc}}$ then the IQC (2.4) holds with*

$$\Psi^* P \Psi := \begin{pmatrix} I_{n_q} & 0 \\ 0 & I_{n_p} \end{pmatrix}^* \begin{pmatrix} P_{11} & P_{12} \\ P_{12}^\top & P_{22} \end{pmatrix} \begin{pmatrix} I_{n_q} & 0 \\ 0 & I_{n_p} \end{pmatrix} \tag{2.54}$$

for $\Delta(q)$ with $k = n_q$.

Proof. Choose any $\tilde{q} \in \mathbb{R}^k$ and suppose that $\delta_i = \frac{\tilde{\Delta}_i(\tilde{q}_i)}{\tilde{q}_i} \in [\alpha_i, \beta_i]$ for all $\tilde{q}_i \neq 0$ and any $\delta_i \in [\alpha_i, \beta_i]$ if $\tilde{q}_i = 0$. We then infer that $\delta_i \tilde{q}_i = \tilde{\Delta}_i(\tilde{q}_i)$ for all $i = 1, \dots, k$, which in turn implies

$$\begin{pmatrix} I \\ \Theta(\delta) \end{pmatrix} \tilde{q} = \begin{pmatrix} I\tilde{q} \\ \Theta(\delta)\tilde{q} \end{pmatrix} = \begin{pmatrix} \tilde{q} \\ \tilde{\Delta}(\tilde{q}) \end{pmatrix}.$$

Hence, we can infer that

$$0 \preccurlyeq \tilde{q}^\top \begin{pmatrix} I \\ \Theta(\delta) \end{pmatrix}^\top P \begin{pmatrix} I \\ \Theta(\delta) \end{pmatrix} \tilde{q} = \begin{pmatrix} \tilde{q} \\ \tilde{\Delta}(\tilde{q}) \end{pmatrix}^\top P \begin{pmatrix} \tilde{q} \\ \tilde{\Delta}(\tilde{q}) \end{pmatrix}.$$

This implies that the IQC (2.4) holds with (2.54) for $\Delta(q)$ with $k = n_q$.
∎

These ideas point towards a multitude of extensions which require only minor modifications of the arguments. Here is a selection of some possibilities:

- All given results remain valid for sector-bounded nonlinearities that explicitly depend on time.
- Full-block multiplier stability results are easy to derive for possibly non-diagonal nonlinearities. One can even further expand to nonlinear feedbacks $\Delta := \mathbb{R}^l \times [0, \infty) \to \mathbb{R}^k$ that satisfy the multi-variable sector constraints

$$\left(\tilde{\Delta}(\tilde{q}, t) - P_{1,j}\tilde{q} \right)^\top \left(P_{2,j}\tilde{q} - \tilde{\Delta}(\tilde{q}, t) \right) \geq 0$$

for all $(\tilde{q}, t) \in \mathbb{R}^l \times [0, \infty)$, $j = 1, \dots, m$ with $P_{1,j} \in \mathbb{R}^{l \times k}$ and $P_{2,j} \in \mathbb{R}^{l \times k}$. Stability can be assured based on the LMI-class of multipliers

$$\mathrm{co} \left\{ \begin{pmatrix} -P_{1,j}^\top P_{2,j} - P_{2,j}^\top P_{1,j} & P_{1,j} + P_{2,j} \\ P_{1,j}^\top + P_{2,j}^\top & -2I \end{pmatrix} : j = 1, \dots, m \right\}.$$

2.6.8.3 Zames-Falb multipliers

As another useful IQC-multiplier for sector-bounded and slope-restricted nonlinearities, we present in this section the Zames-Falb multiplier, which was initially introduced by [209]. It is well-known how to construct an asymptotically exact parameterization of Zames-Falb multipliers for general monotone nonlinearities [26]. On the other hand, for odd-monotone nonlinearities in continuous-time, the parameterization given in [26] has only been shown to be a subset of the set of all possible multipliers. As a the contribution, and as first reported in [198], we can close this gap and show that the given parameterization is actually asymptotically tight. For this purpose, consider the operator $p = \Delta(q)$ defined as $p(t) = \tilde\Delta(q)(t)$ for $t \geq 0$, $q \in \mathscr{L}_2$, where $\tilde\Delta : \mathbb{R} \to \mathbb{R}$ satisfies $\tilde\Delta(0) = 0$ as well as the incremental sector condition

$$\alpha \leq \frac{\tilde\Delta(\tilde q_1) - \tilde\Delta(\tilde q_2)}{\tilde q_1 - \tilde q_2} \leq \sup_{\tilde q_a, \tilde q_b \in \mathbb{R}, \tilde q_a \neq \tilde q_b} \frac{\tilde\Delta(\tilde q_a) - \tilde\Delta(\tilde q_b)}{\tilde q_a - \tilde q_b} < \beta \quad \forall \tilde q_1, \tilde q_2 \in \mathbb{R}, \ \tilde q_1 \neq \tilde q_2 \quad (2.55)$$

with fixed constants $\alpha \leq 0 \leq \beta$. The latter assures that $\tau\tilde\Delta$ shares these properties with $\tilde\Delta$ for all $\tau \in [0,1]$. This leads to the following well-known Zames-Falb multiplier, in which $\hat h$ denotes the Fourier-transform of h [209].

IQC-Multiplier 2.14. *Suppose that $\tilde\Delta : \mathbb{R} \to \mathbb{R}$ with $\tilde\Delta(0) = 0$ satisfies the incremental sector condition (2.55) for $\alpha \leq 0 \leq \beta$. If the non-negative function $h \in \mathscr{L}_1$ and $p_0 \in \mathbb{R}$ are related as $\|h\|_1 < p_0$ then the IQC (2.4) holds with*

$$\Pi(i\omega) := \begin{pmatrix} -\alpha & 1 \\ \beta & -1 \end{pmatrix}^\top \begin{pmatrix} 0 & p_0 - \hat h(i\omega) \\ p_0 - \hat h(i\omega)^* & 0 \end{pmatrix} \begin{pmatrix} -\alpha & 1 \\ \beta & -1 \end{pmatrix} \quad (2.56)$$

for all $\tau\Delta$ with $\tau \in [0,1]$. If δ is odd, (2.4) remains true even if h is not sign constrained.

For a proof we refer the reader to [209].

For computations we parameterize h as follows. We choose A_ν, B_ν as in (2.18) and note that

$$Q_\nu(t) := e^{A_\nu t} B_\nu = e^{\rho t} \varphi_\nu(t), \quad \varphi_\nu(t) = S_\nu(t) f_\nu(t)$$

where

$$S_\nu(t) := \begin{pmatrix} \frac{\mathrm{sgn}(t)^0}{0!} & & & & 0 \\ & \frac{\mathrm{sgn}(t)^1}{1!} & & & \\ & & \frac{\mathrm{sgn}(t)^2}{2!} & & \\ & & & \ddots & \\ 0 & & & & \frac{\mathrm{sgn}(t)^{\nu-1}}{(\nu-1)!} \end{pmatrix} \quad \text{and} \quad f_\nu(t) := \begin{pmatrix} 1 \\ |t| \\ |t|^2 \\ \vdots \\ |t|^{\nu-1} \end{pmatrix}.$$

Subsequently, with the free parameters $p_1, p_2, p_3, p_4 \in \mathbb{R}^{1 \times \nu}$, the impulse response of $h \in \mathscr{L}_1$ is taken as

$$(p_1 - p_2)Q_\nu(-t) \ \text{for} \ t < 0 \quad \text{and} \quad (p_3 - p_4)Q_\nu(t) \ \text{for} \ t \geq 0 \quad (2.57)$$

under the positivity constraints

$$p_1\varphi_\nu(-t) > 0, \quad p_2\varphi_\nu(-t) > 0, \quad p_3\varphi_\nu(t) > 0, \quad p_4\varphi_\nu(t) > 0 \ \text{ for } \ t \geq 0. \tag{2.58}$$

With (2.11) this leads to

$$\hat{h}(i\omega) := \psi_\nu^*(i\omega)\left(0_{1\times 1} \ (p_1 - p_2)\right)^\top + \left(0_{1\times 1} \ (p_3 - p_4)\right)\psi_\nu(i\omega)$$

which is strictly proper. This leads to the following parameterized version of the Zames-Falb multiplier 2.14.

IQC-Multiplier 2.15. *Suppose again that* $\tilde{\Delta} : \mathbb{R} \to \mathbb{R}$ *with* $\tilde{\Delta}(0) = 0$ *satisfies the incremental sector condition* (2.55) *for* $\alpha \leq 0 \leq \beta$ *and let the non-negative function* $h \in \mathscr{L}_1$ *be parameterized as above. If* h *and* $p_0 \in \mathbb{R}$ *are related as* $\|h\|_1 < p_0$ *then the IQC* (2.4) *holds with*

$$\Psi^* P\Psi := \begin{pmatrix} -\alpha\psi_\nu & \psi_\nu \\ \beta\psi_\nu & -\psi_\nu \end{pmatrix}^* \begin{pmatrix} 0 & P_{12} \\ P_{12}^\top & 0 \end{pmatrix} \begin{pmatrix} -\alpha\psi_\nu & \psi_\nu \\ \beta\psi_\nu & -\psi_\nu \end{pmatrix} \tag{2.59}$$

and

$$P_{12} = \begin{pmatrix} p_0 & p_2 - p_1 \\ p_4^\top - p_3^\top & 0 \end{pmatrix}$$

for all $\tau\Delta$ *with* $\tau \in [0,1]$. *If* $\tilde{\Delta}$ *is odd,* (2.4) *remains true even if* h *is not sign constrained.*

Let us now show how the constraint $\|h\|_1 < p_0$ and the polynomial inequalities (2.58) can be turned into LMIs. Since (2.58) implies $p_1 Q_\nu(-t), p_2 Q_\nu(-t) \geq 0$ for $t < 0$, we infer $|h(t)| \leq (p_1 + p_2)Q_\nu(-t)$ for $t < 0$ and, similarly, $|h(t)| \leq (p_3 + p_4)Q_\nu(t)$ for $t \geq 0$. Therefore, the constraint $\|h\|_1 < p_0$ is implied by

$$-(p_1 + p_2 + p_3 + p_4)A_\nu^{-1}B_\nu < p_0, \tag{2.60}$$

which is just a linear constraint in the free parameters p_1, p_2, p_3, p_4. On the other hand, the polynomial inequalities (2.58) on the real positive half-line admit an LMI description (see e.g. [26, 153]). Following [26] we can rewrite (2.58) as

$$\sum_{k=1}^{\nu-1} p_{ij} t^k > 0 \quad \forall t \geq 0$$

where p_{ij} is the jth element of $p_i S_\nu(-t)$ for $i = 1, 2$ and $p_i S_\nu(t)$ for $i = 3, 4$. With the simple variable substitution $t = (-i\omega)(i\omega)$ we can equivalently re-express the latter constraints as

$$\sum_{k=1}^{\nu-1} p_{ij}(-i\omega)^k(i\omega)^k > 0 \ \text{ on } \ \mathbb{C}^0,$$

which clearly persists to hold if and only if

$$\sum_{k=1}^{\nu-1} \frac{p_{ij}(-i\omega)^k(i\omega)^k}{(-i\omega-\rho)^{\nu-1}(i\omega-\rho)^{\nu-1}} > 0 \text{ on } \mathbb{C}^0.$$

It is finally not hard to see that this can be written as a standard FDI

$$\tilde{\psi}_\nu^* P_{i,\nu} \tilde{\psi}_\nu \succ 0 \text{ on } \mathbb{C}^0 \tag{2.61}$$

for

$$P_{i,\nu} = \begin{pmatrix} p_{i1} & & 0 \\ & \ddots & \\ 0 & & p_{i(\nu-1)} \end{pmatrix}, \quad i = 1,2,3,4 \quad \text{and} \quad \tilde{\psi}_\nu = \begin{pmatrix} \frac{i\omega}{(i\omega-\rho)^{\nu-1}} \\ \vdots \\ \frac{(i\omega)^{\nu-1}}{(i\omega-\rho)^{\nu-1}} \end{pmatrix}.$$

It is thus possible to turn the polynomial inequalities (2.58) into finite dimensional LMIs. By choosing the minimal realization

$$\tilde{\psi}_\nu = \left[\begin{array}{c|c} \tilde{A}_\nu & \tilde{B}_\nu \\ \hline \tilde{C}_\nu & \tilde{D}_\nu \end{array} \right] = \left[\begin{array}{cccccccc|c} \rho & 0 & \cdots & \cdots & \cdots & \cdots & 0 & & 1 \\ 1 & \ddots & \ddots & \ddots & \ddots & \ddots & & \vdots & 0 \\ 0 & \ddots & \ddots & \ddots & \ddots & \ddots & & \vdots & \vdots \\ \vdots & \ddots & \ddots & \ddots & \ddots & \ddots & & \vdots & \vdots \\ \vdots & \ddots & \ddots & \ddots & \ddots & \ddots & & \vdots & \vdots \\ \vdots & \ddots & \ddots & \ddots & \ddots & \ddots & 0 & & \vdots \\ 0 & \cdots & \cdots & \cdots & 0 & 1 & \rho & & 0 \\ \hline 0 & \cdots & \cdots & \cdots & \cdots & 0 & c_0^0 & & 0 \\ \vdots & \cdot\cdot & \cdot\cdot & \cdot\cdot & \cdot\cdot & & \vdots & & \vdots \\ \vdots & \cdot\cdot & \cdot\cdot & \cdot\cdot & \cdot\cdot & & \vdots & & \vdots \\ \vdots & \cdot\cdot & \cdot\cdot & \cdot\cdot & \cdot\cdot & & \vdots & & \vdots \\ 0 & c_0^{\nu-3} & \cdots & & c_k^{\nu-3} & \cdots & c_{\nu-3}^{\nu-3} & & \vdots \\ c_0^{\nu-2} & \cdots & & c_k^{\nu-2} & \cdots & \cdots & c_{\nu-2}^{\nu-2} & & 0 \\ c_1^{\nu-1} & \cdots & c_k^{\nu-1} & \cdots & \cdots & \cdots & c_{\nu-1}^{\nu-1} & & 1 \end{array} \right],$$

where

$$c_k^{\nu-n} := \frac{(\nu-n)!}{k!(\nu-n-k)!}(-\rho)^k,$$

we can infer by the KYP-lemma that the FDIs (2.61) are equivalent to the existence of some symmetric matrices $X_{i,\nu} \in \mathbb{S}^{\nu-1}$, $i = 1,2,3,4$ such that the following LMIs hold true:

$$
\begin{pmatrix} I & 0 \\ \tilde{A}_\nu & \tilde{B}_\nu \\ \tilde{C}_\nu & \tilde{D}_\nu \end{pmatrix}^\top \begin{pmatrix} 0 & X_{i,\nu} & 0 \\ X_{i,\nu} & 0 & 0 \\ 0 & 0 & P_{i,\nu} \end{pmatrix} \begin{pmatrix} I & 0 \\ \tilde{A}_\nu & \tilde{B}_\nu \\ \tilde{C}_\nu & \tilde{D}_\nu \end{pmatrix} \succcurlyeq 0, \quad i = 1,2,3,4. \tag{2.62}
$$

This parameterization is equivalent to the one in [26]. However, it is a new insight that it is tight in the following sense.

Theorem 2.2. *There exist a $p_0 > 0$ and an $h \in \mathscr{L}_1$ with $\|h\|_1 < p_0$ that satisfy (2.4) with (2.56) for all $\tau\Delta$, $\tau \in [0,1]$ if and only if there exists a positive integer ν and $p_1, p_2, p_3, p_4 \in \mathbb{R}^{1\times\nu}$ with (2.58) and (2.60) that satisfy (2.4) with (2.59) for all $\tau\Delta$, $\tau \in [0,1]$.*

Proof. For fixed $\rho < 0$ it is well-known that $e^{\rho t} f_{\mathrm{p}}(t)$ with a polynomial f_{p} can approximate functions in $\mathscr{L}_1[0,\infty)$ and $\mathscr{L}_2$ arbitrarily closely [181]. Let us show that non-negative functions in $\mathscr{L}_1[0,\infty)$ can be approximated in this fashion with positive polynomials f_{p}. We owe the proof of this fact to [129].

Lemma 2.5. *Let $h \in \mathscr{L}_1[0,\infty)$ be non-negative, i.e. $h(t) \geq 0$ for $t \geq 0$. Then for all $\epsilon > 0$ there exists a polynomial f_{p} such that $f_{\mathrm{q}}(t) = e^{\rho t} f_{\mathrm{p}}(t)$ satisfies $\|h - f_{\mathrm{q}}\|_1 < \epsilon$ and $f_{\mathrm{p}}(t) > 0$ for all $t \geq 0$.*

Proof. Define the function $f(t) = \sqrt{h(t)}$ which is in $\mathscr{L}_2$. Then there exist some polynomial $\tilde{f}_{\mathrm{p}}$ such that the $\mathscr{L}_2$-norm of $g_-(t) := f(t) - \tilde{f}_{\mathrm{p}}(t)e^{\rho t/2}$ is smaller than $h_-(t) := \sqrt{\epsilon + \|f\|_2^2} - \|f\|_2$. Then the $\mathscr{L}_2$-norm of $g_+(t) := f(t) + \tilde{f}_{\mathrm{p}}(t)e^{\rho t/2}$ is not larger than $h_+(t) := \sqrt{\epsilon + \|f\|_2^2} + \|f\|_2$. Therefore

$$
\int_0^\infty |h(t) - \tilde{p}(t)^2 e^{\rho t}|\,dt = \int_0^\infty |g_-(t)||g_+(t)|\,dt \leq
$$

$$
\leq \sqrt{\int_0^\infty |g_-(t)|^2\,dt} \sqrt{\int_0^\infty |g_+(t)|^2\,dt} < h_-(t)h_+(t) = \epsilon.
$$

For sufficiently small $\tilde{\epsilon}$ this persists to hold for $f_{\mathrm{p}} := \tilde{f}_{\mathrm{p}}^2 + \tilde{\epsilon}$ which is positive.

∎

To continue the proof of Theorem 2.2, we only need to show 'only if'. Given $h \in \mathscr{L}_1$, there exists some $\delta > 0$ such that (2.4) with (2.56) persists to hold for all other functions in $\mathscr{L}_1$, whose Fourier transform deviates by at most δ from h in the $\mathscr{L}_\infty$-norm over frequency. Now define the non-negative functions $h_+(t) = \max\{h(t), 0\}$ and

$h_-(t) = -\min\{0, h(t)\}$ in $\mathscr{L}_1$. We have $h = h_+ - h_-$. Choose $\epsilon > 0$ with $\|h\|_1 + 4\epsilon < p_0$ and $4\epsilon < \delta$. By Lemma 2.5 there exists ν and p_1, p_2, p_3, p_4 with (2.58) such that the $\mathscr{L}_1$-norms of $h_+(t) - p_1 Q_\nu(-t)$, $h_-(t) - p_2 Q_\nu(-t)$ on $(-\infty, 0]$ and those of $h_+(t) - p_3 Q_\nu(t)$, $h_-(t) - p_4 Q_\nu(t)$ on $[0, \infty)$ are smaller than ϵ. Now define H by (2.57). We then infer by the triangle inequality that $\|H - h\|_1 \leq 4\epsilon$. In turn, this implies that the $\mathscr{L}_\infty$-norm of $\hat{H} - \hat{h}$ is bounded by $4\epsilon < \delta$ such that (2.4) with (2.56) still holds for H. Finally, we also have $\|H\|_1 \leq \|H - h\|_1 + \|h\|_1 \leq 4\epsilon + \|h\|_1 < p_0$ which implies that (2.60) is true as well.

∎

As shown in [26], the same result holds with additional constraint $h(t) > 0$ for all $t \in \mathbb{R}$ if setting $p_2 = 0$ and $p_4 = 0$. Both results taken together lead to asymptotically tight convex parameterizations of the set of all Zames-Falb multipliers.

Remark 2.12. For extensions to repeated and/or MIMO monotone nonlinearities we refer the reader to [145, 31, 105, 112]. Other related results are found in e.g. [140], [114, 75] and [81, 83], which respectively address a generalization of the Zames-Falb multiplier for friction nonlinearities, an IQC-multiplier for quasi-concave and hysteresis nonlinearities, and, a scenario where the nominal system is allowed to be unstable.

2.6.9 Popov multipliers

We conclude this section by briefly considering the so-called Popov multipliers. Popov multipliers are IQC-multipliers that are structured as

$$\Psi^* P \Psi := \begin{pmatrix} sI_{n_q} & 0 \\ 0 & I_{n_q} \end{pmatrix}^* \begin{pmatrix} 0 & P_2 \\ P_2^\top & 0 \end{pmatrix} \begin{pmatrix} sI_{n_q} & 0 \\ 0 & I_{n_q} \end{pmatrix}. \qquad (2.63)$$

Here $P_2 \in \mathbb{S}^{n_q}$ is a free matrix variable from some LMIable set $\boldsymbol{P}_2$ and sI_{n_q} a non-proper transfer matrix. Popov multipliers allow us to consider IQCs of the form

$$2 \int_0^\infty \Delta(q)(t)^\top P_2 \dot{q}(t) dt \geq -\epsilon |q_0|^2 \quad \forall q, \dot{q} \in \mathscr{L}_2^{n_q}, \qquad (2.64)$$

which should hold for all $\Delta \in \boldsymbol{\Delta}$ and for some $\epsilon \geq 0$. Here $q_0 := C x_0$ is a given vector defined by the initial condition $x(0) := x_0$ of the nominal system $G := (A, B, C, D)$.

Popov multipliers can be directly combined with standard IQC-multipliers as

$$\Psi^* P \Psi := \begin{pmatrix} \Psi_{11} & \Psi_{12} \\ sI_{n_q} & 0 \\ 0 & I_{n_q} \end{pmatrix}^* \begin{pmatrix} P_1 & 0 & 0 \\ 0 & 0 & P_2 \\ 0 & P_2^\top & 0 \end{pmatrix} \begin{pmatrix} \Psi_{11} & \Psi_{12} \\ sI_{n_q} & 0 \\ 0 & I_{n_q} \end{pmatrix}, \qquad (2.65)$$

with P_1 being a free matrix variable from an LMIable set $\boldsymbol{P}_1$ and $\Psi = \left(\begin{array}{cc} \Psi_{11} & \Psi_{12} \end{array} \right) \in \mathscr{RH}_\infty^{\bullet \times 2n_q}$ being some fixed transfer matrix. This yields the extended IQC

$$IQC\left(\Psi^* P_1 \Psi, q, \Delta(q)\right) + 2 \int_0^\infty \Delta(q)(t)^\top P_2 \dot{q}(t) dt \geq -\epsilon |q_0|^2 \quad \forall q, \dot{q} \in \mathscr{L}_2^{n_q}, \quad (2.66)$$

which should be satisfied for all $P_1 \in \boldsymbol{P}_1$, $P_2 \in \boldsymbol{P}_2$ and $\Delta \in \boldsymbol{\Delta}$ and for some $\epsilon \geq 0$. Note that the outer-factor Ψ of the IQC-multiplier (2.65) is a non-proper transfer matrix. This requires the nominal plant $G \in \mathscr{RH}_\infty^{n_q \times n_p}$ to satisfy the additional property

$$G(\infty) = 0. \quad (2.67)$$

For a proof of the IQC-theorem with Popov-multipliers and non-zero initial conditions, a slight relaxation of the constraint (2.67) and further readings, we refer the reader to [74], [85] and [88, 78] respectively.

Like standard IQC-multipliers, also Popov multipliers are very useful to describe properties of uncertainties and/or nonlinearities. Moreover, they can also easily be combined with other IQC-multipliers as discussed in Remark 2.2 in order to reduce conservatism. Let us briefly explore the possibilities by means of three elementary examples that were taken from [74].

Recall the nonlinear operator $p = \Delta(q)$ defined by $p(t) = \Delta(q)(t) := \tilde{\Delta}(q(t))$ from Section 2.6.8.1, which satisfies the sector constraint (2.52) for fixed constants $\alpha \leq 0 \leq \beta$. Then we can consider the following alternative IQC-multiplier for sector bounded nonlinearities.

IQC-Multiplier 2.16. *The IQC* (2.64) *holds with* (2.63) *for* $\Delta(q)$ *and* $P_2 \in \mathbb{R}$.

Proof. The claim is proved by observing that

$$2P_2 \lim_{\zeta \to \infty} \int_0^\zeta \tilde{\Delta}(q)(t) \dot{q}(t) dt = 2P_2 \lim_{\zeta \to \infty} \int_{q(0)}^{q(\zeta)} \tilde{\Delta}(\xi) d\xi = -2P_2 \int_0^{q_0} \tilde{\Delta}(\xi) d\xi \geq -\epsilon |q_0|^2$$

holds for all $q, \dot{q} \in \mathscr{L}_2^{n_q}$, for the given constants $\alpha \leq 0 \leq \beta$ and for $\epsilon = |\alpha P_2|$ if $P_2 > 0$ and for $\epsilon = |\beta P_2|$ if $P_2 < 0$. Here we exploited the fact that $q, \dot{q} \in \mathscr{L}_2$ imply that $q(t) \to 0$ as $t \to \infty$ (see e.g. [37]).   ■

Let us also recall the class of LTI real diagonally repeated operators $\Delta \in \boldsymbol{\Delta}_{\text{lti,re,dr}}$ from Section 2.6.3.1. Then we can employ the Popov multiplier (2.63) as follows.

IQC-Multiplier 2.17. *The IQC* (2.64) *holds with* (2.63) *for all* $\Delta \in \boldsymbol{\Delta}_{\text{lti,re,dr}}$ *and for all* $P_2 \in \mathbb{S}^{n_q}$.

Proof. Observe that

$$2\delta \int_0^\infty q(t)P_2\dot{q}(t)dt = \delta \cdot \lim_{t\to\infty} \left[q(t)^\top P_2 q(t) \right]_0^t = -\delta q_0^\top P_2 q_0 \geq -\epsilon |q_0|^2 \quad \forall q,\dot{q} \in \mathscr{L}_2^{n_q}$$

holds for all $\delta \in [-1,1]$ and for $\epsilon = \bar{\sigma}(P_2)^2$. Here $\bar{\sigma}(P_2)$ denotes the largest singular value of P_2. Hence, the IQC (2.64) is satisfied for all $\Delta \in \mathbf{\Delta}_{\mathrm{lti,re,dr}}$.

∎

As a final example, let us recall from Section 2.6.3 the class of LTV parametric uncertainties $\Delta \in \mathbf{\Delta}_{\mathrm{ltv,rb}}$ that have a bounded rate-of-variation. Then we can employ the extended Popov multiplier (2.65) as follows.

IQC-Multiplier 2.18. *The IQC (2.66) holds with (2.65) for all $\Delta \in \mathbf{\Delta}_{\mathrm{ltv,rb}}$ if $\mathbb{S}^{n_q} \ni P_2 \succcurlyeq 0$, $P_1 = \mathrm{diag}(\beta P_2, 0_{n_q \times n_q})$ and $\Psi = \left(\begin{array}{cc} \Psi_{11} & \Psi_{12} \end{array} \right) = I_{2n_q}$.*

An improved version of this IQC is found in [76].

Proof. Evaluating the IQC (2.66) for the IQC-multiplier (2.65) yields the inequality

$$\int_0^\infty \beta q(t)^\top P_2 q(t)dt + 2\int_0^\infty \delta(t)q(t)P_2\dot{q}(t)dt = \lim_{t\to\infty} \left[\delta(t)q(t)^\top P_2 q(t) \right]_0^t +$$

$$+ \int_0^\infty (\beta - \dot{\delta}(t))q(t)^\top P_2 q(t)dt \geq \delta(0)q(t)^\top P_2 q(t) \geq \epsilon |q_0|^2 \quad \forall q,\dot{q} \in \mathscr{L}_2^{n_q},$$

which holds for all $\delta \in \Lambda$ and for $\epsilon = \bar{\sigma}(P_2)^2$. We conclude that the IQC (2.66) holds for all $\Delta \in \mathbf{\Delta}_{\mathrm{ltv,rb}}$.

∎

2.7 IQC-multipliers for performance

Apart from the numerous existing IQC-multipliers for uncertainties and nonlinearities there is also a whole variety of performance characteristics that can be expressed in terms of the IQC (2.8). In the following subsections we will present a selected overview of the most relevant ones.

2.7.1 Induced $\mathscr{L}_2$-gain performance

Although elementary, let us start with the most commonly applied IQC-multiplier for induced $\mathscr{L}_2$-gain performance.

IQC-Multiplier 2.19. *Consider the stable system $G_{zw} \in \mathscr{RH}_\infty^{n_z \times n_w}$ and suppose there exists some $\gamma > 0$ and some small $\epsilon > 0$ such that for all trajectories of $z = G_{zw}w$, $w \in \mathscr{L}_2^{n_w}$, the performance IQC (2.8) is satisfied with*

$$\Psi_\mathrm{p}^* P_\mathrm{p} \Psi_\mathrm{p} := \begin{pmatrix} I_{n_z} & 0 \\ 0 & I_{n_w} \end{pmatrix} \begin{pmatrix} \gamma^{-1} I_{n_z} & 0 \\ 0 & -\gamma I_{n_w} \end{pmatrix} \begin{pmatrix} I_{n_z} & 0 \\ 0 & I_{n_w} \end{pmatrix}. \tag{2.68}$$

Then the induced $\mathscr{L}_2$-gain from w to z is less than $\gamma > 0$.

Proof. Let us evaluate the performance IQC (2.8) for the IQC-multiplier (2.68). This directly yields the inequality

$$\gamma^{-1} \|z\|_2^2 \leq (\gamma - \epsilon)\|w\|_2^2 \quad \forall w \in \mathscr{L}_2^{n_w}.$$

Hence, for all trajectories of $z = G_{zw}w$ with $w \in \mathscr{L}_2^{n_w}$, the induced $\mathscr{L}_2$-gain from w to z is less than γ.
∎

Note that one is typically interested in the best achievable $\mathscr{L}_2$-gain of the uncertain system (2.7), while guaranteeing stability for all $\Delta \in \mathbf{\Delta}$. In this case one can simply apply Corollary 2.3 with (2.68) and appropriate IQC-multipliers for $\mathbf{\Delta}$ and minimize γ subject to the LMI constraint (2.15). Here we emphasize that the IQC-multiplier (2.68) is not affine in the parameter γ. In Section 2.7.3 it will be shown how one can easily overcome this issue by applying a simple Schur complement argument.

2.7.2 Passivity performance

In complete analogy to the induced $\mathscr{L}_2$-gain performance specification from the previous section, it is also possible to verify whether or not the map $w \to z$ is passive.

IQC-Multiplier 2.20. *Suppose there exists some $\epsilon > 0$ such that for all trajectories of $z = G_{zw}w$, $w \in \mathscr{L}_2^{n_w}$, the performance IQC (2.8) is satisfied with*

$$\Psi_\mathrm{p}^* P_\mathrm{p} \Psi_\mathrm{p} := \begin{pmatrix} I_n & 0 \\ 0 & I_n \end{pmatrix} \begin{pmatrix} 0 & -I_n \\ -I_n & 0 \end{pmatrix} \begin{pmatrix} I_n & 0 \\ 0 & I_n \end{pmatrix} \tag{2.69}$$

and $n := n_z = n_w$. Then the map $w \to z$ is strictly input passive.

Proof. Let us again evaluate the performance IQC (2.8), but now for (2.69). This directly yields the inequality

$$2\langle z, w \rangle \geq \epsilon \|w\|_2^2 \quad \forall w \in \mathscr{L}_2^{n_w}.$$

Hence, for all trajectories of $z = G_{zw}w$ with $w \in \mathcal{L}_2^{n_w}$, the map $w \to z$ is strictly input passive.

■

2.7.3 Quadratic performance

The previous two IQC-multipliers from Section 2.7.1 and 2.7.2 are special versions of the following more general IQC-multiplier for quadratic performance.

IQC-Multiplier 2.21. *Quadratic performance is achieved if for all trajectories of $z = G_{zw}w$, $w \in \mathcal{L}_2^{n_w}$, there exists some $\epsilon > 0$ such that the IQC (2.8) is satisfied with*

$$\Psi_{\mathrm{p}}^* P_{\mathrm{p}} \Psi_{\mathrm{p}} := \begin{pmatrix} I_{n_z} & 0 \\ 0 & I_{n_w} \end{pmatrix} \begin{pmatrix} T^\top P_{\mathrm{p}11}^{-1} T & P_{\mathrm{p}12} \\ P_{\mathrm{p}12}^\top & P_{\mathrm{p}22} \end{pmatrix} \begin{pmatrix} I_{n_z} & 0 \\ 0 & I_{n_w} \end{pmatrix}. \tag{2.70}$$

Here $P_{\mathrm{p}11}$, $P_{\mathrm{p}12}$ and $P_{\mathrm{p}22}$ are free matrix variables with $P_{\mathrm{p}11} \succ 0$, while T is either zero or a constant matrix with full row rank.

Due to the structure of some important performance IQCs it is convenient to define the left upper block of P_{p} as $T P_{\mathrm{p}11}^{-1} T$. Indeed, with $T = I$, $P_{\mathrm{p}11} = \gamma I$, $P_{\mathrm{p}12} = 0$ and $P_{\mathrm{p}22} = -\gamma I$ we obtain the IQC-multiplier (2.68) for induced $\mathcal{L}_2$-gains and with $T = I$, $P_{\mathrm{p}11} = 0$, $P_{\mathrm{p}12} = -I$ and $P_{\mathrm{p}22} = 0$ we obtain the IQC-multiplier (2.69) for passivity. It is hence possible to verify robust stability and a variety of quadratic performance specifications for the system (2.7) by applying Corollary 2.3 with the IQC-multiplier (2.70). However, note that (2.70) is not affine in the unknown matrix variable $P_{\mathrm{p}11}^{-1}$. In order to overcome this issue, one can proceed as follows: For $\Psi_{\mathrm{p}} = I$ the realization matrices C_2 and D_2 in (2.14) have the structure

$$\begin{pmatrix} C_2 & D_{21} & D_{22} \end{pmatrix} := \begin{pmatrix} \tilde{C}_2 & \tilde{D}_{21} & \tilde{D}_{22} \\ 0 & 0 & I \end{pmatrix}.$$

With (2.70), the LMI (2.15) in Corollary 2.3 hence reads as

$$\begin{pmatrix} I & 0 & 0 \\ A & B_1 & B_2 \\ C_1 & D_{11} & D_{12} \\ \tilde{C}_2 & \tilde{D}_{21} & \tilde{D}_{22} \\ 0 & 0 & I \end{pmatrix}^\top \begin{pmatrix} 0 & X & 0 & 0 & 0 \\ X & 0 & 0 & 0 & 0 \\ 0 & 0 & P & 0 & 0 \\ 0 & 0 & 0 & T^\top P_{\mathrm{p}11}^{-1} T & P_{\mathrm{p}12} \\ 0 & 0 & 0 & P_{\mathrm{p}12}^\top & P_{\mathrm{p}22} \end{pmatrix} \begin{pmatrix} I & 0 & 0 \\ A & B_1 & B_2 \\ C_1 & D_{11} & D_{12} \\ \tilde{C}_2 & \tilde{D}_{21} & \tilde{D}_{22} \\ 0 & 0 & I \end{pmatrix} \prec 0.$$

In order to turn this inequality into a genuine LMI, we first rewrite it as

$$
\Phi_{11}(X, P_{\mathrm{p}12}, P_{\mathrm{p}22}) + \Phi_{12}^{\top} P_{\mathrm{p}11}^{-1} \Phi_{12} :=
$$

$$
\begin{pmatrix} I & 0 & 0 \\ A & B_1 & B_2 \\ C_1 & D_{11} & D_{12} \\ \tilde{C}_2 & \tilde{D}_{21} & \tilde{D}_{22} \\ 0 & 0 & I \end{pmatrix}^{\top}
\begin{pmatrix} 0 & X & 0 & 0 & 0 \\ X & 0 & 0 & 0 & 0 \\ 0 & 0 & P & 0 & 0 \\ 0 & 0 & 0 & 0 & P_{\mathrm{p}12} \\ 0 & 0 & 0 & P_{\mathrm{p}12}^{\top} & P_{\mathrm{p}22} \end{pmatrix}
\begin{pmatrix} I & 0 & 0 \\ A & B_1 & B_2 \\ C_1 & D_{11} & D_{12} \\ \tilde{C}_2 & \tilde{D}_{21} & \tilde{D}_{22} \\ 0 & 0 & I \end{pmatrix} +
$$

$$
+ (\star)^{\top} P_{\mathrm{p}11}^{-1} \left(T\tilde{C}_2 \quad T\tilde{D}_{21} \quad T\tilde{D}_{22} \right) \prec 0.
$$

Hence, since $P_{\mathrm{p}11} \succ 0$, we can apply the Schur complement in order to obtain the inequality

$$
\begin{pmatrix} \Phi_{11}(X, P_{\mathrm{p}12}, P_{\mathrm{p}22}) & \Phi_{12}^{\top} \\ \Phi_{12} & -P_{\mathrm{p}11} \end{pmatrix} \prec 0,
$$

which is now affine in all the variables X, P, $P_{\mathrm{p}11}$, $P_{\mathrm{p}12}$ and $P_{\mathrm{p}22}$. On the other hand, if $T = 0$, we simply obtain $\Phi_{11}(X, P_{\mathrm{p}12}, P_{\mathrm{p}22}) \prec 0$, which is affine in all the variables X, P, $P_{\mathrm{p}12}$ and $P_{\mathrm{p}22}$.

There are many more concrete performance criteria that can be viewed as special cases of (2.70). For example, one could think of generalized $\mathscr{L}_2$-gain performance [32, 33], a variety of $\mathscr{H}_2$- and generalized $\mathscr{H}_2$-performance specifications all with different interpretations [127, 125, 124, 126] and $\mathscr{L}_1$ or peak-to-peak performance [1, 120] to name a few. Moreover, one can as well consider mixed versions. In the two sections that follow, we will respectively discuss $\mathscr{H}_2$ and generalized $\mathscr{L}_2$-gain performance in more detail.

2.7.4 $\mathscr{H}_2$-performance

For stable transfer matrices $G_{zw} \in \mathscr{R}\mathscr{H}_2^{n_z \times n_w}$ the $\mathscr{H}_2$-norm is given by

$$
\|G_{zw}\|_{H_2} = \sqrt{\frac{1}{2\pi}\mathrm{trace} \int_{-\infty}^{\infty} G_{zw}(i\omega)^* G_{zw}(i\omega)d\omega}. \tag{2.71}
$$

In order to turn this norm into computations, one can equivalently characterize $\|G_{zw}\|_{H_2} < \sqrt{\gamma}$ with a slack variable $Y \in \mathbb{S}^{n_w}$ as

$$\frac{1}{2\pi} \int_{-\infty}^{\infty} G_{zw}(i\omega)^* G_{zw}(i\omega) d\omega \prec Y \quad \text{and} \quad \text{trace}(Y) < \gamma. \tag{2.72}$$

We approach the $\mathscr{H}_2$-performance objective as the evaluation of the energy response to an impulsive input at time $t = 0$. This means that we consider inputs $w(0) = \delta_\mathrm{d} e_j$, where e_j is the jth standard unit-vector of the input space $\mathbb{R}^{n_w}$. Here the impulse δ_d, which is not a signal in $\mathscr{L}_2$, is interpreted as imposing a non-zero initial condition. With $\hat{z}_j(i\omega) = G_{zw}(i\omega) e_j$ the performance objective (2.72) can now be written as

$$\frac{1}{2\pi} \sum_{j=1}^{n_w} \int_{-\infty}^{\infty} \hat{z}_j(i\omega)^* \hat{z}_j(i\omega) d\omega < Y \quad \text{and} \quad \text{trace}(Y) < \gamma. \tag{2.73}$$

Let us now show how to verify robust stability and $\mathscr{H}_2$-performance for the system (2.7) by applying Corollary 2.3 with the IQC-multiplier 2.21 for quadratic performance. For this purpose, suppose that $P_{\mathrm{p}11} = T = I$ and that $P_{\mathrm{p}12}$ and $P_{\mathrm{p}22}$ are empty matrices. Further assume that $\mathrm{col}(D_{12}, D_{22}) = 0$ in order to ensure the norm to be finite, and that the non-zero initial condition $w(0)$ is expressed as $x_j(0) = B_2 e_j$, $j = 1, \ldots, n_w$. With (2.70), the LMI (2.15) in Corollary 2.3 then reads as

$$\begin{pmatrix} I & 0 \\ A & B_1 \\ \hline C_1 & D_{11} \\ \hline \tilde{C}_2 & \tilde{D}_{21} \end{pmatrix}^{\mathsf{T}} \begin{pmatrix} 0 & X & 0 & 0 \\ X & 0 & 0 & 0 \\ \hline 0 & 0 & P & 0 \\ \hline 0 & 0 & 0 & I \end{pmatrix} \begin{pmatrix} I & 0 \\ A & B_1 \\ \hline C_1 & D_{11} \\ \hline \tilde{C}_2 & \tilde{D}_{21} \end{pmatrix} \prec 0. \tag{2.74}$$

Robust stability and $\mathscr{H}_2$-performance is now guaranteed up to a level of $\sqrt{\gamma}$ if there exist matrices $X \in \mathbb{S}^\bullet$, $P \in \boldsymbol{P}$ and $Y \in \mathbb{S}^{n_w}$ for which (2.74) and the following LMIs are satisfied:

$$B_2^{\mathsf{T}} X B_2 \prec Y, \quad \text{trace}(Y) < \gamma. \tag{2.75}$$

Remark 2.13. Note that (2.73) is only equivalent to (2.72) if G_{zw} is LTI. For general operators $G_{zw} : \mathscr{H}_2^{n_w} \to \mathscr{H}_2^{n_z}$ (2.73) is only a sufficient condition. For further information and other interpretations of the $\mathscr{H}_2$-norm we refer the reader e.g. to [123, 127, 125, 124, 126, 82].

Proof. Because the proof is analogous to that of Lemma 2.1, we only provide a sketch. First observe that robust stability is implied by (2.74) and the KYP-lemma, just because of

$$\begin{pmatrix} \tilde{C}_2 & \tilde{D}_{21} \end{pmatrix}^{\mathsf{T}} \begin{pmatrix} \tilde{C}_2 & \tilde{D}_{21} \end{pmatrix} \succcurlyeq 0.$$

To prove robust performance, we recall that $x_j(0) = B_2 e_j$ and right- and left-multiply the first inequality in (2.75) with e_j and its transpose. This yields the inequality

$$x_j(0)^{\mathsf{T}} X x_j(0) < e_j^{\mathsf{T}} Y e_j. \tag{2.76}$$

Now take any trajectory of the system (2.7) for $x_j(0) = B_2 e_j$ and right- and left-multiply (2.74) with $\mathrm{col}(x_j(t), p_j(t))$ and its transpose. Here $x_j(t)$ and $p_j(t)$ are the jth evolutions of the state x and the input p. Then, after integration on $[0, \infty)$, exploiting (2.76) and applying Parsevals theorem we obtain

$$\frac{1}{2\pi} \int_{-\infty}^{\infty} \begin{pmatrix} \hat{q}_j(i\omega) \\ \hat{p}_j(i\omega) \end{pmatrix}^* \Pi(i\omega) \begin{pmatrix} \hat{q}_j(i\omega) \\ \hat{p}_j(i\omega) \end{pmatrix} + \hat{z}_j(i\omega)^* \hat{z}_j(i\omega) d\omega \leq e_j^\top Y e_j$$

for all e_j, $j = 1, \ldots, n_w$. Here $\hat{p}_j$, $\hat{q}_j$ and $\hat{z}_j$ are the Fourier transforms of p_j, q_j and z_j respectively. By observing that the first term is non-negative, summing over j leads to (2.73) thanks to trace$(Y) < \gamma$.

■

2.7.5 Dynamic generalized quadratic-performance

In this section we present a generalization of generalized $\mathscr{L}_2$-performance cost as formulated by [32]. Generalized $\mathscr{L}_2$-performance has been introduced in order to be able to handle, for example, independent norm-bounds or unstructured but element-by-element bounded uncertainties. The extension to dynamically weighted signal constraints with a multitude of interesting motivations, such as e.g. arbitrarily tight approximations of the $\mathscr{H}_2$-cost, has been proposed in [33] and will be further generalized here.

Let us consider the stable system $z = G_{zw} w$, with $w \in \mathscr{L}_2^{n_w}$ and $z \in \mathscr{L}_2^{n_z}$, as well as its extension $G \in \mathscr{R}\mathscr{H}_\infty^{n_y \times n_y}$, $n_y := n_z + n_w$, with the partition

$$G = \begin{pmatrix} 0 & G_{zw} \\ 0 & 0 \end{pmatrix}.$$

We are interested in the performance specification

$$\sup_{y \in \mathscr{Y}} \int_{-\infty}^{\infty} \hat{y}(i\omega)^* \left(G(i\omega)^* + G(i\omega) \right) \hat{y}(i\omega) d\omega < \gamma, \tag{2.77}$$

where y is confined to the set

$$\mathscr{Y} = \left\{ y \in \mathscr{L}_2^{n_y} : P\left(\int_{-\infty}^{\infty} [\Psi(i\omega)\hat{y}(i\omega)][\Psi(i\omega)\hat{y}(i\omega)]^* d\omega \right) + P_0 \preccurlyeq 0 \right\}.$$

Here $Y \to P(Y)$ is a linear map that takes symmetric matrices into symmetric matrices and $\Psi \in \mathscr{R}\mathscr{L}_\infty^{\bullet \times n_y}$ is a given transfer matrix with the partition $\Psi = (\Psi_z \ \Psi_w) \in \mathscr{R}\mathscr{L}_\infty^{\bullet \times (n_z + n_w)}$. We can now formulate the following performance IQC-multiplier.

IQC-Multiplier 2.22. *For all trajectories of* $z = G_{zw}w$, $w \in \mathscr{L}_2^{n_w}$, *the performance objective* (2.77) *is achieved and the performance IQC* (2.8) *is satisfied (for some* $\epsilon > 0$*) with the IQC-multiplier*

$$\Pi_{\mathrm{p}}(Y) := \begin{pmatrix} -\Pi_{11}(Y)^{-1} & -\Pi_{11}(Y)^{-1}\Pi_{12}(Y) \\ -\Pi_{12}(Y)^*\Pi_{11}(Y)^{-1} & \Pi_{22}(Y) - \Pi_{12}(Y)^*\Pi_{11}(Y)^{-1}\Pi_{12}(Y) \end{pmatrix}, \quad (2.78)$$

if there exist some symmetric matrix Y *and some* γ *with*

$$Y \preccurlyeq 0, \quad \mathrm{trace}\,(Y P_0) < \gamma \quad \text{and} \quad \Psi^* P^*(Y)\Psi + G^* + G \prec 0 \text{ on } \mathbb{C}^0. \quad (2.79)$$

Here P^* *is the adjoint of* P, *while* $\Pi_{\mathrm{p}}(Y)$ *and* $\Psi^* P^*(Y)\Psi$ *are related to one another as*

$$\Psi^* P^*(Y)\Psi = \begin{pmatrix} \Psi_z & \Psi_w \end{pmatrix}^* P^*(Y) \begin{pmatrix} \Psi_z & \Psi_w \end{pmatrix} = \begin{pmatrix} \Pi_{11}(Y) & \Pi_{12}(Y) \\ \Pi_{12}(Y)^* & \Pi_{22}(Y) \end{pmatrix}. \quad (2.80)$$

The proof is found at the end of this section. Since $P^*(Y)$ is a linear map, we conclude that (2.79) is actually an LMI problem. Hence, we can perform robust stability and performance analysis with (2.77) as cost function. Moreover, this generalizes the results of [32, 33] in the following sense: Consider the partitions $y = \mathrm{col}(y_1, y_2)$ with $y_1 \in \mathscr{L}_2^{n_z}$, $y_2 \in \mathscr{L}_2^{n_w}$ and $\Psi = \mathrm{diag}(\Psi_1, \Psi_2) \in \mathcal{RL}_\infty^{\bullet \times (n_z + n_w)}$ and suppose that $\mathscr{Y}$ is given by

$$\mathscr{Y} := \{y = \mathrm{col}(y_1, y_2) : y_1 \in \mathscr{Y}_1, \quad y_2 \in \mathscr{Y}_2\},$$

where

$$\mathscr{Y}_i := \left\{ y_i \in \mathscr{L}_2^{\bullet} : P_i \left(\int_{-\infty}^{\infty} [\Psi_i(i\omega)\hat{y}_i(i\omega)][\star]^* d\omega \right) + P_{i0} \preccurlyeq 0, \right\}, \quad i = 1, 2. \quad (2.81)$$

Then (2.80) specializes into

$$\Psi^* P^*(Y)\Psi := \begin{pmatrix} \Psi_1 & 0 \\ 0 & \Psi_2 \end{pmatrix}^* \begin{pmatrix} P_1^*(Y_1) & 0 \\ 0 & P_2^*(Y_2) \end{pmatrix} \begin{pmatrix} \Psi_1 & 0 \\ 0 & \Psi_2 \end{pmatrix}$$

and we recover the dynamic generalized $\mathscr{L}_2$-performance criterion from [33]. Furthermore, by choosing $\Psi_1 := I$ and $\Psi_2 := I$, the IQC-multiplier further simplifies to the one discussed in [32]. In view of (2.81), our generalization hence allows for the inclusion of dynamically weighted LMI constraints on the 'cross-correlation'

$$\int_{-\infty}^{\infty} y_1(i\omega)y_2(i\omega)^* d\omega.$$

The extra resulting benefit remains to be explored. Altogether this allows us to consider a diverse class of interesting cost criteria as will be illustrated and discussed next.

Example 2.4. The most elementary example is the computation of the induced $\mathscr{L}_2$-gain

$$\sup_{\|w\|_2^2 \leq 1} \|G_{zw}w\|_2 < \gamma. \tag{2.82}$$

This cost can be equivalently expressed as (2.77) by choosing $P(Y) := \text{trace}(Y)$, $P_0 := -1$ and $\Psi := I$. Subsequently, if we evaluate (2.79) for the particular choices, we obtain the inequalities

$$\mathbb{R} \ni Y < 0, \quad -Y < \gamma \quad \text{and} \quad \begin{pmatrix} YI_{n_z} & G_{zw} \\ G_{zw}^* & YI_{n_w} \end{pmatrix} \prec 0 \text{ on } \mathbb{C}^0,$$

These clearly imply (2.82).

Example 2.5. For many problems it might be more natural to bound the disturbance inputs individually as

$$\|W_i w_i\|_2^2 \leq 1, \quad i = 1, \ldots, n_w. \tag{2.83}$$

Here w_i is the ith entry of $w \in \mathscr{L}_2^{n_w}$ and $W_i \in \mathscr{RH}_\infty$ a given stable weight. Then we can consider the following performance objective

$$\sup_{\|W_i w_i\|_2^2 \leq 1, \ 1 \leq i \leq n_w} \|G_{zw}w\| < \gamma.$$

This can be equivalently expressed as (2.77) by employing (2.81) with $P_1(Y_1) := \text{trace}(Y_1)$, $P_{10} := -1$, $\Psi_1 := 1$ and $P_2(Y_2) := Y_2$, $P_{20} := -I$, $\Psi_2 := \text{diag}(W_1, \ldots, W_{n_w})$. Indeed, if we evaluate (2.79) for the particular choices, we obtain the inequalities

$$\mathbb{R} \ni Y_1 < 0, \quad \mathbb{R}^{n_w \times n_w} \ni Y_2 \prec 0, \quad -Y_1 - \text{trace}(Y_2) < \gamma$$

and

$$\begin{pmatrix} Y_1 I_{n_z} & G_{zw} \\ G_{zw}^* & \Psi_2^* Y_2 \Psi_2 \end{pmatrix} \prec 0 \text{ on } \mathbb{C}^0.$$

One can show that this in turn implies (2.83).

There are many more applications for the performance IQC-multiplier described in this section. One could e.g. think of implicit weights, arbitrarily tight approximations of the $\mathscr{H}_2$-cost and robustness against unstructured element-by-element bounded uncertainties, to name a few. We refer the reader to [32, 33] for further details. Let us conclude the discussion by proving our claim.

Proof. Observe that

$$\int_{-\infty}^{\infty} \hat{y}(i\omega)^* (G(i\omega)^* + G(i\omega))\hat{y}(i\omega)d\omega =$$

$$= \text{trace} \int_{-\infty}^{\infty} (G(i\omega)^* + G(i\omega))\hat{y}(i\omega)\hat{y}(i\omega)^* d\omega$$.

Similarly, with the adjoint P^* and the symmetric matrix $Y \in \mathbb{S}^\bullet$ we have

$$\text{trace}\left(Y \left(P \left(\int_{-\infty}^{\infty} [\Psi(i\omega)\hat{y}(i\omega)][\Psi(i\omega)\hat{y}(i\omega)]^* d\omega \right) + P_0 \right) \right) =$$

$$\text{trace}\left(P^*(Y) \int_{-\infty}^{\infty} [\Psi(i\omega)\hat{y}(i\omega)][\Psi(i\omega)\hat{y}(i\omega)]^* d\omega \right) + \text{trace}(Y P_0) =$$

$$\text{trace} \int_{-\infty}^{\infty} P^*(Y)[\Psi(i\omega)\hat{y}(i\omega)][\Psi(i\omega)\hat{y}(i\omega)]^* d\omega + \text{trace}(Y P_0) =$$

$$\text{trace} \int_{-\infty}^{\infty} \Psi(i\omega)^* P^*(Y)\Psi(i\omega)\hat{y}(i\omega)\hat{y}(i\omega)^* d\omega + \text{trace}(Y P_0).$$

This implies that

$$\int_{-\infty}^{\infty} \hat{y}(i\omega)^* \left(G(i\omega)^* + G(i\omega)\right)\hat{y}(i\omega)d\omega - \gamma +$$

$$+ \text{trace}\left(Y \left(P \left(\int_{-\infty}^{\infty} [\Psi(i\omega)\hat{y}(i\omega)][\Psi(i\omega)\hat{y}(i\omega)]^* d\omega \right) + P_0 \right) \right) =$$

$$= \text{trace}(Y P_0) - \gamma +$$

$$+ \text{trace} \int_{-\infty}^{\infty} (\Psi(i\omega)^* P^*(Y)\Psi(i\omega) + G(i\omega)^* + G(i\omega))\hat{y}(i\omega)\hat{y}(i\omega)^* d\omega. \quad (2.84)$$

Now let us assume that there exists some Y and some γ with (2.79). Then

$$\text{trace}(Y P_0) \leq \gamma - \epsilon$$

persists to hold for some small $\epsilon > 0$. If we choose $y \in \mathscr{Y}$, the right-hand side of (2.84) is non-positive. Hence, the same holds for the left-hand side. Furthermore, due to the definition of $\mathscr{Y}$ we have that

$$\text{trace}\left(Y \left(P \left(\int_{-\infty}^{\infty} [\Psi(i\omega)\hat{y}(i\omega)][\Psi(i\omega)\hat{y}(i\omega)]^* d\omega \right) + P_0 \right) \right) \geq 0.$$

This implies that

$$\int_{-\infty}^{\infty} \hat{y}(i\omega)^* \left(G(i\omega)^* + G(i\omega)\right)\hat{y}(i\omega)d\omega \leq \gamma - \epsilon$$

for all $y \in \mathscr{Y}$, which shows that the performance specification is achieved.

Let us continue by zooming into the structure of the third inequality in (2.79). By inspection we observe that this is identical to

$$\begin{pmatrix} \Pi_{11}(Y) & G_{zw} + \Pi_{12}(Y) \\ \star & \Pi_{22}(Y) \end{pmatrix} = \begin{pmatrix} \Psi_z^* P^*(Y)\Psi_z & G_{zw} + \Psi_z^* P^*(Y)\Psi_w \\ \star & \Psi_w^* P^*(Y)\Psi_w \end{pmatrix} \prec 0 \text{ on } \mathbb{C}^0.$$

Since $\Pi_{11}(Y) \prec 0$, we can apply the Schur complement in order to infer

$$\Pi_{22}(Y) - (G_{zw}^* + \Pi_{12}(Y)^*)\Pi_{11}(Y)^{-1}(G_{zw} + \Pi_{12}(Y)) \prec 0 \text{ on } \mathbb{C}^0.$$

It is not hard to see that this is identical to

$$\begin{pmatrix} G_{zw} \\ I \end{pmatrix}^* \Pi_{\mathrm{p}} \begin{pmatrix} G_{zw} \\ I \end{pmatrix} \prec 0 \text{ on } \mathbb{C}^0 \tag{2.85}$$

for Π_{p} in (2.78). Moreover, since (2.85) is strict, there exists some small $\epsilon > 0$ such that

$$\begin{pmatrix} G_{zw} \\ I \end{pmatrix}^* \Pi_{\mathrm{p}} \begin{pmatrix} G_{zw} \\ I \end{pmatrix} \preccurlyeq -\epsilon I \text{ on } \mathbb{C}^0 \tag{2.86}$$

persists to hold. By left- and right-multiplying (2.86) with $\hat{w}(i\omega)^*$ and $\hat{w}(i\omega)$ respectively, exploiting the system description $z = G_{zw}w$ and integration on $(-\infty, \infty)$, we obtain the IQC (2.8) with Π_{p} replaced by (2.78).   ∎

2.8 Further connections and possibilities

2.8.1 Connection to the μ-theory

It is important to note that the IQC-tools are also related to the well-known and widely applied classical μ-theory [122, 44, 10]. In order to show a connection between the two approaches, let us consider structured complex matrices

$$\Delta_c = \mathrm{diag}\left(\delta_{r_1} I_\bullet, \ldots, \delta_{r_l} I_\bullet, \delta_{c_1} I_\bullet, \ldots, \delta_{c_m} I_\bullet, \Delta_{f_1}, \ldots, \Delta_{f_n}\right) \in \mathbb{C}^{n_p \times n_q},$$

where

1. $\delta_{r_i} \in \mathbb{R}$, $i = 1, \ldots, l$, are real parameters with $|\delta_{r_i}| < 1$;
2. $\delta_{c_j} \in \mathbb{C}$, $j = 1, \ldots, m$, are complex parameters with $|\delta_{c_j}| < 1$;
3. $\Delta_{f_k} \in \mathbb{C}^{\bullet \times \bullet}$, $k = 1, \ldots, n$, are full complex matrices with $\|\Delta_{f_k}\| < 1$.

We denote the set of all these complex matrices as $\boldsymbol{\Delta}_c$. In addition, let us also introduce the definition of the structured singular value.

Definition 2.4 (Structured singular value). The structure singular value of the com-
plex matrix $G_c \in \mathbb{C}^{n_q \times n_p}$ with respect to the set of structured complex matrices $\boldsymbol{\Delta}_c$
is defined as

$$\mu_{\boldsymbol{\Delta}_c}(G_c) = \frac{1}{\sup\{r \geq 0 : \det(I - G_c \Delta_c) \neq 0 \ \forall \Delta_c \in r\boldsymbol{\Delta}_c\}}. \tag{2.87}$$

Then we can state the following well-known result [122].

Theorem 2.3. *Let G_c be a complex matrix and $\boldsymbol{\Delta}_c$ an arbitrary set of (structured)
complex matrices. Then the inequality*

1. *$\mu_{\boldsymbol{\Delta}_c}(G_c) \leq \gamma_u$ holds if and only if $\det(I - G_c \Delta_c) \neq 0$ for all $\Delta_c \in \gamma_u^{-1}\boldsymbol{\Delta}_c$.*
2. *$\mu_{\boldsymbol{\Delta}_c}(G_c) > \gamma_l$ holds if and only if $\det(I - G_c \Delta_c) = 0$ for some $\Delta_c \in \gamma_l^{-1}\boldsymbol{\Delta}_c$.*

Let us now recall the feedback system (2.1) from Section 2.2 and suppose that
Δ is confined to the class of all real rational proper and stable Δ's whose frequency
response takes its values in $\boldsymbol{\Delta}_c$:

$$\boldsymbol{\Delta} := \left\{ \Delta \in \mathcal{RH}_\infty^{n_p \times n_q} : \Delta(i\omega) \in \boldsymbol{\Delta}_c \ \forall \omega \in \mathbb{R} \cup \{\infty\} \right\}. \tag{2.88}$$

Then the feedback interconnection (2.1) is robustly stable for all $\Delta \in \gamma^{-1}\boldsymbol{\Delta}$ if and
only if

$$\mu_{\boldsymbol{\Delta}_c}(G(i\omega)) \leq \gamma \ \forall \omega \in \mathbb{R} \cup \{\infty\}. \tag{2.89}$$

This yields an exact characterization of robust stability for the feedback interconnection (2.1) if Δ is confined to (2.88).

Unfortunately, the computation of the structure singular value is in general a hard
problem, which enforces us to compute lower- and upper-bounds [23, 9]. For example,
it is well-known that $\|G_c\|$ is a crude but tractable upper-bound for $\mu_{\boldsymbol{\Delta}_c}(G_c)$. On the
other hand, it is also possible to introduce structured complex D-scalings

$$D_c = \text{diag}(D_{r_1}, \ldots, D_{r_l}, D_{c_1}, \ldots, D_{c_m}, d_{f_1} I_\bullet, \ldots, d_{f_n} I_\bullet),$$

in accordance with the structure of $\boldsymbol{\Delta}_c$ and where

1. D_{r_i}, $i = 1, \ldots, l$, are complex Hermitian positive definite matrices;
2. D_{c_j}, $j = 1, \ldots, m$, are complex Hermitian positive definite matrices;
3. d_{f_k}, $k = 1, \ldots, n$, are real positive numbers.

We denote the set of all these complex matrices as $\boldsymbol{D}_c$. Note that all matrices $D_c \in \boldsymbol{D}_c$
are nonsingular and satisfy $\Delta_c D_c = D_c \Delta_c$ for all $\Delta_c \in \boldsymbol{\Delta}_c$. This leads to the following
improved upper-bound for the structured singular value [122, 44]:

$$\mu_{\boldsymbol{\Delta}_c}(G_c) \leq \inf_{D_c \in \boldsymbol{D}_c} \|D_c G_c D_c^{-1}\|.$$

Key is that $\|D_c G_c D_c^{-1}\| < \gamma^2$ can be equivalently reformulated as the LMI

$$\begin{pmatrix} G_c \\ I \end{pmatrix}^* \begin{pmatrix} \gamma^{-2} \Pi_c & 0 \\ 0 & -\Pi_c \end{pmatrix} \begin{pmatrix} G_c \\ I \end{pmatrix} \prec 0$$

in the matrix variable $\Pi_c := D_c^* D_c$. Here D_c and Π_c share their structure and their properties, such that also $\Pi_c \in \boldsymbol{D}_c$. We can hence compute upper-bounds for (2.89) at each frequency by using D-scalings. The largest of these upper-bounds is found at the 'critical frequency' and can also be obtained by searching for a frequency dependent scaling $\Pi_c(\omega) \in \boldsymbol{D}_c$ with

$$\begin{pmatrix} G(i\omega) \\ I \end{pmatrix}^* \begin{pmatrix} \gamma^{-2} \Pi_c(\omega) & 0 \\ 0 & -\Pi_c(\omega) \end{pmatrix} \begin{pmatrix} G(i\omega) \\ I \end{pmatrix} \prec 0 \ \ \forall \omega \in \mathbb{R} \cup \{\infty\}.$$

Moreover, we recall from Remark 2.6 that $\Pi_c(\omega)$ can be approximated with arbitrary quality by adopting the IQC-multipliers from Section 2.6.1 as follows:

$$\Psi^* P \Psi := \begin{pmatrix} \star \\ \star \end{pmatrix}^* \begin{pmatrix} P_{11} & 0 \\ 0 & -P_{11} \end{pmatrix} \begin{pmatrix} \gamma^{-1} \psi_\nu \otimes I_{n_q} & 0 \\ 0 & \psi_\nu \otimes I_{n_p} \end{pmatrix}. \tag{2.90}$$

Here the matrix P is confined to

$$P \in \boldsymbol{P}_\mu := \Big\{ P = \mathrm{diag}(P_{11}, -P_{11}) \in \mathbb{S}^{n_q + n_p} : P_{11} \succ 0 \Big\}$$

with

$$P_{11} = \mathrm{diag}\Big(P_{r_1}, \ldots, P_{r_l}, P_{c_1}, \ldots, P_{c_m}, P_{f_1} \otimes I_\bullet, \ldots, P_{f_n} \otimes I_\bullet \Big),$$

where the dimensions are compatible with the block structure of Δ_c. We arrive at the following connection between the IQC-framework and μ-theory: The structured singular value $\mu_{\boldsymbol{\Delta}_c}(G(i\omega))$ is not larger than $\gamma > 0$ for all $\omega \in \mathbb{R} \cup \{\infty\}$ and hence robust stability of the feedback interconnection (2.1) is guaranteed for all $\Delta \in \gamma^{-1} \boldsymbol{\Delta}$ if there exist some $P \in \boldsymbol{P}_\mu$ for which the FDI (2.5) in Theorem 2.1 is satisfied with Π replaced by (2.90). We can hence exploit IQC-theory in order to compute upper-bounds of the structured singular value at the 'critical frequency' of $I - G\Delta$. Let us finally emphasize that [44, 10] introduced the so-called DG-scalings, in order to improve the upper-bound computations for the structured singular value in case Δ_c comprises real parametric uncertainties. DG-scalings, in turn, are directly related to the IQC-multiplier discussed in Section 2.6.3.1. On the other hand, the IQC-tools allow us to even further reduce conservatism by applying full-block multipliers as discussed in Sections 2.6.2.2 and 2.6.3.2.

2.8.2 Alternative computational methods

In Section 2.5.2 we applied the conventional tools for turning the robust stability and performance test of Lemma 2.1 into a finite dimensional convex feasibility test. Unfortunately, this is not always the most efficient method. By applying the KYP-lemma, we introduce $0.5n_G(n_G + 1)$ additional decision variables (where n_G is the state-dimension of state-matrix A in (2.14)). Clearly, for larger state dimensions (say $n_G > 100$), this leads to undesirable extensive computational loads, or even intractability of the optimization problem. In order to overcome this issue, several relatively systematic tools have been developed and proposed in the literature. For example, [96, 91] present two cutting plane algorithms that exploit the structural properties of an Hamiltonian matrix. As illustrated, these algorithms can in certain cases solve IQC-analysis problems substantially more efficiently. As another promising approach, [3, 4] proposes different algorithms based on non-smooth optimization techniques, which can even be employed for robust controller synthesis. We emphasize that, although these alternative approaches offer meaningful advantages for analysis, they might not be quite as useful for robust controller synthesis, especially for those robust controller synthesis problems that can be turned into finite dimensional convex feasibility test through the use of 'conventional' techniques.

2.8.3 Discrete-time systems

Another important advantage of the IQC-tools is that the time-axis can be chosen as all non-negative integers $\mathbb{N}_0$ in order to investigate discrete time systems too. For this purpose, suppose that $l_2^{\bullet}$ denotes the space of all vector-valued sequences with a finite l_2-norm and let $\mathscr{RL}_\infty^{\bullet \times \bullet}$ ($\mathscr{RH}_\infty^{\bullet \times \bullet}$) denote the space of real rational and proper transfer matrices with no poles on the unit circle (in the closed unit disk). Now consider the feedback interconnection (2.7), where at present $G \in \mathscr{RH}_\infty^{(n_q + n_z) \times (n_p + n_w)}$ admits the realization $G(z) := C_G(\frac{1}{z}I - A_G)^{-1}B_G + D_G$, where A has no eigenvalues in the closed unit disk, while $\Delta : l_2^{n_q} \to l_2^{n_p}$ is a bounded and causal operator in some class $\boldsymbol{\Delta}$. Completely analogously to the continuous-time case, we can define the IQC (2.4), which now should hold for all $\Delta \in \boldsymbol{\Delta}$, for all $P \in \boldsymbol{P}$ and for all $q \in l_2^{n_q}$, as well as the IQC (2.8) for performance, which now should hold for all $P_\mathrm{p} \in \boldsymbol{P}_\mathrm{p}$ and for all $w \in l_2^{n_w}$. Here the IQC-multipliers $\Psi^* P \Psi$ and $\Psi_\mathrm{p}^* P_\mathrm{p} \Psi_\mathrm{p}$ are confined to $\mathscr{RL}_\infty^{\bullet \times \bullet}$ with $\Psi, \Psi_\mathrm{p} \in \mathscr{RH}_\infty^{\bullet \times \bullet}$. Robust stability and performance is then achieved if all conditions in Corollary 2.2 are satisfied. The essential difference is that the FDI (2.13) should hold on the complex unit circle and not on the extended imaginary axis. For computations we can again introduce the realization (2.14), where A has no eigenvalues in the closed unit disk, and infer by the discrete-time analogue of the KYP-lemma [139] that the FDI (2.13) is equivalent to the existence of some symmetric matrix X such that the following LMI is satisfied:

$$
\begin{pmatrix} I & 0 & 0 \\ \hline A & B_1 & B_2 \\ \hline C_1 & D_{11} & D_{12} \\ C_2 & D_{21} & D_{22} \end{pmatrix}^{\top} \begin{pmatrix} -X & 0 & 0 & 0 \\ 0 & X & 0 & 0 \\ \hline 0 & 0 & P & 0 \\ 0 & 0 & 0 & P_{\mathrm{p}} \end{pmatrix} \begin{pmatrix} I & 0 & 0 \\ \hline A & B_1 & B_2 \\ \hline C_1 & D_{11} & D_{12} \\ C_2 & D_{21} & D_{22} \end{pmatrix} \prec 0.
$$

Similarly, it is also possible to consider discrete-time versions of the IQC-multipliers presented in the Sections 2.6 and 2.7. The IQC-tools can hence be straightforwardly employed for the robust stability and performance analysis of discrete-time systems. It is even possible to consider a mixture of continuous- and discrete-time, which allows us investigate hybrid systems or systems with jumps (see e.g. [47]).

2.8.4 Further possibilities

Admittedly, the literature on integral quadratic constraints is vast, and it is impossible to address every single aspect, connection and generalization that has been considered. Nevertheless, let us briefly mention a few topics that we did not address. For example, it might be restrictive to consider systems that are described as mappings of signals. This fixes what is considered to be an input or an output of an system. For the class of systems where this is not the case it is possible to extend the IQC-tools to the more general behavioral approach. We refer the reader to [136] and references therein. Secondly, it obviously can be restrictive to consider nominal systems that are linear time-invariant. As an interesting example, we would like to refer the reader to [79, 77, 80] for a study on the robustness analysis of periodic trajectories and periodically forced uncertain feedback systems based on IQC-tools. In fact, there is a vast nonlinear theory that addresses the stability analysis and controller synthesis for nonlinear systems. Finally, it is also relevant to note that there are IQC-tools available that can be exploited for the stability and performance analysis of (and even controller synthesis for) some classes of large scale interconnected and distributed systems (see e.g. [13, 34, 42, 106, 107, 78, 45, 201, 94, 25, 156, 160]).

2.9 Illustrations

In this section we illustrate the IQC-tools by means of two numerical examples. The first one is an elementary estimation problem, which serves to demonstrate the most

important aspects of a proper IQC-analysis. Here we also address the issues related
to the particular choice of the basis-function (2.11). The second example is a little
more realistic helicopter problem from [178], which serves to show the ease with
which we can consider more involved and practical scenarios. We emphasize that
the goal is to illustrate the IQC-analysis tools and not to perform a full robustness
analysis including time-domain simulations, frequency-domain analysis, Monte-Carlo
simulations, etc.

2.9.1 Robustness analysis of an estimator

Consider the standard system interconnection for estimation as depicted in Figure 2.6.
Given the stable linear system $G \in \mathscr{RH}_\infty^{\bullet \times \bullet}$, the goal of estimation is to dynamically
process the measurement signal y_2 with a stable linear filter $E = (A_E, B_E, C_E, D_E) \in$
$\mathscr{RH}_\infty^{\bullet \times \bullet}$ in order to provide an estimate of the signal y_1 such that the induced $\mathscr{L}_2$-gain
from the disturbance input w to the performance output z is minimized.

In this example, G is defined through the realization

$$
\left[\begin{array}{c|c} A & B_w \\ \hline C_1 & D_{1w} \\ C_2 & D_{2w} \end{array} \right] = \left[\begin{array}{cc|cc} 0 & -1 & -2 & 0 \\ 1 & -0.5 & 1 & 0 \\ \hline 1 & 0 & 0 & 0 \\ -1 & 1 & 0 & 0.01 \end{array} \right]
$$

and the estimator E was designed with the standard $\mathscr{H}_\infty$-synthesis tools of Matlab;
it guarantees the induced $\mathscr{L}_2$-gain from w to z to be less than 2.01.

In reality G is actually only the nominal part of the uncertain system $\Sigma(\delta_1, \delta_2, G)$,
which is affected by two parametric uncertainties as follows:

$$
\begin{pmatrix} \dot{x}(t) \\ y_1(t) \\ y_2(t) \end{pmatrix} = \begin{pmatrix} A + B_{p_1} \delta_1 C_{q_1} + B_{p_2} \delta_2(t) C_{q_2} & B_w \\ C_1 & D_{1w} \\ C_2 & D_{2w} \end{pmatrix} \begin{pmatrix} x(t) \\ w(t) \end{pmatrix}.
$$

Here δ_1 is a time-invariant parameter satisfying $|\delta_1| \leq \alpha_1$, for some non-negative con-
stant α_1, and $\delta_2(t)$ is a (continuously differentiable) time-varying parameter satisfying
$|\delta_2(t)| \leq \alpha_2$ and $|\dot{\delta}_2(t)| \leq r_2$, for the non-negative constants α_2 and r_2. The matrices
B_{p_1}, B_{p_2}, C_{q_1} and C_{q_2} are respectively given by $B_{p_1} = \text{col}(1,0)$, $B_{p_2} = \text{col}(0,1)$ and
$C_{q_1} = C_{q_2} = (0 \ 1)$. The nominal system G can hence be retrieved from the uncertain
system $\Sigma(\delta_1, \delta_2, G)$ by setting $\delta_1 = \delta_2 = 0$.

In addition to the parametric uncertainties, the input signal to the estimator is
actually a delayed version of the measurement signal y_2, i.e. $y_{2d}(t) := y_2(t - \beta_3)$ with

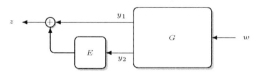

Fig. 2.6 Standard interconnection for nominal estimation

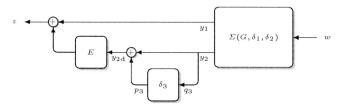

Fig. 2.7 A particular uncertain estimation problem

$\beta_3 \in [0, \alpha_3]$, where α_3 is some non-negative constant. In accordance with the discussion in Section 2.6.5.2, we can pull-out this uncertainty by defining the delay operator $p_3(t) = \delta_3(q_3(t))$, where $\delta_3(q_3)(t) = q_3(t - \beta_3) - q_3(t)$ (with the corresponding Fourier transform given by $\hat{p}_3(i\omega) = \hat{\delta}_3(i\omega)\hat{q}_3(i\omega) = (e^{-i\omega\beta_3} - 1)\hat{q}_3(i\omega))$ and consider the uncertain system interconnection as shown in Figure 2.7. We arrive at the following central question: Does the estimator E, which achieves performance in the nominal case for the configuration shown in Figure 2.6, also achieve robust performance in the presence of the uncertainties δ_1, δ_2 and δ_3 for the configuration shown in Figure 2.7? In order to verify this, we proceed by first obtaining the linear fractional representation

$$
\begin{pmatrix} \dot{x}(t) \\ \dot{x}_E(t) \\ q_1(t) \\ q_2(t) \\ q_3(t) \\ z(t) \end{pmatrix} = \left(\begin{array}{cc|ccc|c} A & 0 & B_{p_1} & B_{p_2} & 0 & B_w \\ B_E C_2 & A_E & 0 & 0 & B_E & B_E D_{2w} \\ \hline C_{q_1} & 0 & 0 & 0 & 0 & 0 \\ C_{q_2} & 0 & 0 & 0 & 0 & 0 \\ C_2 & 0 & 0 & 0 & 0 & D_{2w} \\ \hline C_1 + D_E C_2 & C_E & 0 & 0 & D_E & D_{1w} + D_E D_{2w} \end{array} \right) \begin{pmatrix} x(t) \\ x_E(t) \\ p_1(t) \\ p_2(t) \\ p_3(t) \\ w(t) \end{pmatrix},
$$

$$p_1(t) = \delta_1 q_1(t), \quad p_2(t) = \delta_2 q_2(t), \quad p_3(t) = \delta_3(q_3(t)).$$

Then we can perform a robustness analysis of the uncertain system interconnection as shown in Figure 2.7, by employing the developed IQC-analysis tools.

As a start, we first do an IQC-analysis for different values $\alpha := \alpha_1 = \alpha_2 = 2\alpha_3$, such that $|\delta_1| \le \alpha$, $|\delta_2(t)| \le \alpha$ and $\alpha_3 \le \frac{1}{2}\alpha$ for all $t \in [0, \infty)$. Further we choose the

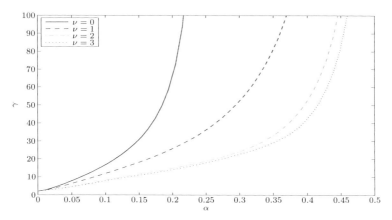

Fig. 2.8 Analysis results for different values of $\alpha \in [0, 0.5]$ and $\nu = \{0, 1, 2, 3\}$.

rate-of-variation r_2 of δ_2 to be bounded by $r_2 = 0.1$, such that $|\dot{\delta}_2(t)| \leq 0.1$ for all $t \in [0, \infty)$. The computations were carried out with Matlab/LMIlab and its Robust Control Toolbox on an average desktop computer (Core2Duo 3GHz CPU, 4GB RAM) by employing the IQC-multipliers 2.5, 2.7, 2.8 and 2.19 from Section 2.6.3.1, 2.6.4, 2.6.5.2 and 2.7.1 respectively, for the time-invariant parametric uncertainty δ_1, the rate-bounded time-varying parametric uncertainty δ_2, the delay uncertainty δ_3 and with induced $\mathscr{L}_2$-gain performance. Here we selected the first basis-function ψ_ν in (2.11) with pole-location $\rho = -1$ and for different McMillan degrees $\nu = \{0, 1, 2, 3\}$, yielding LMI-optimization problems with 28, 112, 254 and 454 decision variables respectively. On average it took 0.45, 1.2, 6.5 and 43 seconds, respectively, to obtain a feasible solution[1].

The computed bounds on the induced $\mathscr{L}_2$-gain are plotted in Figure 2.8. As can be seen, we recover nominal performance (i.e. $\gamma = 2.01$) for $\alpha = 0$, regardless of the pole-location and McMillan degree of the basis-function ψ_ν. For static (non-dynamic) IQC-multipliers (i.e. $\nu = 0$), we see a rapid increase of the guaranteed bound on the induced $\mathscr{L}_2$-gain γ, even for moderate values of α. In addition, stability of $\Sigma(G, \delta_1, \delta_2)$ and, hence, stability of the system interconnection in Figure 2.7 can not be guaranteed for values of α that are larger than approximately 0.245. On the other hand, if we allow for dynamics in the IQC-multipliers (ie. $\nu > 0$), stability is guaranteed for much larger values of α up to 0.49; this at the cost of a higher computational burden.

---

[1] The computation time not only heavily depends on the processor speed of the computer, but also on the particular choice of input parameters (i.e. desired relative accuracy, maximum number of iterations, feasibility radius, initial condition, etc.) to the Matlab command 'mincx'. Therefore, the given computation times should only be interpreted as estimates.

Let us also investigate robust performance in case we only vary one uncertain parameter, while fixing the others. This provides insight in the sensitivity to changes of the individual parameters. In Figure 2.9 the computed bounds on the induced $\mathscr{L}_2$-gains are shown for

1. various values of α_1 and fixed $\alpha_2 = 0.1$, $\alpha_3 = 0.05$ and $r_2 = 0.1$ (upper-row),
2. various values of α_2 and r_2 and fixed $\alpha_1 = 0.1$, $\alpha_3 = 0.05$ (middle-row),
3. various values of α_3 and fixed $\alpha_1 = 0.1$, $\alpha_2 = 0.1$ and $r_2 = 0.1$ (lower-row).

Again the analysis was performed using the first basis-function ψ_ν in (2.11) with pole-location $\rho = -1$ and for different McMillan degrees $\nu = \{0,1,2,3\}$. Once more we see for each scenario that the guaranteed bound on the induced $\mathscr{L}_2$-gain γ is significantly lower if allowing for dynamics in the IQC-multipliers (ie. if $\nu > 0$). Now consider the upper plot of Figure 2.9 and recall the results shown in Figure 2.8. Comparing these, we observe that the interconnection is less sensitive to changes of the uncertain parameter δ_1. For fixed $\alpha_2 = 0.1$, $\alpha_3 = 0.05$ and $r_2 = 0.1$ robust stability is even guaranteed for values of α_1 up to about 0.98. The same can be concluded for the delay operator δ_3. As can be seen in the lower-plot of Figure 2.9, the guaranteed bound on the induced $\mathscr{L}_2$-gain only increased by approximately 3.2 if increasing α_3 from 0 to 0.5. Note that this is not very surprising since the delay is not located in a feedback loop. In contrast to the uncertainties δ_1 and δ_3, the interconnection is more sensitive to changes of the uncertain time-varying parameter δ_2. The second row of plots in Figure 2.9 display a rapid increase of the computed bounds on the induced $\mathscr{L}_2$-gains for increasing values of α_2 and fixed $r_2 = \{0,0.1,0.2,0.3\}$. Finally, taking into account the maximum rate-of-variation r_2 of the uncertain time-varying parameter δ_2 allows us to obtain significantly lower guaranteed bounds on the induced $\mathscr{L}_2$-gain. This is clearly visible in the second row of plots in Figure 2.9, as the dashed blue line, the dashed-dotted green line and the dotted black line all move to towards the solid red line, if increasing r_2 from 0 to 0.3. Here the solid red line denotes the guaranteed level of robust performance for arbitrarily fast variations of δ_2.

As discussed in Remark 2.6, there is no general recipe for choosing an appropriate basis-function ψ_ν. Therefore, it is important to provide some numerical insights for our particular choices. For this purpose, we recompute the bounds on the induced $\mathscr{L}_2$-gains of the first IQC-analysis problem of this section for various $\alpha \in [0,0.2]$ and $\nu = \{0,1,2,3\}$, for both basis-functions (2.11) and for pole-locations ranging from -100 to -0.001. In Figure 2.10 the results are depicted for the first basis-function in (2.11). Here the particular values on the vertical axis denote the lowest guaranteed bound on the induced $\mathscr{L}_2$-gains for some pole-location ρ and with the corresponding values of α as given in Table 2.1. Clearly, for $\nu = 0$ the pole-location ρ does not affect the computations. On the other hand, for $\nu > 0$, the guaranteed bound on the induced $\mathscr{L}_2$-gain drops as ρ increases until approximately -0.1. From there, the guaranteed bound on the induced $\mathscr{L}_2$-gain increases again, if ρ approaches the imaginary axis. In addition, for $\nu = 2,3$, this leads to numerical problems for pole-locations close to the

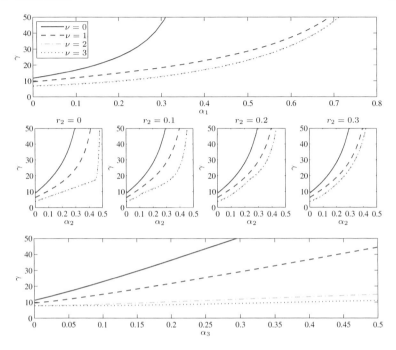

Fig. 2.9 Analysis results for various bounds on one uncertain parameter, while fixing the others.

nr.	α	nr.	α	nr.	α	nr.	α	nr.	α
1	0.001	5	0.043	9	0.085	13	0.127	17	0.169
2	0.011	6	0.053	10	0.095	14	0.137	18	0.179
3	0.022	7	0.064	11	0.106	15	0.148	19	0.190
4	0.032	8	0.074	12	0.116	16	0.158	20	0.200

Table 2.1 Various values of α.

imaginary axis. In contrast, this is not the case for the second basis-function in (2.11). As can be seen in Figure 2.11, the computations remain numerically well behaved, even for pole-locations close to the imaginary axis. Of course the latter plots only give us insights for this particular example and it might very well happen that the first basis-function in (2.11) behaves numerically better in other examples, if compared to the second one. From the numerical results, as shown in the Figures 2.10 and 2.11, we conclude that for the given basis-functions (2.11) ρ should be chosen somewhere between -1 and -0.1. Hence, our initial guess in the first IQC-analysis problem of

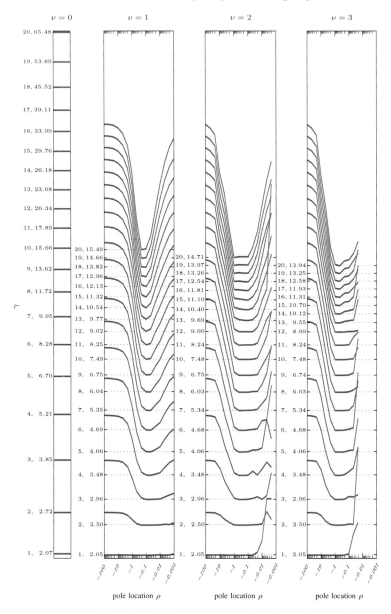

Fig. 2.10 Best achievable induced $\mathscr{L}_2$-gains γ obtained with the first basis-function in (2.11) for various $\rho \in [-100, -0.001]$ and for increasing $\alpha \in [0, 0.2]$.

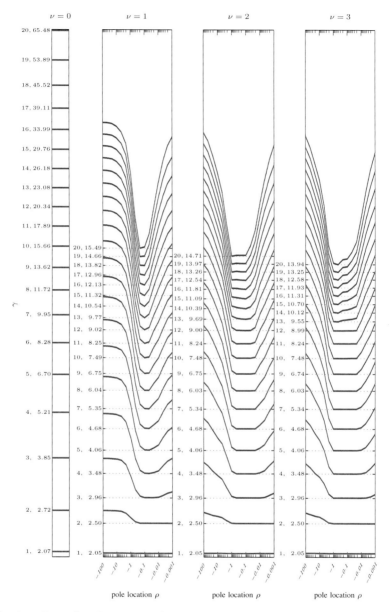

Fig. 2.11 Best achievable induced $\mathscr{L}_2$-gains γ obtained with the second basis-function in (2.11) for various $\alpha \in [0, 0.2]$ and $\rho \in [-100, -0.001]$.

this section seems reasonable. We emphasize that the behavior of these particular basis-functions has been observed in many problems and usually lead to stable computations if ρ is chosen properly. However, how to choose the basis-function optimally remains an open research question.

2.9.2 Robustness analysis of a helicopter

Let us now turn our attention towards a somewhat more realistic example. The book [178] describes the design of a nominal $\mathcal{H}_\infty$-controller for a Westland Lynx twin engine multi-purpose military helicopter. The starting point for this study was an 8th order model $G = (A, B, C, 0)$ given by the realization[2]

$$
\left[\begin{array}{c|c} A & B \\ \hline C & 0 \end{array}\right] =
\left[\begin{array}{cccccccc|cccc}
0 & 0 & 0 & 1.00 & 0.05 & 0 & 0 & 0 & 0 & 0 & 0 & 0 \\
0 & 0 & 1.00 & 0 & 0.06 & 0 & 0 & 0 & 0 & 0 & 0 & 0 \\
0 & 0 & -11.57 & -2.54 & -0.06 & 0.11 & -0.10 & 0.01 & 0.12 & 0.08 & -2.75 & -0.02 \\
0 & 0 & 0.44 & -2.00 & 0 & 0.02 & 0.02 & 0 & -0.03 & 0.48 & 0.01 & 0 \\
0 & 0 & -2.04 & -0.46 & -0.74 & 0.02 & -0.01 & 0 & 0.30 & 0.02 & -0.50 & -0.21 \\
-32.10 & 0 & -0.50 & 2.30 & 0 & -0.02 & -0.02 & 0.02 & 0.29 & -0.54 & -0.02 & 0 \\
0.10 & 32.06 & -2.35 & -0.50 & 0.83 & 0.02 & -0.04 & 0 & -0.02 & 0.02 & -0.54 & 0.24 \\
-1.91 & 1.71 & 0 & -0.06 & 0 & 0.01 & 0 & -0.29 & -4.82 & 0 & 0 & 0 \\
\hline
0 & 0 & 0 & 0 & 0 & 0.06 & 0.05 & -1.00 & 0 & 0 & 0 & 0 \\
1 & 0 & 0 & 0 & 0 & 0 & 0 & 0 & 0 & 0 & 0 & 0 \\
0 & 1 & 0 & 0 & 0 & 0 & 0 & 0 & 0 & 0 & 0 & 0 \\
0 & 0 & 0 & -0.05 & 1.00 & 0 & 0 & 0 & 0 & 0 & 0 & 0 \\
0 & 0 & 1.00 & 0 & 0 & 0 & 0 & 0 & 0 & 0 & 0 & 0 \\
0 & 0 & 0 & 1.00 & 0 & 0 & 0 & 0 & 0 & 0 & 0 & 0 \\
\end{array}\right],
$$

with respectively

1. the pitch attitude, roll attitude, roll rate (body axis), pitch rate (body axis), yaw rate, forward velocity, lateral velocity and vertical velocity as states;
2. the heave velocity, pitch attitude, roll attitude and heading rate as the four primary controlled outputs, as well as the roll and pitch rate as two additional measurement outputs;
3. the main rotor collective, longitudinal cyclic, lateral cyclic and tail rotor collective as blade angle control inputs.

The authors propose a disturbance rejection design in order to cope with the atmospheric turbulence. The design is based on the weighted open-loop plant

$$
\begin{pmatrix} z_1 \\ z_2 \\ y \end{pmatrix} =
\begin{pmatrix} W_{z_1} W_{w_1} & -W_{z_1} G_d W_{w_2} & -W_{z_1} G \\ 0 & 0 & W_{z_2} \\ W_{w_1} & -G_d W_{w_2} & -G \end{pmatrix}
\begin{pmatrix} w_1 \\ w_2 \\ u \end{pmatrix},
$$

---

[2] The values in this realization were rounded of to 2 digits in order to fit the model on this page. This changes the eigenvalues of A only very slightly. The exact model can be found on http://www.nt.ntnu.no/users/skoge/book

in accordance with the system interconnection depicted in Figure 2.12. Here the disturbance model G_d is obtained from G, by assuming that the atmospheric turbulence can be modeled as gust velocity components that perturb the forward, lateral and vertical velocity states. Then G_d is given by $G_d = (A, B_d, C, 0)$, where B_d is defined by the last three columns of A (i.e. $B_d := A \cdot \mathrm{col}(0_{5 \times 3}, I_3)$). The performance weights W_{z_1}, W_{z_2}, W_{w_1} and W_{w_2} are given by

$$\begin{aligned}
W_{z_1} &= \mathrm{diag}\left(\tfrac{0.5(s+12)}{(s+0.012)}, \ \tfrac{0.89(s+2.81)}{(s+0.005)} \cdot I_2, \ \tfrac{0.5(s+10)}{(s+0.01)}, \ \tfrac{2s}{(s+4)(s+4.5)} \cdot I_2 \right), \\
W_{z_2} &= \tfrac{(s+0.0001)}{(s+10)} \cdot I_4, \\
W_{w_1} &= \mathrm{diag}\left(I_4, \ 0.1 \cdot I_2 \right), \\
W_{w_2} &= 30 \cdot I_3,
\end{aligned}$$

and the controller K was designed with the standard $\mathscr{H}_\infty$-synthesis tools of Matlab and guarantees the induced $\mathscr{L}_2$-gain from $w = \mathrm{col}(w_1, w_2)$ to $z = \mathrm{col}(z_1, z_2)$ to be less than 2.34. With this controller, the resulting weighted closed-loop system has a McMillan degree of 39. For further details on the nominal design we refer the reader to [178].

It is obvious at this point that many (robustness) analysis tests can be performed in order to verify whether or not the design satisfies the desired specifications. As an illustration of the IQC-tools, we investigate how robust $\mathscr{L}_2$-gain performance degrades if the measurement channels are affected by delay uncertainties. For this purpose, we proceed in accordance with the discussion in Section 2.6.5.2 and pull-out the delay uncertainties by defining the delay operator $\Delta_{\mathrm{td}} := (e^{-i\omega\beta} - 1)I_{n_q}$ for some $\beta \in [0, \alpha]$, $\alpha > 0$. Then we obtain the uncertain system interconnection as depicted in Figure 2.13, where W_p and W_q are constant weights that select the measurement channels that are assumed to be affected by time-delay uncertainties (e.g. if all channels are affected then $W_p = I$ and $W_q = I$). The corresponding linear fractional representation is given by

$$\begin{pmatrix} q \\ z_1 \\ z_2 \\ y \end{pmatrix} = \begin{pmatrix} 0 & 0 & W_q G_d W_{w_2} & W_q G \\ -W_{z_1} W_p & W_{z_1} W_{w_1} & -W_{z_1} G_d W_{w_2} & -W_{z_1} G \\ 0 & 0 & 0 & W_{z_2} \\ -W_p & W_{w_1} & -G_d W_{w_2} & -G \end{pmatrix} \begin{pmatrix} p \\ w_1 \\ w_2 \\ u \end{pmatrix}, \quad \begin{matrix} p = \Delta_{\mathrm{td}} q, \\ u = Ky. \end{matrix}$$

Let us first consider the IQC-analysis for the case that all measurement channels are subject to the same maximum time-delay α. The computations were carried out with Matlab/LMIlab and its Robust Control Toolbox on an average desktop computer (Core2Duo 3GHz CPU, 4GB RAM) by employing the IQC-multipliers 2.8 and 2.19 from Section 2.6.5.2 and 2.7.1 respectively, for the delay uncertainty Δ_{td} and induced $\mathscr{L}_2$-gain performance. Here we used the first basis-function ψ_ν in (2.11) with pole-location $\rho = -1$ and for different McMillan degrees $\nu = \{0, 1, 2\}$, yielding LMI-

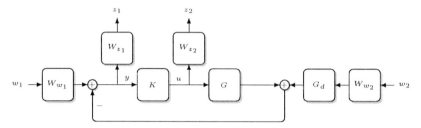

Fig. 2.12 Weighted system interconnection for a helicopter control

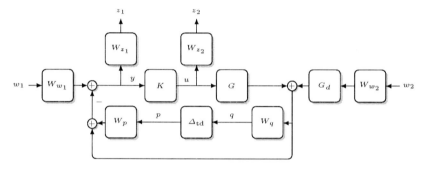

Fig. 2.13 Uncertain weighted system interconnection for a helicopter control

optimization problems with 1369, 2194 and 3271 decision variables respectively. On average it took about 3, 20 and 140 minutes respectively to obtain a feasible solution. In order to compare the results, we also performed a μ-analysis by employing the weighting function (2.42) which is also used in the delay IQC-multiplier (2.44). On average these computations took about 50 seconds[3]. The computed bounds on the induced $\mathscr{L}_2$-gains are plotted in Figure 2.14. Again, we recover nominal performance (i.e. $\gamma = 2.34$) for $\alpha = 0$, regardless of the McMillan degree of the basis-function ψ_ν. As opposed to the first example, there is no extra benefit if the IQC-multipliers are chosen to be dynamic (i.e. if $\nu > 0$). On the other hand, if compared to the μ-analysis, the IQC-tools yield lower guaranteed bounds on the induced $\mathscr{L}_2$-gain for time-delays larger than 0.05 seconds. This is due to the extra constraint (2.48) which is imposed by the delay IQC-multiplier (2.44) and which refines the standard approach used in μ-analysis. We further conclude that robust performance is lost for time-delays larger

---

[3] The robust performance margin was computed with Matlab using the command 'robustperf'. This yields an induced $\mathscr{L}_2$-gain of $1/r$ for time-delay uncertainties $\beta \in [0, r)$. In order to compute the robust stability margin for time-delay uncertainties $\beta \in [0, \alpha)$, we re-scaled the computations appropriately by means of a bi-section based algorithm.

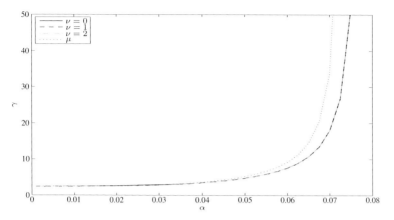

Fig. 2.14 IQC- and μ-analysis results for different values of $\alpha \in [0, 0.08]$ and $\nu = \{0, 1, 2\}$.

than approximately 0.05 seconds; the closed-loop system is guaranteed to remain stable for time-delays not larger than approximately 0.075 seconds.

Because a maximum allowable time-delay of 0.075 seconds might be critical, one can also investigate the effect of time-delay in each individual measurement channel separately. For this purpose, we repeat the previous IQC- and μ-analysis six times with W_p and W_q being the standard unit-vector (i.e. $W_p = W_q^\top = \text{col}(0, \ldots, 0, 1, 0, \ldots, 0)$, where the 1 is located at the ith entry corresponding to the ith measurement channel). The computed bounds on the induced $\mathscr{L}_2$-gains are plotted in Figure 2.15. We observe for the first four measurement channels that the analysis results are identical to the results obtained in the previous experiment. However, only in the first measurement channel the maximum allowable time-delay is about 0.075 seconds. In the other channels we can tolerate significantly larger time-delays up to 0.7 seconds. This is also the case for the 5th and 6th measurement channel. However here the analysis results are a little different. First of all, the benefit of dynamics in the IQC-multiplier (i.e. a higher McMillan degree of ψ_ν, $\nu > 0$), is substantial. And, secondly, the μ-analysis results coincide with the IQC-analysis results for $\nu > 0$. Hence, the extra constraint (2.48) which is imposed by the delay IQC-multiplier (2.44) and which refines the standard approach used in the μ-analysis does not contribute to the IQC-analysis.

Remark 2.14. As we have seen, the computation-time rapidly increases as the number of decision variables grow; up to 140 minutes for 3271 decision variables if the computations are carried out with Matlab/LMIlab. Alternatively, one might consider solvers such as Sedumi [109] or SDPT3 [184, 187], which can reduce the computation-time with about a factor of 10 and sometimes even more. However, it is the author's experience that LMIlab yields the most reliable results.

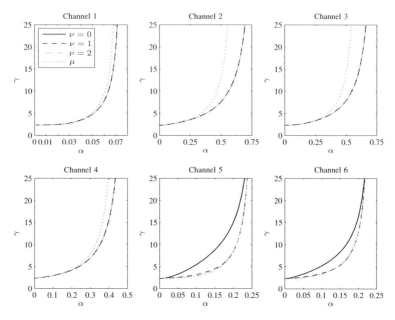

Fig. 2.15 IQC- and μ-analysis results for each measurement channel separately, for different values of $\alpha \in [0, 0.75]$ and $\nu = \{0, 1, 2\}$.

2.10 An alternative proof of the IQC-theorem

The last part of this chapter is a little more technical and is intended for the interested reader. As we have illustrated, the IQC-framework is a powerful tool for the stability and performance analysis of complex system interconnections. On the other hand, another central notion in systems theory is dissipativity [205, 206]. Roughly speaking, a dissipative dynamical system is characterized by the property that, at any time, the amount of energy that is stored in the system never exceeds the amount of energy that has been supplied to the system. This can be formalized for a general class of systems by means of the so-called dissipation inequality which involves a storage function and a supply rate. An appealing aspect for the analysis of linear dissipative systems is that LMIs very naturally emerge. This makes the framework attractive from a computational point-of-view. In fact, one can often relatively easily generalize the notions from linear to time-varying, uncertain and/or nonlinear dissipative systems.

It is relevant to note that there is a common interest in finding a connection between the IQC-framework and the dissipativity approach [22, 73, 171, 6, 182, 71, 207, 113, 177, 176] (see also [46] for another link between the IQC and the multiplier approach [209, 37]). A connection between the two approaches is of particular importance, because it would open the way to merge frequency-domain techniques with time-domain conditions known from Lyapunov-theory. This could lead to generalizations that would be hard to obtain directly in the state-space. So far, a link has only been established for the special case of static and a rather restrictive class of dynamic IQC-multipliers that are satisfied on all finite time horizons. In case of general dynamic IQCs, which only need to hold on infinite time horizons, it remains unclear how to proceed. Here we would like to emphasize that [177] contains a technical glitch and that [176] has been submitted at a later date.

As a contribution, we now provide a novel reformulation of the IQC-theorem which is of independent interest. We will exploit the result for the main contribution of this Section, which is a dissipativity based proof of the IQC-theorem. For a rather general class of dynamic IQC-multipliers it will be shown that, once the conditions of the IQC-theorem are satisfied, it is possible to construct a non-negative Lyapunov function that satisfies a dissipation inequality. The proof relies on (i) the reformulation of the IQC-theorem; (ii) a symmetric Wiener-Hopf factorization [57], which is guaranteed to exist and can be easily constructed through the solution of an algebraic Riccati equation (ARE) [156] (see also [118]); (iii) the gluing lemma, which describes how certain operations on frequency-domain conditions can be performed in the state-space [159]; and (iv) on dissipativity theory. We emphasize that these results have already (almost literally) appeared in [196].

2.10.1 Preparations

Unfortunately it is not possible to prove the IQC-theorem in full generality if we rely on dissipation arguments. As the main restricting hypothesis we need to assume that the IQC-multiplier Π is confined to the set

$$\Pi \subset \left\{ \Pi = \Pi^* \in \mathscr{RL}_\infty^{(n_q+n_p) \times (n_q+n_p)} : \Pi_{11} \succcurlyeq 0, \ \Pi_{22} \preccurlyeq 0 \text{ on } \mathbb{C}^0 \right\}.$$

If this hypothesis is satisfied, then Theorem 2.1 simplifies as follows.

Corollary 2.4. *Assume that*

1. *the feedback interconnection of G and Δ is well-posed;*
2. *the IQC-multiplier $\Pi \in \Pi$ is satisfied for Δ;*
3. *the following FDI is satisfied:*

$$\begin{pmatrix} G \\ I \end{pmatrix}^* \Pi \begin{pmatrix} G \\ I \end{pmatrix} \prec 0 \text{ on } \mathbb{C}^0. \tag{2.91}$$

Then the feedback interconnection (2.1) *is stable.*

We recall that the main results in [116] have been proved by means of a homotopy method, without any inertia properties on Π_{11} and Π_{22}. For this reason, the first two conditions in Theorem 2.1 need to be satisfied with $\tau\Delta$ for all $\tau \in [0,1]$, which implies that $\Pi_{11} \succcurlyeq 0$ on $\mathbb{C}^0$. In addition, if we assume that $\Pi_{22} \prec 0$ on $\mathbb{C}^0$ then

$$IQC(\Pi, q, \Delta(q)) \geq 0 \quad \forall q \in \mathscr{L}_2^{n_q}$$

automatically implies that

$$IQC(\Pi, q, \tau\Delta(q)) \geq 0 \quad \forall q \in \mathscr{L}_2^{n_q} \quad \text{and} \quad \forall \tau \in [0,1).$$

Although this might seem restrictive, we still cover most practical IQC-multipliers that are found in the literature. In fact, apart from the full-block multiplier relaxation schemes in Section 2.6.2.2 that rely on partial convexity, Pólya and sum-of-square (SOS) arguments, all IQC-multipliers presented in this thesis satisfy the above-mentioned inertia constraints. Without too much loss of generality we can hence assume that $\Pi \in \boldsymbol{\Pi}$. Further note that we only require the feedback interconnection of G and $\tau\Delta$ to be well-posed for $\tau = 1$. Consequently, it is possible to consider more general uncertainty sets, and even allow for singletons.

Moving towards robust stability analysis, and as discussed in Section 2.5.1, we parameterize a whole family of IQC-multipliers as

$$\boldsymbol{\Pi} = \{\Psi^* P \Psi : P \in \boldsymbol{P}\}$$

with an LMIable set of symmetric matrices $\boldsymbol{P}$ and with some fixed and typically tall transfer matrix $\Psi \in \mathscr{RH}_\infty^{\bullet \times (n_q + n_p)}$ such that

$$IQC(\Psi^* P \Psi, q, \Delta(q)) \geq 0 \quad \forall q \in \mathscr{L}_2^{n_q}$$

is satisfied for all $P \in \boldsymbol{P}$ and for all $\Delta \in \boldsymbol{\Delta}$. Robust stability for our specialized class of multipliers $\boldsymbol{\Pi}$ can now be guaranteed as follows.

Corollary 2.5. *Suppose that the first two conditions in Corollary 2.4 hold for all* $\Delta \in \boldsymbol{\Delta}$ *with* $\boldsymbol{\Pi}$ *replaced by* $\Psi^* \boldsymbol{P} \Psi$. *Then the feedback interconnection* (2.1) *is robustly stable if there exists some* $P \in \boldsymbol{P}$ *for which the following FDI holds:*

$$\begin{pmatrix} G \\ I \end{pmatrix}^* \Psi^* P \Psi \begin{pmatrix} G \\ I \end{pmatrix} \prec 0 \text{ on } \mathbb{C}^0. \tag{2.92}$$

In the sequel we will provide a proof this version of the IQC-theorem. We recall that, if $\boldsymbol{P}$ is LMIable, condition (2.92) can be easily checked numerically by exploiting Lemma 2.2. Indeed, with

$$\Psi \begin{pmatrix} G \\ I \end{pmatrix} = \left[\begin{array}{c|c} A & B \\ \hline C & D \end{array} \right] \tag{2.93}$$

and $\mathrm{eig}(A) \subset \mathbb{C}^-$, (2.92) is equivalent to the existence of some matrix $X = X^\top$ for which the following LMI holds true:

$$\begin{pmatrix} I & 0 \\ A & B \\ C & D \end{pmatrix}^\top \begin{pmatrix} 0 & X & 0 \\ X & 0 & 0 \\ 0 & 0 & P \end{pmatrix} \begin{pmatrix} I & 0 \\ A & B \\ C & D \end{pmatrix} \prec 0. \tag{2.94}$$

2.10.2 Key difficulties to resolve

Let us briefly discuss the main difficulties, which occur when trying to prove Corollary 2.5 with dissipation arguments. For this purpose, consider the open sets $V \subset \mathbb{R}^\bullet$, $U \subset \mathbb{R}^\bullet$, $Y \subset \mathbb{R}^\bullet$ and define the sufficiently smooth mappings $f : V \times U \to \mathbb{R}^\bullet$, $g : V \times U \to Y$ and $s : U \times Y \to \mathbb{R}$. Then we can formally introduce the following definition of dissipativity [205].

Definition 2.5 (Dissipativity). The system $\dot{x} = f(x, u)$, $y = g(x, u)$ is dissipative with respect to the supply rate $s(\cdot, \cdot)$ if there exists a storage function $X : V \to \mathbb{R}$ such that

$$X(x(t_2)) + \int_{t_1}^{t_2} s(u(t), y(t)) dt \le X(x(t_1)) \tag{2.95}$$

holds for all trajectories with $x(t) \in V$, $u(t) \in U$, $y(t) \in Y$ for $t \in [t_1, t_2]$ and for all $t_1 < t_2$.

Let us first consider the special case of static IQC-multipliers (i.e. $\Psi = I$). For any trajectory of the feedback interconnection (2.1) with initial condition $x(0)$ we can right and left-multiply (2.94) with $\mathrm{col}(x(t), p(t))$ and its transpose. Then, after integration on $[0, T]$, we obtain

$$x(T)^\top X x(T) + \int_0^T y(t)^\top P y(t) dt \le x(0)^\top X x(0) \quad \forall T \ge 0. \tag{2.96}$$

Here $t \to y(t)^\top P y(t)$ is the supply function with

$$y = \begin{pmatrix} Gp \\ p \end{pmatrix} = \begin{pmatrix} q \\ \Delta(q) \end{pmatrix}.$$

We observe that the left-upper block of P in the partition $\mathrm{col}(G, I)$ is non-negative just because $\Pi = \Psi^* P \Psi = P \in \boldsymbol{\Pi}$. Therefore, (2.94) implies that $X \succ 0$ if and only if

$\mathrm{eig}(A) \subset \mathbb{C}^-$. Further note that

$$\int_0^T y(t)^\top P y(t) dt \geq 0 \qquad (2.97)$$

holds with $y = \mathrm{col}(q, \Delta(q))$ for all $T > 0$, again, just because $\Pi = \Psi^* P \Psi = P \in \boldsymbol{\Pi}$. These two properties allow us to prove Corollary 2.92 for the special case of static IQC-multipliers by classical arguments.

On the other hand, for the general case of dynamic IQC-multipliers (i.e. Ψ is an arbitrary tall and stable transfer matrix), one cannot directly relate the trajectories of (2.93) to the trajectories of the feedback interconnection (2.1), since Ψ is not stably invertible. Moreover, P is in general an indefinite matrix. Hence, $\mathrm{eig}(A) \subset \mathbb{C}^-$ neither implies, nor is implied by positive definiteness of X. Further note that (2.97) is in general only non-negative for $T = \infty$. All this prevents us to prove stability as for static multipliers. In the sequel we will show how to resolve these difficulties.

2.10.3 A novel reformulation of the IQC-theorem

As a preparation for the proof of Corollary 2.4, we will first reformulate the feedback interconnection (2.1) as well as the conditions in Corollary 2.4. For this purpose, let us introduce the signals $q = q_1 = p_1$ and $p = p_2 = q_2$ with $p_1, q_1, p_2, q_2 \in \mathscr{L}_{2e}^\bullet$. Then (2.1) can be expressed as

$$\underbrace{\begin{pmatrix} q_1 \\ q_2 \end{pmatrix}}_{q_\mathrm{e}} = \underbrace{\begin{pmatrix} -I & 2G \\ 0 & I \end{pmatrix}}_{G_\mathrm{e}} \underbrace{\begin{pmatrix} p_1 \\ p_2 \end{pmatrix}}_{p_\mathrm{e}} + \underbrace{\begin{pmatrix} 2\mu_1 \\ 0 \end{pmatrix}}_{\mu_\mathrm{s}},$$

$$\underbrace{\begin{pmatrix} p_1 \\ p_2 \end{pmatrix}}_{p_\mathrm{e}} = \underbrace{\begin{pmatrix} I & 0 \\ 2\Delta & -I \end{pmatrix}}_{\Delta_\mathrm{e}(q_\mathrm{e})} \underbrace{\begin{pmatrix} q_1 \\ q_2 \end{pmatrix}}_{} + \underbrace{\begin{pmatrix} 0 \\ 2\eta_2 \end{pmatrix}}_{\eta_\mathrm{s}}.$$

Moreover, by replacing the structured external disturbances μ_s and η_s by the unstructured external disturbances $\mu_\mathrm{e} := \mathrm{col}(\mu_1, \mu_2)$ and $\eta_\mathrm{e} := \mathrm{col}(\eta_1, \eta_2)$ respectively, we obtain the following extended feedback interconnection:

$$\begin{aligned} q_\mathrm{e} &= G_\mathrm{e} p_\mathrm{e} + \mu_\mathrm{e}, \\ p_\mathrm{e} &= \Delta_\mathrm{e}(q_\mathrm{e}) + \eta_\mathrm{e}. \end{aligned} \qquad (2.98)$$

If we also introduce the extended IQC-multipliers

$$\Pi_{\mathrm{e}} := \begin{pmatrix} 0 & \Pi_{\epsilon} \\ \Pi_{\epsilon} & 0 \end{pmatrix}, \Pi_{\epsilon} := \begin{pmatrix} \Pi_{11\epsilon} & \Pi_{12} \\ \Pi_{12}^{*} & \Pi_{22} \end{pmatrix} = \begin{pmatrix} \Pi_{11} & \Pi_{12} \\ \Pi_{12}^{*} & \Pi_{22} \end{pmatrix} + \begin{pmatrix} \epsilon I & 0 \\ 0 & 0 \end{pmatrix}$$

for $\epsilon \geq 0$, we can state the following result.

Theorem 2.4.

1. *The feedback interconnection* (2.1) *is stable if and only if the feedback interconnection* (2.98) *is stable.*
2. *For all* $\epsilon \geq 0$ *the IQC defined by* Π_{ϵ} *is satisfied for* Δ *if and only if the IQC defined by* Π_{e} *is satisfied for* Δ_{e}.
3. *The FDI* (2.91) *is satisfied if and only if there exists a small* $\epsilon > 0$ *such that*

$$\begin{pmatrix} G_{\mathrm{e}} \\ I \end{pmatrix}^{*} \Pi_{\mathrm{e}} \begin{pmatrix} G_{\mathrm{e}} \\ I \end{pmatrix} \prec 0 \ \text{ on } \ \mathbb{C}^{0}. \tag{2.99}$$

Proof. Let us start by proving the first statement. For this purpose, let us introduce the notation

$$\mathcal{I}(\Sigma_1, \Sigma_2) := \begin{pmatrix} I & -\Sigma_1 \\ -\Sigma_2 & I \end{pmatrix},$$

and observe that $\mathcal{I}(G_{\mathrm{e}}, \Delta_{\mathrm{e}})$ can be written as

$$\mathcal{I}(G_{\mathrm{e}}, \Delta_{\mathrm{e}}) = T_2 T_3 T_4 = T_2 T_3 T_5 T_5 T_4,$$

with $T_1 = \mathrm{diag}(-2I, 2I)$, $T_3 = \mathrm{diag}(\mathcal{I}(G, \Delta), -I)$, $T_5 = \mathrm{diag}(I, -I)$,

$$T_2 = \begin{pmatrix} G_{\mathrm{e}} T_1 & -G_{\mathrm{e}} \\ 0 & I \end{pmatrix} \ \text{ and } \ T_4 = \begin{pmatrix} I & 0 \\ \Delta_{\mathrm{e}} & -I \end{pmatrix}.$$

Hence, if $\mathcal{I}(G, \Delta)$ has a causal inverse on $\mathscr{L}_{2e}$, we have that

$$\mathcal{I}(G_{\mathrm{e}}, \Delta_{\mathrm{e}})^{-1} = T_4^{-1} T_3^{-1} T_2^{-1} = (T_5 T_4)^{-1} (T_3 T_5)^{-1} T_2^{-1}.$$

Also note that there exist constants $\alpha_2, \alpha_{2i}, \alpha_4, \alpha_{4i} > 0$ with

$$\|T_2\|_{\mathrm{i}2} \leq \alpha_2, \ \ \|T_2^{-1}\|_{\mathrm{i}2} \leq \alpha_{2i}, \ \ \|T_4\|_{\mathrm{i}2} \leq \alpha_4, \ \ \|T_4^{-1}\|_{\mathrm{i}2} \leq \alpha_{4i}.$$

Hence, if we assume that the feedback interconnection (2.1) is well-posed and that there exists some $\alpha_1 > 0$ such that $\|\mathcal{I}(G, \Delta)^{-1}\|_{\mathrm{i}2} \leq \alpha_1$, then

$$\|\mathcal{I}(G,\Delta)^{-1}\|_{\text{i2}} + 1 \leq \alpha_1 + 1$$
$$\Rightarrow \quad \|(T_5 T_3)^{-1}\|_{\text{i2}} \leq \alpha_1 + 1$$
$$\Rightarrow \quad \|T_2^{-1}\|_{\text{i2}} \|(T_5 T_3)^{-1}\|_{\text{i2}} \|(T_5 T_4)^{-1}\|_{\text{i2}} \leq (\alpha_1 + 1)\alpha_{2i}\alpha_{4i}$$
$$\Rightarrow \quad \|(T_5 T_4)^{-1}(T_3 T_5)^{-1} T_2^{-1}\|_{\text{i2}} \leq (\alpha_1 + 1)\alpha_{2i}\alpha_{4i}$$
$$\Rightarrow \quad \|\mathcal{I}(G_{\text{e}},\Delta_{\text{e}})^{-1}\|_{\text{i2}} \leq (\alpha_1 + 1)\alpha_{2i}\alpha_{4i}.$$

Therefore, stability of the feedback interconnection (2.98) is implied by stability of the feedback interconnection (2.1). Conversely, suppose that the feedback interconnection (2.98) is well-posed and that there exists some $\alpha_3 > 0$ such that $\|\mathcal{I}(G_{\text{e}},\Delta_{\text{e}})^{-1}\|_{\text{i2}} \leq \alpha_3$, then

$$\|T_4^{-1} T_3^{-1} T_2^{-1}\|_{\text{i2}} \leq \alpha_3$$
$$\Rightarrow \quad \|T_4\|_{\text{i2}} \|T_4^{-1} T_3^{-1} T_2^{-1}\|_{\text{i2}} \|T_2\|_{\text{i2}} \leq \alpha_2 \alpha_3 \alpha_4$$
$$\Rightarrow \quad \|T_3^{-1}\|_{\text{i2}} \leq \alpha_2 \alpha_3 \alpha_4$$
$$\Rightarrow \quad \|\mathcal{I}(G,\Delta)^{-1}\|_{\text{i2}} - 1 \leq \alpha_2 \alpha_3 \alpha_4$$
$$\Rightarrow \quad \|\mathcal{I}(G,\Delta)^{-1}\|_{\text{i2}} \leq \alpha_2 \alpha_3 \alpha_4 + 1.$$

Therefore, stability of the feedback interconnection (2.1) is implied by stability of the feedback interconnection (2.98).

In order to prove the second statement, we need to show that, for all $\epsilon \geq 0$,

$$IQC(\Pi_\epsilon, q, \Delta(q)) \geq 0 \quad \forall q \in \mathscr{L}_2 \tag{2.100}$$

is equivalent to

$$IQC(\Pi_{\text{e}}, q_{\text{e}}, \Delta_{\text{e}}(q_{\text{e}})) \geq 0 \quad \forall q_{\text{e}} \in \mathscr{L}_2. \tag{2.101}$$

For this purpose, let us introduce the Fourier-transforms $\hat{q}_1$ and $\widehat{\Delta(q_1)}$ of the signals $q_1 = q$ and $p_2 = \Delta(q_1)$, respectively, and observe that (2.100) is equivalent to

$$\int_{-\infty}^{\infty} \hat{q}_1^* \Pi_{11\epsilon} \hat{q}_1 + 2\text{Re}\left(\hat{q}_1^* \Pi_{12} \widehat{\Delta(q_1)}\right) + \widehat{\Delta(q_1)}^* \Pi_{22} \widehat{\Delta(q_1)} d\omega \geq 0. \tag{2.102}$$

Since $\Pi_{22} \preccurlyeq 0$ on $\mathbb{C}^0$ we infer that

$$\left(\widehat{\Delta(q_1)} - \hat{q}_2\right)^* \Pi_{22} \left(\widehat{\Delta(q_1)} - \hat{q}_2\right) \leq 0 \quad \text{a.e. on } \mathbb{C}^0$$

for all $q_1, q_2 \in \mathscr{L}_2$. Hence for all $q_1, q_2 \in \mathscr{L}_2$ it holds that

$$\widehat{\Delta(q_1)}^* \Pi_{22} \widehat{\Delta(q_1)} \leq 2\text{Re}\left(\hat{q}_2^* \Pi_{22} \widehat{\Delta(q_1)}\right) - \hat{q}_2^* \Pi_{22} \hat{q}_2 \quad \text{a.e. on } \mathbb{C}^0$$

Therefore (2.102) implies

$$\int_{-\infty}^{\infty} \hat{q}_1^* \Pi_{11\epsilon}\hat{q}_1 + 2\text{Re}\left(\hat{q}_1^* \Pi_{12}\widehat{\Delta(q_1)}\right) + 2\text{Re}\left(\hat{q}_2^* \Pi_{22}\widehat{\Delta(q_1)}\right) - \hat{q}_2^* \Pi_{22}\hat{q}_2 d\omega \geq 0$$

and thus

$$\text{Re}\int_{-\infty}^{\infty} \begin{pmatrix} \hat{q}_1 \\ \hat{q}_2 \end{pmatrix}^* \begin{pmatrix} \Pi_{11\epsilon} & \Pi_{12} \\ \Pi_{12}^* & \Pi_{22} \end{pmatrix} \begin{pmatrix} \hat{q}_1 \\ 2\widehat{\Delta(q_1)} - \hat{q}_2 \end{pmatrix} d\omega \geq 0.$$

By defining the Fourier transforms $\hat{q}_\text{e} = \text{col}(\hat{q}_1, \hat{q}_2)$ and $\hat{p}_\text{e} = \text{col}(\hat{p}_1, \hat{p}_2)$ of the signals $q_\text{e} = \text{col}(q_1, q_2)$ and $p_\text{e} = \text{col}(p_1, p_2)$, respectively, and observing that $\widehat{\Delta_\text{e}(q_\text{e})} = \text{col}(\hat{q}_1, 2\widehat{\Delta(q_1)} - \hat{q}_2)$ is nothing but the Fourier transform of $p_\text{e} = \Delta_\text{e}(q_\text{e})$, we can infer that (2.101) is satisfied for all $\epsilon \geq 0$.

In order to show that (2.101) implies (2.100), set $\tilde{q}_\text{e} = \text{col}(q, \Delta(q))$ for $q \in \mathcal{L}_2$. By the definition of Δ_e we have $\tilde{p}_\text{e} = \Delta_\text{e}(\tilde{q}_\text{e}) = \text{col}(q, \Delta(q)) = \tilde{q}_\text{e}$. Hence, for $q_\text{e} = \tilde{q}_\text{e}$ the lifted IQC (2.101) simplifies into

$$\Sigma(\Pi_\text{e}, \tilde{q}_\text{e}, \tilde{q}_\text{e}) \geq 0, \quad \tilde{q}_\text{e} = \Delta_\text{e}(\tilde{q}_\text{e}), \quad \tilde{q}_\text{e} \in \mathcal{L}_2.$$

Direct inspection reveals that this is nothing but (2.100).

It remains show the third statement. For this purpose, observe that the FDI (2.91) can be written as

$$G^* \Pi_{11} G + G^* \Pi_{12} + \Pi_{12}^* G + \Pi_{22} \prec 0 \text{ on } \mathbb{C}^0.$$

Clearly there exists some small $\epsilon > 0$ such that

$$G^* (\Pi_{11} + \epsilon I) G + G^* \Pi_{12} + \Pi_{12}^* G + \Pi_{22} \prec 0 \text{ on } \mathbb{C}^0 \tag{2.103}$$

persists to hold. With $\Pi_{11\epsilon} = \Pi_{11} + \epsilon I$ we can then apply the Schur complement in order to obtain

$$\begin{pmatrix} -\Pi_{11\epsilon} & \Pi_{11\epsilon}G \\ G^* \Pi_{11\epsilon} & G^* \Pi_{12} + \Pi_{12}^* G + \Pi_{22} \end{pmatrix} \prec 0 \text{ on } \mathbb{C}^0,$$

and this can be expressed as

$$\text{He}\left(\begin{pmatrix} \Pi_{11\epsilon} & \Pi_{12} \\ \Pi_{12}^* & \Pi_{22} \end{pmatrix} \begin{pmatrix} -I & 2G \\ 0 & I \end{pmatrix}\right) \prec 0 \text{ on } \mathbb{C}^0,$$

which is (2.99). The reverse direction directly follows from pre- and post-multiplying (2.99) with $\text{col}(G, I)^*$ and $\text{col}(G, I)$ respectively.

∎

Remark 2.15. Note that Theorem 2.4 is an auxiliary result of independent interest. For example, IQC-multipliers of the form

$$\Pi = \begin{pmatrix} 0 & \Pi_{12} \\ \Pi_{12}^* & 0 \end{pmatrix}$$

have emerged in a number of gain-scheduled and distributed controller synthesis problems [156, 158]. Since the augmented IQC-multiplier Π_e possesses the very same structure, Theorem 2.4 might be useful for generalizations to larger classes of dynamic IQC-multipliers. Further, we will exploit the result in order to proof Corollary 2.4.

2.10.4 The proof

As discussed in Section 2.5.1 and 2.10.1, one of the key features of the IQC-framework involves the computational search of a symmetric matrix $P \in \boldsymbol{P}$ that satisfies the FDI in Corollary 2.5 in order to verify whether or not the feedback interconnection (2.1) is robustly stable for a certain class $\boldsymbol{\Delta}$. It hence makes sense to prove Corollary 2.4 by taking Corollary 2.5 as a starting point. For this reason, let us assume that the first two conditions in Corollary 2.4 hold for all $\Delta \in \boldsymbol{\Delta}$, with $\boldsymbol{\Pi}$ replaced by $\Psi^* \boldsymbol{P} \Psi$, and that there exists some $P \in \boldsymbol{P}$ for which the FDI (2.92) is satisfied. Let us also assume, for notational simplicity, that $\Pi_{11} \succ 0$ on $\mathbb{C}^0$ such that ϵ in Theorem 2.4 can be set to zero. By Theorem 2.4 the FDI (2.92) is then equivalent to

$$\begin{pmatrix} \Psi G_e \\ \Psi \end{pmatrix}^* \begin{pmatrix} 0 & P \\ P & 0 \end{pmatrix} \begin{pmatrix} \Psi G_e \\ \Psi \end{pmatrix} \prec 0 \text{ on } \mathbb{C}^0. \tag{2.104}$$

If we assume Ψ and G_e to admit the realizations $(A_\Psi, B_\Psi, C_\Psi, D_\Psi)$ and (A_e, B_e, C_e, D_e), with $\mathrm{eig}(A_\Psi) \subset \mathbb{C}^-$ and $\mathrm{eig}(A_e) \subset \mathbb{C}^-$, we can define the realization

$$\begin{pmatrix} \Psi G_e \\ \Psi \end{pmatrix} = \left[\begin{array}{ccc|c} A_\Psi & 0 & B_\Psi C_e & B_\Psi D_e \\ 0 & A_\Psi & 0 & B_\Psi \\ 0 & 0 & A_e & B_e \\ \hline C_\Psi & 0 & D_\Psi C_e & D_\Psi D_e \\ 0 & C_\Psi & 0 & D_\Psi \end{array} \right] .$$

By the KYP-lemma (2.104) is equivalent to the existence of a matrix $X_e = X_e^\top$ for which the following LMI is satisfied:

$$
\begin{pmatrix} \star \\ \hline \star \\ \star \\ \star \\ \hline \star \\ \star \\ \star \end{pmatrix}^{\top}
\begin{pmatrix}
\begin{array}{cc|cc} 0 & X_{\mathrm{e}} & 0 & 0 \\ \hline X_{\mathrm{e}} & 0 & 0 & 0 \\ \hline 0 & 0 & 0 & P \\ 0 & 0 & P & 0 \end{array}
\end{pmatrix}
\begin{pmatrix}
\begin{array}{ccc|c} I & & & 0 \\ \hline A_\Psi & 0 & B_\Psi C_{\mathrm{e}} & B_\Psi D_{\mathrm{e}} \\ 0 & A_\Psi & 0 & B_\Psi \\ 0 & 0 & A_{\mathrm{e}} & B_{\mathrm{e}} \\ \hline C_\Psi & 0 & D_\Psi C_{\mathrm{e}} & D_\Psi D_{\mathrm{e}} \\ 0 & C_\Psi & 0 & D_\Psi \end{array}
\end{pmatrix} \prec 0. \qquad (2.105)
$$

Here X_{e} is assumed to be partitioned as

$$
X_{\mathrm{e}} = \begin{pmatrix} X_{11} & X_{12} & X_{13} \\ X_{12}^{\top} & X_{22} & X_{23} \\ X_{13}^{\top} & X_{23}^{\top} & X_{33} \end{pmatrix},
$$

were X_{11}, X_{22} and X_{33} have compatible dimensions with A_Ψ, A_Ψ and A_{e} respectively.

Now observe that V_{e} is in general an indefinite matrix. As discussed in Section 2.10.2, this is one of the main problems that prevented us from concluding stability of the feedback interconnection (2.1). In order to overcome this difficulty we need to factorize the IQC-multiplier $\Psi^* P \Psi$ as $\tilde{\Psi}^* \tilde{P} \tilde{\Psi}$, with $\tilde{P} = \tilde{P}^{\top}$, $\tilde{\Psi}, \tilde{\Psi}^{-1} \in \mathscr{RH}_\infty$. For this reason, let us collect some existing insights from the literature [156].

Lemma 2.6. *Let $\Pi \in \mathscr{RL}_\infty^{\bullet \times \bullet}$. Then Π satisfies $\Pi = \Pi^*$ if and only if it admits a realization with the structure*

$$
\Pi = \left[\begin{array}{cc|c} -A_\Pi^{\top} & E_\Pi & C_\Pi^{\top} \\ 0 & A_\Pi & B_\Pi \\ \hline -B_\Pi^{\top} & C_\Pi & D_\Pi \end{array} \right], \qquad (2.106)
$$

with $D_\Pi = D_\Pi^{\top}$, $E_\Pi = E_\Pi^{\top}$ and $\mathrm{eig}(A_\Pi) \subset \mathbb{C}^-$.

Lemma 2.7. *Let $G \in \mathscr{RH}_\infty$ and $\Pi = \Pi^* \in \mathscr{RL}_\infty$ satisfy*

$$
\mathrm{He}(\Pi G) \prec 0 \text{ on } \mathbb{C}^0.
$$

Then there exists a $\tilde{\Psi} \in \mathscr{RH}_\infty$ with $\tilde{\Psi}(\infty) = I$, $\tilde{\Psi}^{-1} \in \mathscr{RH}_\infty$ such that

$$
\Pi = \tilde{\Psi}^* \Pi(\infty) \tilde{\Psi}.
$$

If Π is realized as in (2.106), then D_Π is invertible and the ARE

$$
A_\Pi^{\top} Z + Z A_\Pi + E_\Pi - \left(B_\Pi^{\top} Z + C_\Pi \right)^{\top} D_\Pi^{-1} \left(B_\Pi^{\top} Z + C_\Pi \right) = 0 \qquad (2.107)
$$

has a unique stabilizing solution $Z = Z^{\top}$, i.e. $A_\Pi - B_\Pi \tilde{C}_\Pi$ is Hurwitz for

$$\tilde{C}_\Pi := D_\Pi^{-1}(B_\Pi^\top Z + C_\Pi).$$

A symmetric canonical Wiener-Hopf factorization is given by

$$\Pi = \tilde{\Psi}^* \Pi(\infty)\tilde{\Psi} \quad with \quad \tilde{\Psi} = (A_\Pi, B_\Pi, \tilde{C}_\Pi, I).$$

Let us observe that

$$\Psi^* P\Psi = \left[\begin{array}{cc|c} -A_\Psi^\top & C_\Psi^\top P C_\Psi & C_\Psi^\top P D_\Psi \\ 0 & A_\Psi & B_\Psi \\ \hline -B_\Psi^\top & D_\Psi^\top P C_\Psi & D_\Psi^\top P D_\Psi \end{array}\right] \qquad (2.108)$$

has precisely the structure of (2.106). Hence, by Lemma 2.7, feasibility of (2.105) implies the existence of the stabilizing solution Z of the ARE (2.107) corresponding to realization (2.108) which can be expressed as

$$\begin{pmatrix} I & 0 \\ A_\Psi & B_\Psi \\ C_\Psi & D_\Psi \end{pmatrix}^\top \begin{pmatrix} 0 & Z & 0 \\ Z & 0 & 0 \\ 0 & 0 & P \end{pmatrix} \begin{pmatrix} I & 0 \\ A_\Psi & B_\Psi \\ C_\Psi & D_\Psi \end{pmatrix} = \begin{pmatrix} \tilde{C}_\Psi^\top \\ I \end{pmatrix} \tilde{P} \begin{pmatrix} \tilde{C}_\Psi & I \end{pmatrix}, \qquad (2.109)$$

We can thus identify G and Π in Lemma 2.7 with G_{e} and $\Psi^* P\Psi$ respectively, and define the realization $\tilde{\Psi} = (A_\Psi, B_\Psi, \tilde{C}_\Psi, I)$ such that

$$\Psi^* P\Psi = \tilde{\Psi}^* \tilde{P}\tilde{\Psi},$$

with

$$\tilde{P} = D_\Psi^\top P D_\Psi, \quad \tilde{C}_\Psi = (D_\Psi^\top P D_\Psi)^{-1}(B_\Psi^\top Z + D_\Psi^\top P C_\Psi) \quad \text{and} \quad \tilde{\Psi}, \tilde{\Psi}^{-1} \in \mathscr{RH}_\infty.$$

It is now possible to merge (2.109) and (2.105) by applying the gluing-lemma [159] (See Appendix F.1). This yields the inequality

$$\begin{pmatrix} \star \\ \hline \star \\ \star \\ \hline \star \\ \hline \star \\ \star \end{pmatrix}^\top \begin{pmatrix} 0 & \tilde{X}_{\mathrm{e}} & 0 & 0 \\ \hline \tilde{X}_{\mathrm{e}} & 0 & 0 & 0 \\ \hline 0 & 0 & 0 & \tilde{P} \\ 0 & 0 & \tilde{P} & 0 \end{pmatrix} \begin{pmatrix} I & 0 \\ \hline A_\Psi & 0 & B_\Psi C_{\mathrm{e}} & B_\Psi D_{\mathrm{e}} \\ 0 & A_\Psi & 0 & B_\Psi \\ 0 & 0 & A_{\mathrm{e}} & B_{\mathrm{e}} \\ \hline \tilde{C}_\Psi & 0 & C_{\mathrm{e}} & D_{\mathrm{e}} \\ 0 & \tilde{C}_\Psi & 0 & I \end{pmatrix} \prec 0 \qquad (2.110)$$

for

$$\tilde{X}_{\mathrm{e}} := \begin{pmatrix} X_{11} & X_{12} - Z & X_{13} \\ X_{12}^\top - Z & X_{22} & X_{23} \\ X_{13}^\top & X_{23}^\top & X_{33} \end{pmatrix}.$$

The simple congruence transformation with

$$\begin{pmatrix} I & 0 & 0 & 0 \\ 0 & I & 0 & 0 \\ 0 & 0 & I & 0 \\ 0 & -\tilde{C}_\Psi & 0 & I \end{pmatrix}^\top \quad (2.110) \quad \begin{pmatrix} I & 0 & 0 & 0 \\ 0 & I & 0 & 0 \\ 0 & 0 & I & 0 \\ 0 & -\tilde{C}_\Psi & 0 & I \end{pmatrix} \prec 0$$

then reveals that (2.110) is equivalent to

$$\text{He}\begin{pmatrix} \tilde{X}_e \tilde{A}_e & \tilde{X}_e \tilde{B}_e \\ \tilde{P}\tilde{C}_e & \tilde{P}\tilde{D}_e \end{pmatrix} \prec 0, \qquad (2.111)$$

where

$$\left[\begin{array}{c|c} \tilde{A}_e & \tilde{B}_e \\ \hline \tilde{C}_e & \tilde{D}_e \end{array}\right] = \left[\begin{array}{ccc|c} A_\Psi & -B_\Psi D_e \tilde{C}_\Psi & B_\Psi C_e & B_\Psi D_e \\ 0 & A_\Psi - B_\Psi \tilde{C}_\Psi & 0 & B_\Psi \\ 0 & -B_e \tilde{C}_\Psi & A_e & B_e \\ \hline \tilde{C}_\Psi & -D\tilde{C}_\Psi & C_e & D_e \end{array}\right] = \tilde{\Psi} G_e \tilde{\Psi}^{-1} =: \tilde{G}_e. \quad (2.112)$$

Clearly, since A_Ψ, $A_\Psi - B_\Psi \tilde{C}_\Psi$ and A_e are all Hurwitz we can conclude the same for $\tilde{A}_e$. Moreover, the left-upper block of (2.111) reads as $\text{He}(\tilde{X}_e \tilde{A}_e) \prec 0$ which implies that $\tilde{X}_e \succ 0$.

Now recall from the second statement in Theorem 2.4 that

$$\Sigma(\Pi, q, \Delta(q)) \geq 0 \quad \text{is equivalent to} \quad \Sigma(\Pi_e, q_e, \Delta_e(q_e)) \geq 0.$$

With $\Pi = \tilde{\Psi}^* \tilde{P} \tilde{\Psi}$, $\tilde{q}_e = \tilde{\Psi} q_e$ and $\tilde{p}_e = \tilde{\Psi} p_e$, we obtain

$$\Sigma(\tilde{P}, \tilde{q}_e, \tilde{\Delta}_e(\tilde{q}_e)) \geq 0,$$

where $\tilde{\Delta}_e = \tilde{\Psi} \Delta_e \tilde{\Psi}^{-1}$ defines a bounded and causal operator on $\mathscr{L}_{2e}$. We conclude that

$$\int_0^\infty \tilde{q}_e(t)^\top \tilde{P} \tilde{\Delta}_e(\tilde{q}_e)(t) dt \geq 0 \quad \forall \tilde{q}_e \in \mathscr{L}_2.$$

Moreover, for any $\tilde{q}_e \in \mathscr{L}_{2e}$ we infer that the truncated signal (i.e. $\tilde{q}_{e_T} = \tilde{q}_e$ on $[0, T]$ and $\tilde{q}_{e_T} = 0$ on (T, ∞)) satisfies $\tilde{q}_{e_T} \in \mathscr{L}_2$. By causality of $\tilde{\Delta}_e$ we conclude

$$\int_0^\infty \tilde{q}_{e_T}(t)^\top \tilde{P} \tilde{\Delta}_e(\tilde{q}_{e_T})(t) dt \geq 0 \ \Rightarrow \ \int_0^\infty \tilde{q}_{e_T}(t)^\top \tilde{P} \tilde{\Delta}_e(\tilde{q}_{e_T})_T(t) dt \geq 0 \ \Rightarrow$$

$$\Rightarrow \int_0^\infty \tilde{q}_{e_T}(t)^\top \tilde{P} \tilde{\Delta}_e(\tilde{q}_e)_T(t) dt \geq 0 \ \Rightarrow \ \int_0^T \tilde{q}_e(t)^\top \tilde{P} \tilde{\Delta}_e(\tilde{q}_e)(t) dt \geq 0 \ \forall \ T \geq 0.$$

$$(2.113)$$

In contrast to what we saw in Section 2.10.2, it is now nice to see that (2.113) does satisfy the nonnegativity property for all $T \geq 0$. This resolves the last obstacle for proving stability of (2.1). In fact, all previous steps were a preparation in order to transform the augmented feedback interconnection (2.98) into

$$\begin{aligned}
\tilde{q}_e &= \tilde{G}_e \tilde{p}_e + \tilde{\mu}_e \\
\tilde{p}_e &= \tilde{\Delta}_e(\tilde{q}_e) + \tilde{\eta}_e,
\end{aligned} \tag{2.114}$$

with the external disturbances related by $\tilde{\mu}_e = \tilde{\Psi} \mu_e$ and $\tilde{\eta}_e = \tilde{\Psi} \eta_e$. As illustrated in Figure 2.16, observe that we proceeded in a very classical fashion [209, 37]. The key twist is to perform all arguments in the state-space, starting with the original parametrization of the IQC-multiplier $\Pi = \Psi^* P \Psi$. Note that stability of (2.98) is equivalent to the stability of (2.114). Indeed, let $\tilde{\Psi}_e = \operatorname{diag}(\tilde{\Psi}, \tilde{\Psi})$ and suppose that there exists a $\gamma_1 > 0$ such that $\|\mathcal{I}(\tilde{G}_e, \tilde{\Delta}_e)^{-1}\|_{i2} \leq \gamma_1$. With $\gamma_1 = \gamma_2 (\|\tilde{\Psi}_e\|_{i2} \|\tilde{\Psi}_e^{-1}\|_{i2})^{-1}$ we can then infer that

$$\|\mathcal{I}(\tilde{G}_e, \tilde{\Delta}_e)^{-1}\|_{i2} \|\tilde{\Psi}_e\|_{i2} \|\tilde{\Psi}_e^{-1}\|_{i2} \leq \|\tilde{\Psi}_e \mathcal{I}(\tilde{G}_e, \tilde{\Delta}_e)^{-1} \tilde{\Psi}_e^{-1}\|_{i2} = \|\mathcal{I}(G_e, \Delta_e)^{-1}\|_{i2} \leq \gamma_2.$$

The converse statement follows from identical arguments.

We are now ready to prove Corollary 2.5. We need to show that there exists a $\gamma > 0$ such that $\|\mathcal{I}(\tilde{G}_e, \tilde{\Delta}_e)^{-1}\|_{i2} \leq \gamma$ for all $\Delta \in \boldsymbol{\Delta}$. For this purpose, let us define a linear fractional representation of (2.114) as follows:

$$\begin{pmatrix} \tilde{q}_e \\ \tilde{q}_e \\ \tilde{p}_e \end{pmatrix} = \begin{pmatrix} \tilde{G}_e & I & \tilde{G}_e \\ \tilde{G}_e & I & \tilde{G}_e \\ I & 0 & I \end{pmatrix} \begin{pmatrix} \tilde{y}_e \\ \tilde{\mu}_e \\ \tilde{\eta}_e \end{pmatrix}, \quad \tilde{y}_e = \tilde{\Delta}_e(\tilde{q}_e).$$

By recalling the realization (2.112) it is straightforward to see that

$$\begin{pmatrix} \tilde{G}_e & I & \tilde{G}_e \\ \tilde{G}_e & I & \tilde{G}_e \\ I & 0 & I \end{pmatrix} = \left[\begin{array}{c|ccc} \tilde{A}_e & \tilde{B}_e & 0 & \tilde{B}_e \\ \hline \tilde{C}_e & \tilde{D}_e & I & \tilde{D}_e \\ \tilde{C}_e & \tilde{D}_e & I & \tilde{D}_e \\ 0 & I & 0 & I \end{array} \right]. \tag{2.115}$$

Moreover, it is not hard to show that there exists a sufficiently large $\gamma > 0$ such that the inequality (2.111), corresponding to the system (2.112), can be augmented with the performance channels $\operatorname{col}(\tilde{\mu}_e, \tilde{\eta}_e) \to \operatorname{col}(\tilde{q}_e, \tilde{p}_e)$ as

$$\operatorname{He} \begin{pmatrix} \tilde{X}_e \tilde{A}_e & \tilde{X}_e \tilde{B}_e & 0 & \tilde{X}_e \tilde{B}_e \\ \tilde{P} \tilde{C}_e & \tilde{P} \tilde{D}_e & \tilde{P} & \tilde{P} \tilde{D}_e \\ 0 & 0 & -\frac{\gamma}{2} I & 0 \\ 0 & 0 & 0 & -\frac{\gamma}{2} I \end{pmatrix} + \frac{1}{\gamma} (\star)^\top \begin{pmatrix} \tilde{C}_e & \tilde{D}_e & I & \tilde{D}_e \\ 0 & I & 0 & I \end{pmatrix} \prec 0, \tag{2.116}$$

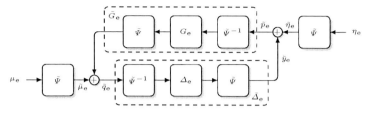

Fig. 2.16 Transformed feedback interconnection

corresponding to the open-loop plant (2.115). For some small $\epsilon > 0$ we can now replace
the right-hand side of (2.116) by $-\epsilon I$, without violating (2.116). For any trajectory
of (2.114) with initial condition zero, we can right- and left-multiply the resulting
inequality with $\mathrm{col}(x(t), \tilde{y}_e(t), \tilde{\mu}_e(t), \tilde{\eta}_e(t))$ and its transpose in order to obtain

$$\frac{d}{dt} x(t)^\top \tilde{X}_e x(t) + 2\tilde{q}_e(t)^\top \tilde{P} \tilde{y}_e(t) + \frac{1}{\gamma}(\|\tilde{q}_e(t)\|^2 + \|\tilde{p}_e(t)\|^2)$$
$$+ \epsilon(\|x(t)\|^2 + \|\tilde{y}_e(t)\|^2) \le (\gamma - \epsilon)(\|\tilde{\mu}_e(t)\|^2 + \|\tilde{\eta}_e(t)\|^2).$$

After integration on $[0,T]$ this in turn yields the inequality

$$x(T)^\top \tilde{X}_e x(T) + 2\int_0^T \tilde{q}_e(t)^\top \tilde{P}\tilde{y}_e(t)dt + \frac{1}{\gamma}\int_0^T \|\tilde{q}_e(t)\|^2 + \|\tilde{p}_e(t)\|^2 dt +$$
$$+ \epsilon \int_0^T \|x(t)\|^2 + \|\tilde{y}_e(t)\|^2 dt \le (\gamma - \epsilon)\int_0^T \|\tilde{\mu}_e(t)\|^2 + \|\tilde{\eta}_e(t)\|^2 dt \quad \forall\, T \ge 0. \quad (2.117)$$

In view of $\tilde{y}_e = \tilde{\Delta}_e(\tilde{q}_e)$, we can now exploit (2.113) to get

$$\frac{1}{\gamma}\int_0^T \|\tilde{q}_e(t)\|^2 + \|\tilde{p}_e(t)\|^2 dt + \epsilon \int_0^T \|x(t)\|^2 + \|\tilde{\Delta}(\tilde{q}_e)(t)\|^2 dt +$$
$$x(T)^\top \tilde{X}_e x(T) \le (\gamma - \epsilon)\int_0^T \|\tilde{\mu}_e(t)\|^2 + \|\tilde{\eta}_e(t)\|^2 dt \quad \forall\, T \ge 0.$$

Since the right-hand size is bounded for $T \to \infty$ we can infer the same (due to $\tilde{X}_e \succ 0$)
for the left-hand side. This shows that $\tilde{q}_e, \tilde{p}_e, x, \tilde{\Delta}(\tilde{q}_e) \in \mathscr{L}_2$. Robust stability of the
feedback interconnection (2.1) is finally proven by observing that for all $\tilde{\mu}_e, \tilde{\eta}_e \in \mathscr{L}_2$
we have

$$\frac{1}{\gamma}\int_0^\infty \|\tilde{q}_e(t)\|^2 + \|\tilde{p}_e(t)\|^2 dt \le \gamma \int_0^\infty \|\tilde{\mu}_e(t)\|^2 + \|\tilde{\eta}_e(t)\|^2 dt.$$

Clearly this shows that $\|\mathcal{I}(\tilde{G}_e, \tilde{\Delta}_e)^{-1}\|_{i2} \le \gamma$ for all $\Delta \in \boldsymbol{\Delta}$ and the proof is finished.

2.10.5 Outline and possibilities

Let us conclude this section by briefly discussing the results. We have established a connection between the IQC-framework and the dissipativity approach. This not only shows that IQCs can be interpreted as dynamic supply rates, but also opens the way to merge frequency-domain techniques with time-domain conditions known from Lyapunov-theory. For further readings and some preliminary applications such as robustness analysis and synthesis with parameter-varying IQCs and robustness analysis for delayed nonlinear and parameter-varying systems we refer the reader to [171, 132, 131, 202] and [130] respectively. Further one could think of incorporation hard time-domain constraints, such as $\mathscr{L}_1$- or generalized $\mathscr{H}_2$- constraints, which could only be applied if the IQCs were valid on all finite horizons. This might lead to nice applications in the field of anti-windup compensator design (see e.g. [19] and references therein). Moreover, it might even be possible to establish a connection with parameter dependent Lyapunov functions. We refer the reader to [177, 176] for some more concrete applications. As an major obstacle towards new applications, we emphasize that the inertia assumption on the storage function in [177, Theorem 3] is by no means true in general. As shown in Section 2.10.4 it is required to perform a shifting operation on the storage function through the solution of an algebraic Riccati equation (ARE). Therefore, the storage function does not solely consist of original data, which causes difficulties in finding any of the listed generalizations. This could be resolved by means of an LMI characterization of the ARE as was done in [6] for the special case of dynamic D-scalings.

2.11 Chapter summary

In this preparatory chapter we have developed the well-known IQC-framework for the analysis of uncertain dynamical systems. The tools enable us to efficiently and systematically investigate the stability and performance properties of a rather extensive class of uncertain dynamical systems via finite dimensional linear matrix inequalities. On the one hand, this chapter can be viewed as a stand-alone survey on the stability analysis with integral quadratic constraints. On the other hand, it forms the foundation on which we build our synthesis results in the subsequent chapters.

Along with the general setup, we have provided an extensive overview on the formulation and parametrization of concrete IQC-multipliers, which capture the properties of various types of uncertainties and nonlinearities as well as a diverse class of performance specifications. This includes e.q. IQCs for uncertain linear time-invariant dynamics, time-invariant and (rate-bounded) time-varying parametric uncertainties, time-delay uncertainties, passive and norm-bounded uncertainties and/or nonlinearities, sector bounded and slope-restricted nonlinearities, induced $\mathscr{L}_2$-gain performance, passivity performance, quadratic performance, $\mathscr{H}_2$-performance and dynamic generalized quadratic performance.

The developed framework has been illustrated by means of two comprehensive numerical examples. Here we demonstrated the most important aspects of a proper IQC-analysis and how the IQC-tools can be applied for both simple academical as well as for more realistic robustness analysis problems. Also we have addressed some important aspects related to the implementation of the tools and the numerical computations.

We concluded the chapter, with an alternative proof of the IQC-stability theorem based on dissipativity arguments. This opens the way to merge frequency-domain techniques with time-domain conditions known from Lyapunov-theory. The full potential of this result remains to be explored.

Chapter 3
From analysis to synthesis

3.1 Introduction

In this chapter we move on from a long introduction on integral quadratic constraint (IQC) analysis towards the more challenging IQC-synthesis problem. As we have seen in Chapter 2, the IQC-analysis problem can be solved via finite dimensional linear matrix inequalities (LMIs) and convex optimization techniques. This enables us to efficiently perform robust stability and performance analysis for a diverse class of uncertain dynamical systems. Although this is an extraordinary fact on its own, one might not just want to verify whether or not a given controller achieves robust stability and performance. Instead, control engineers often prefer to design one.

Unfortunately, and as we will see in the subsequent chapters, the IQC-synthesis problem is in general much harder to solve. In other words, this means that the IQC-synthesis problem in its full generality cannot be efficiently solved via finite dimensional LMIs without introducing severe conservatism, i.e. without sacrificing too much robust performance. Luckily, it is possible to consider an extensive class of specialized robust control synthesis problems and efficiently solve these with LMI-techniques, without giving up to much on robust performance. It is the main goal of this thesis to reveal, for a number of interesting design questions, how this is possible.

The goal of this chapter is twofold. Being aware of the vastness of Chapter 2, the first goal of this chapter is to provide an introduction for those readers that are already familiar with the IQC-analysis tools. This includes a short outline of topics that will be discussed in chapter 4, 5 and 6 respectively. Secondly, this is again a preparatory chapter in which we formulate the central and most general design question that we will consider in this thesis: the so-called robust gain-scheduled controller synthesis problem. This is convenient and insightful, since all other design questions that will be addressed at a later stage can derived from this central problem by specialization.

We will briefly illustrate this by recalling the solutions of two important synthesis problems: the $\mathscr{H}_\infty$-synthesis problem [41, 48, 72] and the more general nominal gain-scheduling synthesis problem [121, 174, 152]. These well-known results and the applied techniques form a basis for all other synthesis results that will be considered in the subsequent chapters. Apart from the exposition, we emphasize that there are no novel technical contributions in this chapter.

This chapter is organized as follows. First, in Section 3.2 and 3.3 we will introduce all the required ingredients in order to formulate the central design question of this thesis in Section 3.4: the robust gain-scheduled controller synthesis problem. Then, after a few preparations in Section 3.5, we proceed by briefly recalling the solutions of the $\mathscr{H}_\infty$- and gain-scheduling synthesis problem in Section 3.6 and 3.7 respectively. Subsequently, in Section 3.8, we will give a short outline of further possibilities that will be addressed in detail in the following chapters, and we conclude the chapter with a summary in Section 3.9.

3.2 Robust controller synthesis

Recall the standard feedback interconnection for robust stability and performance analysis in Figure 2.2 from Section 2.4.2 and consider its extension with a control channel in Figure 3.1. The corresponding dynamical system is now defined through the linear fractional representation

$$\begin{pmatrix} q \\ z \\ y \end{pmatrix} = \underbrace{\begin{pmatrix} G_{qp} & G_{qw} & G_{qu} \\ G_{zp} & G_{zw} & G_{zu} \\ G_{yp} & G_{yw} & G_{yu} \end{pmatrix}}_{G} \begin{pmatrix} p \\ w \\ u \end{pmatrix}, \quad p = \Delta(q), \tag{3.1}$$

where $p \to q$ represents the uncertainty channel, $w \to z$ the performance channel and $u \to y$ the control channel. The open-loop uncontrolled plant $G \in \mathscr{R}^{(n_q+n_z+n_y)\times(n_p+n_w+n_u)}$ is a real rational and proper transfer matrix and the trouble making component $\Delta : \mathscr{L}_{2e}^{n_q} \to \mathscr{L}_{2e}^{n_p}$ is a bounded and causal operator, which we allow to vary in a given class Δ that satisfies Assumption 2.1 (see Section 2.4.1). We assume that the linear fractional representation (3.1) is well-posed for all $\Delta \in \Delta$, which means that $I - G_{qp}\Delta$ has a causal inverse for all $\Delta \in \Delta$.

The main goal in robust control is to design an LTI controller $u = Ky$ such that, for all $\Delta \in \Delta$, the system (3.1) is rendered stable, while some performance objective is achieved on the channel $w \to z$.

In a first step to render this problem computational, we close the loop with the controller $u = Ky$. This yields the closed-loop system

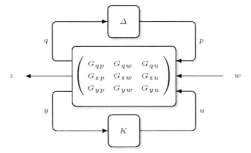

Fig. 3.1 Standard feedback interconnection for robust controller synthesis

$$
\begin{pmatrix} q \\ z \end{pmatrix} = \underbrace{\begin{pmatrix} \mathcal{G}_{qp} & \mathcal{G}_{qw} \\ \mathcal{G}_{zp} & \mathcal{G}_{zw} \end{pmatrix}}_{\mathcal{G}} \begin{pmatrix} p \\ w \end{pmatrix}, \quad p = \Delta(q), \tag{3.2}
$$

where $\mathcal{G}$ is defined through the linear fractional transformation

$$
\begin{pmatrix} \mathcal{G}_{qp} & \mathcal{G}_{qw} \\ \mathcal{G}_{zp} & \mathcal{G}_{zw} \end{pmatrix} := \left(\begin{array}{cc|c} G_{qp} & G_{qw} & G_{qu} \\ G_{zp} & G_{zw} & G_{zu} \\ \hline G_{yp} & G_{yw} & G_{yu} \end{array} \right) \star K.
$$

Note that, apart from stabilization and achieving a particular performance objective, the controller also needs to render (3.2) well-posed for all $\Delta \in \boldsymbol{\Delta}$, which means that $I - \mathcal{G}_{qp}\Delta$ and, hence, $I - G_{yu}K$ must have causal inverses for all $\Delta \in \boldsymbol{\Delta}$. As discussed in Chapter 2, this is typically assumed and sometimes guaranteed for special cases.

Completely analogously to the discussion in Section 2.5, it is possible to embed the robust controller synthesis problem in the IQC-framework. For this purpose, we assume that the IQC

$$
IQC(\Psi^* P\Psi, q, \Delta(q)) \geq 0 \quad \forall q \in \mathscr{L}_2^{n_q} \tag{3.3}
$$

is satisfied for all $\Delta \in \boldsymbol{\Delta}$ and for all $P \in \boldsymbol{P}$. For the precise definition of this expression we refer the reader to Section 2.3. We recall that $\boldsymbol{P}$ is an LMIable set of symmetric matrices $P \in \mathbb{S}^\bullet$ and that $\Psi = (\Psi_1 \ \Psi_2) \in \mathscr{R}\mathscr{H}_\infty^{\bullet \times (n_q + n_p)}$ is a fixed and typically tall transfer matrix. In addition, we assume the induced $\mathscr{L}_2$-gain on the channel $w \to z$ as the desired performance objective. As discussed in Section 2.7.1, this criterion can be expressed through an IQC as

$$
IQC(P_\gamma, z, w) \leq -\epsilon \|w\|_2^2, \tag{3.4}
$$

where $P_\gamma := \operatorname{diag}(\gamma^{-1}I_{n_z}, -\gamma I_{n_w})$. We can now state the following analysis result.

Theorem 3.1. *Let the controller K be given and assume that*

1. *K internally stabilizes G;*
2. *the system (3.2) is well-posed for all $\Delta \in \mathbf{\Delta}$;*
3. *the IQC (3.3) is satisfied for all $\Delta \in \mathbf{\Delta}$ and for all $P \in \mathbf{P}$;*
4. *there exists some $P \in \mathbf{P}$ for which the following FDI holds:*

$$\begin{pmatrix} \Psi_1 \mathcal{G}_{qp} + \Psi_2 & \Psi_1 \mathcal{G}_{qw} \\ \hline \mathcal{G}_{zp} & \mathcal{G}_{zw} \\ 0 & I \end{pmatrix}^* \begin{pmatrix} P & 0 \\ \hline 0 & P_\gamma \end{pmatrix} \begin{pmatrix} \Psi_1 \mathcal{G}_{qp} + \Psi_2 & \Psi_1 \mathcal{G}_{qw} \\ \hline \mathcal{G}_{zp} & \mathcal{G}_{zw} \\ 0 & I \end{pmatrix} \prec 0 \text{ on } \mathbb{C}^0.$$

$$(3.5)$$

Then, for all $\Delta \in \mathbf{\Delta}$, the controller K (internally) stabilizes (3.2) and guarantees the induced $\mathscr{L}_2$-gain from w to z to be less than $\gamma > 0$.

The proof is identical to the one of Lemma 2.1. Here we note that K internally stabilizes G if and only if

$$\begin{pmatrix} q \\ z \\ y \end{pmatrix} = \begin{pmatrix} G_{qp} & G_{qw} & G_{qu} \\ G_{zp} & G_{zw} & G_{zu} \\ G_{yp} & G_{yw} & G_{yu} \end{pmatrix} \begin{pmatrix} p \\ w \\ u \end{pmatrix}, \quad u = Kv + v_1, \quad v = y + v_2$$

defines a proper transfer matrix $\operatorname{col}(p, w, v_1, v_2) \to \operatorname{col}(q, z, u, v)$ that is stable (see e.g. [210, 43]). We conclude that after applying a simple Schur complement with respect to the $\gamma^{-1}I$ block of P_γ, we obtain a semi-infinite convex feasibility test for robust stability and performance analysis. As discussed in Section 2.5.2, this test can be easily turned into a genuine convex feasibility problem by introducing realizations for the transfer matrices G, K and Ψ and exploiting the KYP-lemma (i.e. Lemma 2.2 in Section 2.5.2).

On the other hand, in the case of synthesis, the goal is to find K which, for all $\Delta \in \mathbf{\Delta}$, internally stabilizes G and which renders the induced $\mathscr{L}_2$-gain from w to z less than some a priori given $\gamma > 0$. Unfortunately, in doing so, we evidently loose convexity, since (3.5) is now bilinear in the unknowns P and K.

3.3 Gain-scheduled controller synthesis

Before we proceed with the robust controller synthesis problem, let us first address another important control configuration which is regularly addressed in the literature [121, 174, 152] and which we will frequently consider in this thesis. As an essential observation, it is often very natural to assume that the dynamical system under consideration is not affected by genuine uncertainties, but by a so-called scheduling block Δ_{s} which is assumed to depend linearly on time-varying parameters that are known in real-time. Given such a time-varying dynamical system the goal is then to feed the extra information to a so-called gain-scheduled controller $K_{\mathrm{gs}}(\Delta_{\mathrm{s}})$, which depends (in a possibly nonlinear fashion) on the very same scheduling block Δ_{s}. The configuration is depicted in Figure 3.2. The essential difference with the general robust controller synthesis problem from the previous section is that we would like to exploit the additional information coming from the scheduling block Δ_{s} in order to design a gain-scheduled controller $K_{\mathrm{gs}}(\Delta_{\mathrm{s}})$ which automatically adapts to the changes in the environment. For an overview of application scenarios we refer the reader e.g. to [169, 141, 203, 191, 199, 12, 188, 147, 189, 68] and references therein.

Let us be more precise about the definitions and consider the corresponding dynamical system, which is now defined through the linear fractional representation

$$\begin{pmatrix} q_{\mathrm{s}} \\ z \\ y \end{pmatrix} = \underbrace{\begin{pmatrix} G_{ss} & G_{sw} & G_{su} \\ G_{zs} & G_{zw} & G_{zu} \\ G_{ys} & G_{yw} & G_{yu} \end{pmatrix}}_{G} \begin{pmatrix} p_{\mathrm{s}} \\ w \\ u \end{pmatrix}, \quad p_{\mathrm{s}} = \Delta_{\mathrm{s}} q_{\mathrm{s}} \qquad (3.6)$$

that is well-posed for all $\Delta_{\mathrm{s}} \in \boldsymbol{\Delta}_{\mathrm{s}}$. Here $p_{\mathrm{s}} \to q_{\mathrm{s}}$ represents the scheduling channel, while $w \to z$ and $u \to y$, as before, denote the performance and the control channel respectively. The open-loop uncontrolled plant $G \in \mathscr{R}^{(n_{q_{\mathrm{s}}}+n_z+n_y) \times (n_{p_{\mathrm{s}}}+n_w+n_u)}$ is again assumed to be represented by a real rational and proper transfer matrix and the scheduling block $\Delta_{\mathrm{s}} := \hat{\Delta}_{\mathrm{s}} \circ \delta$ is a linear function of the online measurable time-varying parameter vector $\delta : [0,\infty) \to \Lambda$. Here the map $\hat{\Delta}_{\mathrm{s}} : \mathbb{R}^k \to \mathbb{R}^{n_{q_{\mathrm{s}}} \times n_{p_{\mathrm{s}}}}$ is defined by

$$\hat{\Delta}_{\mathrm{s}}(\delta) := \sum_{i=1}^{k} \delta_i \Omega_i$$

for some fixed matrices $\Omega_i \in \mathbb{R}^{n_{q_{\mathrm{s}}} \times n_{p_{\mathrm{s}}}}$, $\delta \in \mathbb{R}^k$ and we assume that δ takes its values in the compact polytope

$$\Lambda := \mathrm{co}\left\{ \delta^1, \ldots, \delta^m \right\} \subseteq \mathbb{R}^k$$

with $\delta^j = (\delta_1^j, \ldots, \delta_k^j)$, $j \in \{1, \ldots, m\}$ as generator points. Without loss of generality Λ contains the origin. Then Δ_{s} is contained in the set

$$\boldsymbol{\Delta}_{\mathrm{s}} := \left\{ \hat{\Delta}_{\mathrm{s}} \circ \delta : \delta \in PC([0,\infty), \Lambda) \right\} \qquad (3.7)$$

and defines the operator $p_{\mathrm{s}}(t) = \hat{\Delta}_{\mathrm{s}}(\delta(t))q_{\mathrm{s}}(t)$. Here $PC([0,\infty), \Lambda)$ denotes the space of piecewise continuous functions $[0,\infty) \to \Lambda$.

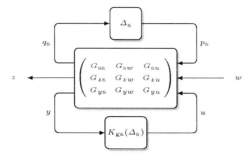

Fig. 3.2 Standard feedback interconnection for nominal gain-scheduling controller synthesis

The goal is to design a gain-scheduled controller $u = K_{gs}(\Delta_s)y$ which, for all $\Delta_s \in \mathbf{\Delta}_s$, stabilizes (3.6) and which renders the induced $\mathscr{L}_2$-gain from w to z less than some a priori given $\gamma > 0$. Here $K_{gs}(\Delta_s)$ is defined through an LTI system K of dimension $(n_u + n_{q_c}) \times (n_y + n_{p_c})$ and a so-called scheduling-function $\hat{\Delta}_c : \mathbb{R}^{n_{p_s} \times n_{q_s}} \to \mathbb{R}^{n_{p_c} \times n_{q_c}}$ as

$$\begin{pmatrix} u \\ q_c \end{pmatrix} = \underbrace{\begin{pmatrix} K_{11} & K_{12} \\ K_{21} & K_{22} \end{pmatrix}}_{K} \begin{pmatrix} y \\ p_c \end{pmatrix}, \quad p_c = \hat{\Delta}_c(\Delta_s)q_c. \tag{3.8}$$

For any $\Delta_s \in \mathbf{\Delta}_s$ we emphasize that the action of the operator $\hat{\Delta}_c(\Delta_s)$ is defined through $p_c(t) = \hat{\Delta}_c(\hat{\Delta}_s(\delta(t)))q_c(t)$.

In a first step to render this problem computational, we reformulate the closed-loop specifications with the static full-block multiplier from Section 2.6.2.2 and as in [116, 150]. For this purpose, we describe the closed-loop system $(\Delta_s \star G) \star K_{gs}(\Delta_s)$ in a standard fashion as

$$\begin{pmatrix} q_e \\ z \end{pmatrix} = \underbrace{\begin{pmatrix} \mathcal{G}_{ee} & \mathcal{G}_{ew} \\ \mathcal{G}_{ze} & \mathcal{G}_{zw} \end{pmatrix}}_{\mathcal{G}} \begin{pmatrix} p_e \\ w \end{pmatrix}, \quad p_e = \Delta_e(\Delta_s)q_e. \tag{3.9}$$

Here the so-called extended scheduling block $\Delta_e(\Delta_s) = \Delta_e \circ \Delta_s$ is defined through the map

$$\Delta_e(\Delta) := \begin{pmatrix} \Delta & 0 \\ 0 & \hat{\Delta}_c(\Delta) \end{pmatrix} \quad \text{for } \Delta \in \mathbb{R}^{n_{p_s} \times n_{q_s}}, \tag{3.10}$$

while $\mathcal{G}$ denotes the interconnection of G and the LTI part of the controller K as given by

$$\underbrace{\left(\begin{array}{c|c} \mathcal{G}_{ee} & \mathcal{G}_{ew} \\ \hline \mathcal{G}_{ze} & \mathcal{G}_{zw} \end{array}\right)}_{\mathcal{G}} = \underbrace{\left(\begin{array}{cc|cc|c} G_{ss} & 0 & G_{sw} & G_{su} & 0 \\ 0 & 0 & 0 & 0 & I_{n_{q_c}} \\ \hline G_{zs} & 0 & G_{zw} & G_{zu} & 0 \\ G_{ys} & 0 & G_{yw} & G_{yu} & 0 \\ 0 & I_{n_{p_c}} & 0 & 0 & 0 \end{array}\right)}_{G_{\mathrm{aug}}} \star \underbrace{\left(\begin{array}{cc} K_{11} & K_{12} \\ K_{21} & K_{22} \end{array}\right)}_{K},$$

Note that $\Delta_{\mathrm{e}}(\Delta_{\mathrm{s}})$ takes its values in the set of matrices

$$\hat{\Delta}_{\mathrm{e}} := \left\{ \hat{\Delta}_{\mathrm{e}}(\delta) = \begin{pmatrix} \hat{\Delta}_{\mathrm{s}}(\delta) & 0 \\ 0 & \hat{\Delta}_{\mathrm{c}}(\hat{\Delta}_{\mathrm{s}}(\delta)) \end{pmatrix} : \delta \in \Lambda \right\}. \tag{3.11}$$

Following [150], the LMI constraint

$$\begin{pmatrix} I \\ \hat{\Delta}_{\mathrm{e}} \end{pmatrix}^{\top} \underbrace{\begin{pmatrix} Q_{\mathrm{e}} & S_{\mathrm{e}} \\ S_{\mathrm{e}}^{\top} & R_{\mathrm{e}} \end{pmatrix}}_{P_{\mathrm{e}} = P_{\mathrm{e}}^{\top}} \begin{pmatrix} I \\ \hat{\Delta}_{\mathrm{e}} \end{pmatrix} \succ 0, \quad Q_{\mathrm{e}} \succ 0, \quad R_{\mathrm{e}} \prec 0, \tag{3.12}$$

defines the so-called set of extended full-block multipliers

$$\mathbf{P}_{\mathrm{e}} = \left\{ P_{\mathrm{e}} = P_{\mathrm{e}}^{\top} : P_{\mathrm{e}} \text{ satisfies (3.12) for all } \hat{\Delta}_{\mathrm{e}} \in \hat{\Delta}_{\mathrm{e}} \right\} \tag{3.13}$$

and guarantees the IQC

$$IQC(P_{\mathrm{e}}, q_{\mathrm{e}}, \tau \Delta_{\mathrm{e}}(\Delta_{\mathrm{s}})q_{\mathrm{e}}) \geq 0 \quad \forall q_{\mathrm{e}} \in \mathscr{L}_2^{n_{q_s}+n_{q_c}}, \quad \forall \tau \in [0,1] \tag{3.14}$$

to be satisfied for all $\Delta_{\mathrm{s}} \in \mathbf{\Delta}_{\mathrm{s}}$ and for all $P_{\mathrm{e}} \in \mathbf{P}_{\mathrm{e}}$. The desired induced $\mathscr{L}_2$-gain performance objective is again assumed to be expressed through the IQC (3.4). This leads to the following analysis result.

Theorem 3.2. *Let the well-posed controller $K_{\mathrm{gs}}(\Delta_{\mathrm{s}}) = K \star \hat{\Delta}_{\mathrm{c}}(\Delta_{\mathrm{s}})$ be given and assume that*

1. *K internally stabilizes G_{aug};*
2. *the IQC (3.14) is satisfied for all $\Delta_{\mathrm{s}} \in \mathbf{\Delta}_{\mathrm{s}}$ and for all $P_{\mathrm{e}} \in \mathbf{P}_{\mathrm{e}}$;*
3. *there exists some $P_{\mathrm{e}} \in \mathbf{P}_{\mathrm{e}}$ for which the following FDI holds:*

$$\begin{pmatrix} \mathcal{G}_{ee} & \mathcal{G}_{ew} \\ \hline I & 0 \\ \hline \mathcal{G}_{ze} & \mathcal{G}_{zw} \\ \hline 0 & I \end{pmatrix}^{*} \begin{pmatrix} P_{\mathrm{e}} & 0 \\ \hline 0 & P_{\gamma} \end{pmatrix} \begin{pmatrix} \mathcal{G}_{ee} & \mathcal{G}_{ew} \\ \hline I & 0 \\ \hline \mathcal{G}_{ze} & \mathcal{G}_{zw} \\ \hline 0 & I \end{pmatrix} \prec 0 \text{ on } \mathbb{C}^0. \tag{3.15}$$

Then, for all $\Delta_s \in \mathbf{\Delta}_s$, the system (3.9) is well-posed and the controller $K_{gs}(\Delta_s)$ stabilizes (3.9) and guarantees the induced $\mathscr{L}_2$-gain from w to z to be less than $\gamma > 0$.

If the scheduling function $\hat{\Delta}_c(\cdot)$ has the right structure for $\boldsymbol{P}_e$ to be LMIable, then we obtain again a semi-infinite convex feasibility test for robust stability and performance analysis, which can be easily turned into computations. On the other hand, in the case of synthesis, the goal is to search for K which internally stabilizes G_{aug} as well as a scheduling function $\hat{\Delta}_c(\cdot)$ such that there exists some $P_e \in \mathbf{P}_e$ in the corresponding class of extended multipliers (3.13) for which the FDI (3.15) holds true. Again we observe that the FDI (3.15) becomes bilinear in the unknowns P_e and K.

3.4 Robust gain-scheduled controller synthesis

Naturally, it is straightforward to merge the robust controller synthesis problem from Section 3.2 with the gain-scheduling synthesis problem from the previous section. This leads to the most general configuration of this thesis, where we distinguish between linear time-varying parameters, which are available online as scheduling parameters for the controller, and genuine uncertainties not necessarily time-varying, parametric or linear, that are not available online. The general configuration is shown in Figure 3.3. The corresponding dynamical system is now defined through the linear fractional representation

$$\begin{pmatrix} q_s \\ q \\ z \\ y \end{pmatrix} = \underbrace{\begin{pmatrix} G_{ss} & G_{sp} & G_{sw} & G_{su} \\ G_{qs} & G_{qp} & G_{qw} & G_{qu} \\ G_{zs} & G_{zp} & G_{zw} & G_{zu} \\ G_{ys} & G_{yp} & G_{yw} & G_{yu} \end{pmatrix}}_{G} \begin{pmatrix} p_s \\ p \\ w \\ u \end{pmatrix}, \quad \begin{pmatrix} p_s \\ p \end{pmatrix} = \begin{pmatrix} \Delta_s q_s \\ \Delta(q) \end{pmatrix}, \quad (3.16)$$

where $p_s \to q_s$, $p \to q$, $w \to z$ and $u \to y$ denote the scheduling, uncertainty, performance and control channel respectively, and where the open-loop uncontrolled generalized plant $G \in \mathscr{R}^{(n_{q_s}+n_q+n_z+n_y) \times (n_{p_s}+n_q+n_w+n_u)}$ is assumed to be a real rational and proper transfer matrix. As defined previously, the genuine uncertainties Δ are contained in the set $\boldsymbol{\Delta}$ and the scheduling block Δ_s is confined to $\mathbf{\Delta}_s$. We assume (3.16) to be well-posed for all $\Delta \in \boldsymbol{\Delta}$ and for all $\Delta_s \in \mathbf{\Delta}_s$.

Completely analogously to the previous discussions, the goal is to design a robust gain-scheduled controller $u = K_{rgs}(\Delta_s)y$ which, for all $\Delta \in \boldsymbol{\Delta}$ and for all $\Delta_s \in \mathbf{\Delta}_s$, stabilizes (3.16) and which renders the induced $\mathscr{L}_2$-gain from w to z less than some a priori given $\gamma > 0$. Here $K_{rgs}(\Delta_s)$ is again defined through an LTI system K and a scheduling-function $\hat{\Delta}_c : \mathbb{R}^{n_{p_s} \times n_{q_s}} \to \mathbb{R}^{n_{p_c} \times n_{q_c}}$ as in (3.8).

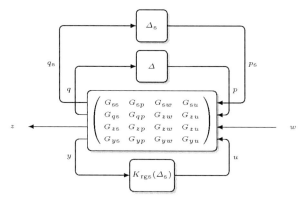

Fig. 3.3 Standard feedback interconnection for nominal gain-scheduling controller synthesis

As before, we reformulate the closed-loop specifications by exploiting static full-block multipliers. This yields the closed-loop system

$$
\begin{pmatrix} q_e \\ q \\ z \end{pmatrix} = \underbrace{\begin{pmatrix} \mathcal{G}_{ee} & \mathcal{G}_{ep} & \mathcal{G}_{ew} \\ \mathcal{G}_{qe} & \mathcal{G}_{qp} & \mathcal{G}_{qw} \\ \mathcal{G}_{ze} & \mathcal{G}_{zp} & \mathcal{G}_{zw} \end{pmatrix}}_{\mathcal{G}} \begin{pmatrix} p_e \\ p \\ w \end{pmatrix}, \quad \begin{pmatrix} p_e \\ p \end{pmatrix} = \begin{pmatrix} \Delta_e(\Delta_s)q_e \\ \Delta(q) \end{pmatrix}, \qquad (3.17)
$$

with the extended scheduling block $\Delta_e(\Delta_s)$, and where $\mathcal{G}$ denotes the interconnection of G and the LTI part of the controller K as given by

$$
\underbrace{\begin{pmatrix} \mathcal{G}_{ee} & \mathcal{G}_{ep} & \mathcal{G}_{ew} \\ \mathcal{G}_{qe} & \mathcal{G}_{qp} & \mathcal{G}_{qw} \\ \mathcal{G}_{ze} & \mathcal{G}_{zp} & \mathcal{G}_{zw} \end{pmatrix}}_{\mathcal{G}} = \underbrace{\begin{pmatrix} G_{ss} & 0 & G_{sp} & G_{sw} & G_{su} & 0 \\ 0 & 0 & 0 & 0 & 0 & I_{n_{q_c}} \\ G_{qs} & 0 & G_{qp} & G_{qw} & G_{qu} & 0 \\ G_{zs} & 0 & G_{zp} & G_{zw} & G_{zu} & 0 \\ G_{ys} & 0 & G_{yp} & G_{yw} & G_{yu} & 0 \\ 0 & I_{n_{p_c}} & 0 & 0 & 0 & 0 \end{pmatrix}}_{G_{\mathrm{aug}}} \star \underbrace{\begin{pmatrix} K_{11} & K_{12} \\ K_{21} & K_{22} \end{pmatrix}}_{K}.
$$

We arrive at the third and final analysis result of this chapter.

Theorem 3.3. *Let the well-posed controller $K_{\mathrm{rgs}}(\Delta_s) = K \star \hat{\Delta}_c(\Delta_s)$ be given and assume that*

1. K *internally stabilizes* G_{aug};
2. *the system* (3.17) *is well-posed for all* $\Delta \in \boldsymbol{\Delta}$ *and for all* $\Delta_{\mathrm{s}} \in \boldsymbol{\Delta}_{\mathrm{s}}$;
3. *the IQC* (3.3) *is satisfied for all* $\Delta \in \boldsymbol{\Delta}$ *and for all* $P \in \boldsymbol{P}$;
4. *the IQC* (3.14) *is satisfied for all* $\Delta_{\mathrm{s}} \in \boldsymbol{\Delta}_{\mathrm{s}}$ *and for all* $P_{\mathrm{e}} \in \mathbf{P}_{\mathrm{e}}$;
5. *there exist* $P \in \boldsymbol{P}$ *and* $P_{\mathrm{e}} \in \mathbf{P}_{\mathrm{e}}$ *for which the following FDI holds:*

$$
\begin{pmatrix} \star \\ \star \\ \star \\ \star \\ \star \end{pmatrix}^{*}
\begin{pmatrix} P_{\mathrm{e}} & 0 & 0 \\ \hline 0 & P & 0 \\ \hline 0 & 0 & P_{\gamma} \end{pmatrix}
\begin{pmatrix} \mathcal{G}_{ee} & \mathcal{G}_{ep} & \mathcal{G}_{ew} \\ I & 0 & 0 \\ \hline \Psi_1 \mathcal{G}_{qe} & \Psi_1 \mathcal{G}_{qp} + \Psi_2 & \Psi_1 \mathcal{G}_{qw} \\ \mathcal{G}_{ze} & \mathcal{G}_{zp} & \mathcal{G}_{zw} \\ 0 & 0 & I \end{pmatrix} \prec 0 \text{ on } \mathbb{C}^0. \quad (3.18)
$$

Then, for all $\Delta \in \boldsymbol{\Delta}$ *and for all* $\Delta_{\mathrm{s}} \in \boldsymbol{\Delta}_{\mathrm{s}}$, $K_{\mathrm{rgs}}(\Delta_{\mathrm{s}})$ *stabilizes* (3.17) *and guarantees the induced* $\mathcal{L}_2$*-gain from* w *to* z *to be less than* $\gamma > 0$.

Although the latter analysis result can be straightforwardly turned into an LMI problem, the corresponding synthesis problem translates the search for K that internally stabilizes G_{aug} and a scheduling function $\hat{\Delta}_{\mathrm{c}}(\cdot)$ such that there exists some $P \in \boldsymbol{P}$ and $P_{\mathrm{e}} \in \mathbf{P}_{\mathrm{e}}$ for which the FDI (3.18). Unfortunately, the simultaneous search for P, P_{e}, K and $\hat{\Delta}_{\mathrm{c}}(\cdot)$ yields a non-convex optimization problem, and to date no convexifying procedure is known for the general robust gain-scheduled controller synthesis problem. Luckily, for specialized problems it is often possible to exploit additional structure in order to transform the original non-convex problem in to a convex one. It is the main goal of this thesis to reveal in which cases this can be achieved. In the remainder of this preparatory chapter, we briefly recall two essential synthesis results from the literature and give a short outlook of further possibilities and generalizations, in order to guide the reader towards our contributions in the subsequent chapters.

3.5 Preparations

As a preparation, let us introduce the realizations

$$
K = \left[\begin{array}{c|c} A_K & B_K \\ \hline C_K & D_K \end{array} \right] \quad (3.19)
$$

with $A_K \in \mathbb{R}^{n_K \times n_K}$ and

$$
G = \begin{pmatrix} G_{ss} & G_{sp} & G_{sw} & G_{su} \\ G_{qs} & G_{qp} & G_{qw} & G_{qu} \\ G_{zs} & G_{zp} & G_{zw} & G_{zu} \\ G_{ys} & G_{yp} & G_{yw} & G_{yu} \end{pmatrix} = \left[\begin{array}{c|cccc} A & B_s & B_p & B_w & B_u \\ \hline C_s & D_{ss} & D_{sp} & D_{sw} & D_{su} \\ C_q & D_{qs} & D_{qp} & D_{qw} & D_{qu} \\ C_z & D_{zs} & D_{zp} & D_{zw} & D_{zu} \\ C_y & D_{ys} & D_{yp} & D_{yw} & D_{yu} \end{array} \right] \tag{3.20}
$$

with $A \in \mathbb{R}^{n_G \times n_G}$ and where the pairs (A, B_u) and (A, C_y) are assumed to be stabilizable and detectable respectively and where, without loss of generality, $G_{yu}(\infty) = D_{yu}$ is assumed to be zero. A realization of the closed-loop system $\mathcal{G}$ is then given by

$$
\mathcal{G} = \begin{pmatrix} \mathcal{G}_{ee} & \mathcal{G}_{ep} & \mathcal{G}_{ew} \\ \mathcal{G}_{qe} & \mathcal{G}_{qp} & \mathcal{G}_{qw} \\ \mathcal{G}_{ze} & \mathcal{G}_{zp} & \mathcal{G}_{zw} \end{pmatrix} = \left[\begin{array}{c|ccc} \mathcal{A} & \mathcal{B}_e & \mathcal{B}_p & \mathcal{B}_w \\ \hline \mathcal{C}_e & \mathcal{D}_{ee} & \mathcal{D}_{ep} & \mathcal{D}_{ew} \\ \mathcal{C}_q & \mathcal{D}_{qe} & \mathcal{D}_{qp} & \mathcal{D}_{qw} \\ \mathcal{C}_z & \mathcal{D}_{ze} & \mathcal{D}_{zp} & \mathcal{D}_{zw} \end{array} \right], \tag{3.21}
$$

where, as well-known,

$$
\left(\begin{array}{c|c|cc} \mathcal{A} & \mathcal{B}_e & \mathcal{B}_p & \mathcal{B}_w \\ \hline \mathcal{C}_e & \mathcal{D}_{ee} & \mathcal{D}_{ep} & \mathcal{D}_{ew} \\ \hline \mathcal{C}_q & \mathcal{D}_{qe} & \mathcal{D}_{qp} & \mathcal{D}_{qw} \\ \mathcal{C}_z & \mathcal{D}_{ze} & \mathcal{D}_{zp} & \mathcal{D}_{zw} \end{array} \right) = \left(\begin{array}{cc|cc|cc} A & 0 & B_s & 0 & B_p & B_w \\ 0 & 0 & 0 & 0 & 0 & 0 \\ C_s & 0 & D_{ss} & 0 & D_{sp} & D_{sw} \\ 0 & 0 & 0 & 0 & 0 & 0 \\ \hline C_q & 0 & D_{qs} & 0 & D_{qp} & D_{qw} \\ C_z & 0 & D_{zs} & 0 & D_{zp} & D_{zw} \end{array} \right) + \left(\begin{array}{cc|c} 0 & B_u & 0 \\ I & 0 & 0 \\ 0 & D_{su} & 0 \\ 0 & 0 & I \\ \hline 0 & D_{qu} & 0 \\ 0 & D_{zu} & 0 \end{array} \right) \times
$$

$$
\times \begin{pmatrix} A_K & B_K \\ C_K & D_K \end{pmatrix} \left(\begin{array}{cc|cc|cc} 0 & I & 0 & 0 & 0 & 0 \\ \hline C_y & 0 & D_{ys} & 0 & D_{yp} & D_{yw} \\ 0 & 0 & 0 & I & 0 & 0 \end{array} \right). \tag{3.22}
$$

Here $\mathrm{eig}(\mathcal{A}) \subset \mathbb{C}^-$, just because we assumed K to be a given internally stabilizing controller. If we also introduce the minimal realizations $\Psi_i = (A_i, B_i, C_i, D_i)$, $i = 1, 2$, such that

$$
\Psi = \begin{pmatrix} \Psi_1 & \Psi_2 \end{pmatrix} = \left[\begin{array}{cc|cc} A_1 & 0 & B_1 & 0 \\ 0 & A_2 & 0 & B_2 \\ \hline C_1 & C_2 & D_1 & D_2 \end{array} \right] =: \left[\begin{array}{c|cc} A_\Psi & B_{\Psi 1} & B_{\Psi 2} \\ \hline C_\Psi & D_{\Psi 1} & D_{\Psi 2} \end{array} \right] \tag{3.23}
$$

with $A_\Psi \in \mathbb{R}^{n_\Psi \times n_\Psi}$ and $\mathrm{eig}(A_\Psi) \subset \mathbb{C}^-$, then it is not hard to verify that the right-hand outer-factor of (3.15) admits the stable realization

$$
\begin{pmatrix}
\mathcal{G}_{ee} & \mathcal{G}_{ep} & \mathcal{G}_{ew} \\
I & 0 & 0 \\
\hline
\Psi_1 \mathcal{G}_{qe} & \Psi_1 \mathcal{G}_{qp} + \Psi_2 & \Psi_1 \mathcal{G}_{qw} \\
\hline
\mathcal{G}_{ze} & \mathcal{G}_{zp} & \mathcal{G}_{zw} \\
0 & 0 & I
\end{pmatrix}
=
\begin{bmatrix}
A_\Psi & B_{\Psi 1}\mathcal{C}_q & B_{\Psi 1}\mathcal{D}_{qe} & B_{\Psi 1}\mathcal{D}_{qp} + B_{\Psi 2} & B_{\Psi 1}\mathcal{D}_{qw} \\
0 & \mathcal{A} & \mathcal{B}_e & \mathcal{B}_p & \mathcal{B}_w \\
0 & \mathcal{C}_e & \mathcal{D}_{ee} & \mathcal{D}_{ep} & \mathcal{D}_{ew} \\
0 & 0 & I & 0 & 0 \\
\hline
C_\Psi & D_{\Psi 1}\mathcal{C}_q & D_{\Psi 1}\mathcal{D}_{qe} & D_{\Psi 1}\mathcal{D}_{qp} + D_{\Psi 2} & D_{\Psi 1}\mathcal{D}_{qw} \\
0 & \mathcal{C}_z & \mathcal{D}_{ze} & \mathcal{D}_{zp} & \mathcal{D}_{zw} \\
0 & 0 & 0 & 0 & I
\end{bmatrix}.
$$

By the KYP-lemma (i.e. Lemma 2.2 in Section 2.5.2) we can now infer that the FDI (3.18) is equivalent to the existence of a symmetric matrix $\mathcal{X} \in \mathbb{S}^{n_\Psi + n_G + n_K}$ for which the following LMI is satisfied:

$$
\begin{pmatrix} \star \\ \star \\ \star \\ \star \\ \star \\ \star \\ \star \\ \star \end{pmatrix}^\top
\begin{pmatrix}
0 & \mathcal{X} & 0 & 0 & 0 \\
\mathcal{X} & 0 & 0 & 0 & 0 \\
0 & 0 & P_e & 0 & 0 \\
0 & 0 & 0 & P & 0 \\
0 & 0 & 0 & 0 & P_\gamma
\end{pmatrix}
\begin{pmatrix}
I & 0 & 0 & 0 \\
A_\Psi & B_{\Psi 1}\mathcal{C}_q & B_{\Psi 1}\mathcal{D}_{qe} & B_{\Psi 1}\mathcal{D}_{qp} + B_{\Psi 2} & B_{\Psi 1}\mathcal{D}_{qw} \\
0 & \mathcal{A} & \mathcal{B}_e & \mathcal{B}_p & \mathcal{B}_w \\
0 & \mathcal{C}_e & \mathcal{D}_{ee} & \mathcal{D}_{ep} & \mathcal{D}_{ew} \\
0 & 0 & I & 0 & 0 \\
C_\Psi & D_{\Psi 1}\mathcal{C}_q & D_{\Psi 1}\mathcal{D}_{qe} & D_{\Psi 1}\mathcal{D}_{qp} + D_{\Psi 2} & D_{\Psi 1}\mathcal{D}_{qw} \\
0 & \mathcal{C}_z & \mathcal{D}_{ze} & \mathcal{D}_{zp} & \mathcal{D}_{zw} \\
0 & 0 & 0 & 0 & I
\end{pmatrix} \prec 0.
$$

(3.24)

Here $\mathcal{X}$ is assumed to be partitioned as

$$
\mathcal{X} = \begin{pmatrix} X_{11} & X_{12} \\ X_{12}^\top & X_{22} \end{pmatrix} = \begin{pmatrix} X_{11} & X_{12} & U_1 \\ X_{12}^\top & X_{22} & U_2 \\ U_1^\top & U_2^\top & \bar{X} \end{pmatrix} = \begin{pmatrix} X & U \\ U^\top & \bar{X} \end{pmatrix}
\tag{3.25}
$$

where X_{11}, X_{22} and $\bar{X}$ have compatible dimensions with A_Ψ, A and A_K respectively; as indicated by the partition lines, this implies the same for $\mathcal{X}_{22}$ and $\mathcal{A}$ as well as for X and $\mathrm{diag}(A_\Psi, A)$.

We are now ready to recall two important synthesis results from the literature. As we will see in the sequel, these synthesis problems admit a solution in terms of LMIs and the applied techniques form a basis for the main results of this thesis.

3.6 Nominal controller synthesis

As one of the prominent achievements in the control field, it has been shown in the early nineties that nominal controller synthesis can be turned into an LMI-problem

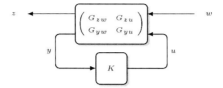

Fig. 3.4 Standard feedback interconnection for nominal controller synthesis

[41, 48, 72]. The goal of this section is to briefly recall the results and the applied techniques. For this purpose, consider the system interconnection shown in Figure 3.4, which is derived from the system interconnection as depicted in Figure 3.3 by setting $\Delta = 0$ and $\Delta_s = 0$. Note that this is possible since $0 \in \Delta$ and $0 \in \Delta_s$. The uncertainty channel $p \to q$ and the scheduling channel $p_s \to q_s$ are now void and direct inspection reveals that the matrix inequality (3.24) simplifies into

$$
\begin{pmatrix} I & 0 \\ \mathcal{A} & \mathcal{B}_w \\ \mathcal{C}_z & \mathcal{D}_{zw} \\ 0 & I \end{pmatrix}^{\top}
\begin{pmatrix} 0 & \mathcal{X} & 0 & 0 \\ \mathcal{X} & 0 & 0 & 0 \\ 0 & 0 & \gamma^{-1}I_{n_z} & 0 \\ 0 & 0 & 0 & -\gamma I_{n_w} \end{pmatrix}
\begin{pmatrix} I & 0 \\ \mathcal{A} & \mathcal{B}_w \\ \mathcal{C}_z & \mathcal{D}_{zw} \\ 0 & I \end{pmatrix} \prec 0. \qquad (3.26)
$$

As one of the essential requirements for the synthesis of an internally stabilizing controller K, we need to find matrices A_K, B_K, C_K and D_K such that $\mathcal{A}$ is Hurwitz. Fortunately, for nominal synthesis it is well-known that this can be equivalently characterized with the constraint $\mathcal{X} \succ 0$. Indeed, (3.26) implies $\mathrm{He}(\mathcal{X}\mathcal{A}) \succ 0$, which directly proves the claim.

Remark 3.1. Let us stress that this elementary characterization for internal stability is by no means true for the general robust gain-scheduled controller synthesis problem as presented in Section 3.4. Indeed, since $P \in \mathbf{P}$ is in general an indefinite matrix, stability of $\mathcal{A}$ neither implies nor is implied by positivity of $\mathcal{X}$. We will further address this issue in detail in Chapter 4.

In order to formulate the results, we also need to consider a so-called dual LMI as obtained by Lemma C.1 in Appendix C. This yields the equivalent matrix inequality

$$
\begin{pmatrix} -\mathcal{A}^{\top} & -\mathcal{C}_z^{\top} \\ I & 0 \\ 0 & I \\ -\mathcal{B}_w^{\top} & -\mathcal{D}_{zw}^{\top} \end{pmatrix}^{\top}
\begin{pmatrix} 0 & \mathcal{Y} & 0 & 0 \\ \mathcal{Y} & 0 & 0 & 0 \\ 0 & 0 & \gamma I_{n_z} & 0 \\ 0 & 0 & 0 & -\gamma^{-1}I_{n_w} \end{pmatrix}
\begin{pmatrix} -\mathcal{A}^{\top} & -\mathcal{C}_z^{\top} \\ I & 0 \\ 0 & I \\ -\mathcal{B}_w^{\top} & -\mathcal{D}_{zw}^{\top} \end{pmatrix} \succ 0. \quad (3.27)
$$

Here we assume $\mathcal{Y} = \mathcal{X}^{-1}$ to be partitioned as

$$\mathcal{Y} = \left(\begin{array}{c|c} Y & V \\ \hline V^\top & \bar{Y} \end{array} \right), \tag{3.28}$$

where Y and $\bar{Y}$ have compatible dimensions with A and A_K respectively.

Remark 3.2. Note that the latter dualization step is highly nontrivial for the general robust gain-scheduled controller synthesis problem. This is caused by the outer-factors of the IQC-multiplier $\Psi^* P \Psi$ which are generally tall and cannot be inverted; it prevents us from applying the dualization Lemma C.1 for the general matrix inequality (3.24). We will also further address this issue in detail in Chapter 4.

Now observe that (3.26) and (3.27) can be rendered affine in γ by applying a simple Schur complement. This, respectively, yields the inequalities

$$\text{He} \begin{pmatrix} \mathcal{X}\mathcal{A} & \mathcal{X}\mathcal{B}_w & 0 \\ 0 & -\frac{\gamma}{2}I_{n_w} & 0 \\ \mathcal{C}_z & \mathcal{D}_{zw} & -\frac{\gamma}{2}I_{n_z} \end{pmatrix} \prec 0 \tag{3.29}$$

and

$$\text{He} \begin{pmatrix} -\mathcal{Y}\mathcal{A}^\top & -\mathcal{Y}\mathcal{C}_z^\top & 0 \\ 0 & \frac{\gamma}{2}I_{n_z} & 0 \\ -\mathcal{B}_w^\top & -\mathcal{D}_{zw}^\top & \frac{\gamma}{2}I_{n_w} \end{pmatrix} \succ 0. \tag{3.30}$$

Controller synthesis requires to design $K = (A_K, B_K, C_K, D_K)$ in order to render (3.29) satisfied for $\mathcal{X} \succ 0$ or equivalently (3.30) for $\mathcal{Y} \succ 0$. As formulated this problem is non-convex. At this point [48, 72] suggest to eliminate the realization matrices of K in the analysis inequalities (3.29) and (3.30) by applying Lemma C.2 in Appendix C. This involves the basis matrices

$$T_y \quad \text{of} \quad \ker \begin{pmatrix} C_y & D_{yw} & 0_{n_y \times n_z} \end{pmatrix} \quad \text{and} \quad T_u \quad \text{of} \quad \ker \begin{pmatrix} B_u^\top & D_{zu}^\top & 0_{n_u \times n_w} \end{pmatrix}.$$

We can now state the following well-known result for the synthesis of $\mathscr{H}_\infty$-controllers [41, 48, 72].

Theorem 3.4. *The following two statements are equivalent:*

1. *There exist matrices A_K, B_K, C_K, D_K and $\mathcal{X}$ for which (3.26) holds with $\mathcal{X} \succ 0$.*
2. *There exist symmetric matrices X, Y for which the following LMIs hold:*

$$T_y^\top \left[\text{He} \begin{pmatrix} XA & XB_w & 0 \\ 0 & -\frac{\gamma}{2}I_{n_w} & 0 \\ C_z & D_{zw} & -\frac{\gamma}{2}I_{n_z} \end{pmatrix} \right] T_y \prec 0, \tag{3.31}$$

$$
T_u^\top \left[\operatorname{He} \begin{pmatrix} -YA^\top & -YC_z^\top & 0 \\ 0 & \frac{\gamma}{2} I_{n_z} & 0 \\ -B_w^\top & -D_{zw}^\top & \frac{\gamma}{2} I_{n_w} \end{pmatrix} \right] T_u \succ 0. \tag{3.32}
$$

$$
\begin{pmatrix} X & I \\ I & Y \end{pmatrix} \succ 0, \tag{3.33}
$$

Once the LMIs (3.31)-(3.33) in the variables X and Y are feasible, one can construct an $\mathcal{X}$ and a controller $K = (A_K, B_K, C_K, D_K)$ of McMillan degree n_G such that (3.26) is satisfied with $\mathcal{X} \succ 0$. Then K internally stabilizes G and renders the induced $\mathscr{L}_2$-gain from w to z less than $\gamma > 0$. For a proof and further details we refer the reader to [41, 48, 72].

This celebrated result covers the most simplified version of the general robust gain-scheduled controller synthesis problem as discussed in Section 3.4; the case that the dynamical system under consideration is not affected by any uncertainty, nonlinearity or scheduling parameter. It can be viewed as the starting point from which researchers have tried to find convex solutions to related more general synthesis problems and the applied techniques form the basis for most of the generalizations that have been proposed in the literature; they are essential as well for our generalizations that will be presented in the subsequent chapters. One prominent generalization of the nominal controller synthesis problem is the nominal gain-scheduled controller synthesis problem, which we considered in Section 3.3 and as will be further discussed next.

3.7 Nominal gain-scheduled controller synthesis

It is interesting to note that also the nominal gain-scheduling configuration, as considered in Section 3.3, admits a convex solution in term of LMIs. This has been shown about two decades ago in [121, 174, 152] and significantly generalizes the $\mathscr{H}_\infty$-synthesis problem from the previous section. Indeed, let us again consider the system interconnection shown in Figure 3.2, which can also be derived from the configuration as depicted in Figure 3.3, by setting $\Delta = 0$. The uncertainty channel $p \to q$ is now void and direct inspection reveals that the matrix inequality (3.24) simplifies into

$$
\begin{pmatrix} \star \\ \star \\ \star \\ \star \\ \star \\ \star \end{pmatrix}^\top \left(\begin{array}{ccccc} 0 & \mathcal{X} & 0 & 0 & 0 \\ \mathcal{X} & 0 & 0 & 0 & 0 \\ 0 & 0 & P_e & 0 & 0 \\ 0 & 0 & 0 & \gamma^{-1} I_{n_z} & 0 \\ 0 & 0 & 0 & 0 & -\gamma I_{n_w} \end{array} \right) \begin{pmatrix} I & 0 & 0 \\ \mathcal{A} & \mathcal{B}_e & \mathcal{B}_w \\ \mathcal{C}_e & \mathcal{D}_{ee} & \mathcal{D}_{ew} \\ 0 & I & 0 \\ \mathcal{C}_z & \mathcal{D}_{ze} & \mathcal{D}_{zw} \\ 0 & 0 & I \end{pmatrix} \prec 0. \tag{3.34}
$$

Analogously to the nominal $\mathscr{H}_\infty$-synthesis problem, we also formulate the dual LMI problem as obtained by Lemma C.1 in Appendix C. Indeed, any $P_e \in \mathbf{P}_e$ is nonsingular. Moreover, with $\tilde{P}_e = P_e^{-1}$ and by applying Lemma C.1, the LMI constraint (3.12) is equivalent to

$$
\begin{pmatrix} -\hat{\Delta}_e^\top \\ I \end{pmatrix}^\top \underbrace{\begin{pmatrix} \tilde{Q}_e & \tilde{S}_e \\ \tilde{S}_e^\top & \tilde{R}_e \end{pmatrix}}_{\tilde{P}_e = \tilde{P}_e^\top} \begin{pmatrix} -\hat{\Delta}_e^\top \\ I \end{pmatrix} \prec 0, \quad \tilde{Q}_e \succ 0, \quad \tilde{R}_e \prec 0. \tag{3.35}
$$

This motivates to introduce the class of so-called dual full-block multipliers $\tilde{\mathbf{P}}_e := \mathbf{P}_e^{-1}$ as given by

$$
\tilde{\mathbf{P}}_e = \left\{ \tilde{P}_e = \tilde{P}_e^\top : \tilde{P}_e \text{ satisfies (3.35) for all } \hat{\Delta}_e \in \hat{\mathbf{\Delta}}_e \right\}. \tag{3.36}
$$

Lemma C.1 can as well be applied to the matrix inequality (3.34) in order to infer its equivalence to the matrix inequality

$$
\begin{pmatrix} \star \\ \star \\ \star \\ \star \\ \star \\ \star \end{pmatrix}^\top \begin{pmatrix} 0 & \mathcal{Y} & 0 & 0 & 0 \\ \mathcal{Y} & 0 & 0 & 0 & 0 \\ 0 & 0 & \tilde{P}_e & 0 & 0 \\ 0 & 0 & 0 & \gamma I_{n_z} & 0 \\ 0 & 0 & 0 & 0 & -\gamma^{-1} I_{n_w} \end{pmatrix} \begin{pmatrix} -\mathcal{A}^\top & -\mathcal{C}_e^\top & -\mathcal{C}_z^\top \\ I & 0 & 0 \\ 0 & I & 0 \\ -\mathcal{B}_e^\top & -\mathcal{D}_{ee}^\top & -\mathcal{D}_{ze}^\top \\ 0 & 0 & I \\ -\mathcal{B}_w^\top & -\mathcal{D}_{ew}^\top & -\mathcal{D}_{zw}^\top \end{pmatrix} \succ 0. \tag{3.37}
$$

Here $\mathcal{Y} := \mathcal{X}^{-1}$ is again assumed to be partitioned as in (3.28).

Now observe that (3.34) and (3.37) can be rendered affine in γ by applying a simple Schur complement. This, respectively, yields the inequalities

$$
\mathrm{He} \begin{pmatrix} \mathcal{X}\mathcal{A} & \mathcal{X}\mathcal{B}_e & \mathcal{X}\mathcal{B}_w & 0 \\ 0 & 0 & 0 & 0 \\ 0 & 0 & -\frac{\gamma}{2} I_{n_w} & 0 \\ \mathcal{C}_z & \mathcal{D}_{ze} & \mathcal{D}_{zw} & -\frac{\gamma}{2} I_{n_z} \end{pmatrix} + \begin{pmatrix} \star \\ \star \\ \star \end{pmatrix}^\top P_e \begin{pmatrix} \mathcal{C}_e & \mathcal{D}_{ee} & \mathcal{D}_{ew} & 0 \\ 0 & I & 0 & 0 \end{pmatrix} \prec 0, \tag{3.38}
$$

and

$$
\mathrm{He} \begin{pmatrix} -\mathcal{Y}\mathcal{A}^\top & -\mathcal{Y}\mathcal{C}_e^\top & -\mathcal{Y}\mathcal{C}_z^\top & 0 \\ 0 & 0 & 0 & 0 \\ 0 & 0 & \frac{\gamma}{2} I_{n_z} & 0 \\ -\mathcal{B}_w^\top & -\mathcal{D}_{ew}^\top & -D_{zw}^\top & \frac{\gamma}{2} I_{n_w} \end{pmatrix} + \begin{pmatrix} \star \\ \star \end{pmatrix}^\top \tilde{P}_e \begin{pmatrix} 0 & I & 0 & 0 \\ -\mathcal{B}_e^\top & -\mathcal{D}_{ee}^\top & -\mathcal{D}_{ze}^\top & 0 \end{pmatrix} \succ 0. \tag{3.39}
$$

Controller synthesis now requires to design $K = (A_K, B_K, C_K, D_K)$ and the scheduling function $\Delta_c(\cdot)$ in order to render (3.38) satisfied for $\mathcal{X} \succ 0$ and $P_e \in \boldsymbol{P}_e$ or equivalently (3.39) for $\mathcal{Y} \succ 0$ and $\tilde{P}_e \in \tilde{\boldsymbol{P}}_e$. Again we observe that this problem, as formulated, is non-convex. Analogously to the nominal controller synthesis problem also [121, 174, 152] suggest to eliminate the realization matrices of K in the analysis inequalities (3.38) and (3.39) by applying Lemma C.2 in Appendix C. This involves the basis matrices

$$T_y \quad \text{of} \quad \ker\left(C_y \; D_{ys} \; D_{yw} \; 0_{n_y \times n_z} \right)$$

and

$$T_u \quad \text{of} \quad \ker\left(B_u^\top \; D_{su}^\top \; D_{zu}^\top \; 0_{n_u \times n_w} \right).$$

It is crucial to observe that, after elimination, the resulting conditions do not involve the complete multiplier P_e or its inverse $\tilde{P}_e$ any more, but that they can be expressed in terms of the set of primal scalings

$$\boldsymbol{P}_s := \left\{ P_s = P_s^\top : (\star)^\top P_s \begin{pmatrix} I \\ \hat{\Delta}_s(\delta^j) \end{pmatrix} \succ 0, \quad (\star)^\top P_s \begin{pmatrix} 0 \\ I \end{pmatrix} \prec 0 \;\; \forall j = 1, \ldots, m \right\}$$

and dual scalings

$$\tilde{\boldsymbol{P}}_s := \left\{ \tilde{P}_s = \tilde{P}_s^\top : (\star)^\top \tilde{P}_s \begin{pmatrix} -\hat{\Delta}_s(\delta^j)^\top \\ I \end{pmatrix} \prec 0, \quad (\star)^\top \tilde{P}_s \begin{pmatrix} I \\ 0 \end{pmatrix} \succ 0 \;\; \forall j = 1, \ldots, m \right\}.$$

We stress that these two sets are described by a finite number of LMIs directly in terms of the scheduling function $\Delta_s(\cdot)$ of the open-loop system and the generators of the parameter set Λ; in particular they do not involve the to-be-constructed controller scheduling function $\Delta_c(\cdot)$. We have now introduce all the necessary ingredients in order to state the following results of [121, 174, 152] for the synthesis of nominal gain-scheduled controllers.

Theorem 3.5. *The following two statements are equivalent:*

1. *There exists a controller scheduling function $\Delta_c(\cdot)$ and matrices A_K, B_K, C_K, D_K, $\mathcal{X}$, $P_e \in \boldsymbol{P}_e$ for which (3.34) holds with $\mathcal{X} \succ 0$.*
2. *There exist symmetric matrices X, Y, $P_s \in \boldsymbol{P}_s$, $\tilde{P}_s \in \tilde{\boldsymbol{P}}_s$ for which the following LMIs hold:*

$$T_y^\top \left[\text{He} \begin{pmatrix} XA & XB_s & XB_w & 0 \\ 0 & 0 & 0 & 0 \\ 0 & 0 & -\frac{\gamma}{2}I_{n_w} & 0 \\ C_z & D_{zs} & D_{zw} & -\frac{\gamma}{2}I_{n_z} \end{pmatrix} + \begin{pmatrix} \star \\ \star \end{pmatrix}^\top P_s \begin{pmatrix} C_s & D_{ss} & D_{sw} & 0 \\ 0 & I & 0 & 0 \end{pmatrix} \right] T_y \prec 0,$$

$$\tag{3.40}$$

$$T_u^\top \left[\mathrm{He} \left(\begin{pmatrix} -YA^\top & -YC_{\mathrm s}^\top & -YC_z^\top & 0 \\ 0 & 0 & 0 & 0 \\ 0 & 0 & \frac{\gamma}{2}I_{n_z} & 0 \\ -B_w^\top & -D_{\mathrm{sw}}^\top & -D_{zw}^\top & \frac{\gamma}{2}I_{n_w} \end{pmatrix} + \begin{pmatrix} \star \\ \star \\ \star \end{pmatrix}^\top \tilde P_{\mathrm s} \begin{pmatrix} 0 & I & 0 & 0 \\ -B_{\mathrm s}^\top & -D_{\mathrm{ss}}^\top & -D_{z\mathrm s}^\top & 0 \end{pmatrix} \right) \right] T_u \succ 0.$$

(3.41)

$$\begin{pmatrix} X & I \\ I & Y \end{pmatrix} \succ 0,$$ (3.42)

Once the LMIs (3.40)-(3.42) in the variables X, Y, $P_{\mathrm s}$ and $\tilde P_{\mathrm s}$ are feasible, one can construct an $\mathcal X$, an extended multiplier $P_{\mathrm e}$, a controller $K = (A_K, B_K, C_K, D_K)$ of McMillan degree n_G and, last but not least, a controller scheduling function $\hat\Delta_{\mathrm c}(\cdot)$ such that (3.34) is satisfied with $\mathcal X \succ 0$ and $P_{\mathrm e}$. Then K internally stabilizes G_{aug} and renders the induced $\mathscr L_2$-gain from w to z less than $\gamma > 0$. For a proof and further details we refer the reader to [121, 174, 152].

Observe that the $\mathscr H_\infty$-controller synthesis problem from Section 3.6 can be obtained from Theorem 3.4 as a special case, by assuming $\Delta_{\mathrm s} = 0$. Theorem 3.4 hence significantly generalizes the synthesis results from Section 3.6. In summary we have presented two important design problems that can be handled by solving a standard semi-definite program; the nominal $\mathscr H_\infty$-synthesis and nominal gain-scheduled controller synthesis problem. We emphasize that both design problems do not take into account the uncertainty block Δ. This scenario can only be handled for some specialized problems as will be discussed next.

3.8 Outline of further possible configurations

To conclude this chapter, let us give an outline of further possible synthesis configurations that are covered. The abstracts in the following subsections briefly sketch the possibilities as well as the topics that will be addressed in the coming chapters.

3.8.1 Robust gain-scheduled estimation

One of the most well-known robust synthesis problems that admits an LMI-solution is the robust estimation problem (see e.g. [51, 180, 165] and references therein). Unfortunately, most of the available algorithms only allow for genuine uncertainties $\Delta \in \Delta$ and do not cover the inclusion of a scheduling block $\Delta_{\mathrm s} \in \Delta_{\mathrm s}$. For plants that depend on both genuine uncertainties $\Delta \in \Delta$ and scheduling parameters $\Delta_{\mathrm s} \in \Delta_{\mathrm s}$ solutions are available, however, only if the uncertainties are parametric and if the dependency on $\Delta \in \boldsymbol{\Delta}$ and $\Delta_s \in \boldsymbol{\Delta}_s$ is affine (see e.g. [20]). On the other hand, in the case of more general uncertainties and a general linear fractional dependency on $\Delta \in \boldsymbol{\Delta}$ and $\Delta_s \in \boldsymbol{\Delta}_s$, it is unknown how to design robust gain-scheduled estimators. As the main theme of Chapter 4, and in accordance with our contributions in the papers [194] and [200], we will formulate sufficient existence conditions for constructing such robust gain-scheduled estimators with guaranteed $\mathscr{L}_2$-gain bounds in terms of finite dimensional LMIs.

3.8.2 Another general synthesis framework

In Chapter 5 we will continue with a more general framework for the systematic synthesis of robust gain-scheduled controllers by convex optimization techniques. Also here we distinguish between linear time-varying parameters, which are assumed to be available online as scheduling parameters for the controller, and genuine uncertainties, not necessarily time-varying, parametric or linear, that are not available online. Under the rough hypothesis that the control channel is not affected by the unmeasurable uncertainties and that the properties of the uncertainties and scheduling variables are captured by suitable families of IQC-multipliers, we will reveal in this chapter how controller synthesis can be turned into a genuine semi-definite program. This design framework, which was initially introduced in our paper [197], is shown to encompass a rich class of concrete scenarios, such as robust estimator or observer design, robust feedforward control, generalized l_2-synthesis, gain-scheduled output-feedback control with uncertain performance weights and robust control with unstable weights, among others.

3.8.3 General IQC-synthesis

Analogously to the existing μ-synthesis tools [143, 40, 11, 28, 146, 144, 63, 208], we propose in Chapter 6 an alternative algorithm for the systematic design of robust controllers based on an iteration of nominal controller synthesis and IQC-analysis. The suggested algorithm, which we first presented in our papers [193] and [198], enables us to systematically perform robust controller synthesis for a significantly larger class of uncertainties if compared to the existing methods. Indeed, while the classical approaches are restricted to the use of real/complex time-invariant or arbitrarily fast time-varying parametric uncertainties, it has been extensively illustrated in Chapter 2 that the IQC-framework offers the possibility to efficiently handle much more general classes of uncertainties. Secondly, in contrast to the classical approaches, the proposed techniques completely avoid gridding and curve-fitting. We present new insights that allow us to reformulate the robust IQC-analysis LMIs into a standard quadratic performance problem. This enables us to generate suitable initial conditions for each subsequent iteration step. Depending on the size of the problem this can significantly speed up the synthesis process.

3.9 Chapter summary

In this transitional chapter we moved on from IQC-analysis towards the more challenging IQC-synthesis problem. We formulated the central and most general design question of this thesis: the robust gain-scheduled controller synthesis problem. Although this problem cannot be easily translated into an LMI problem, we argued that it is possible to consider a large class of robust controller synthesis problems that can be derived from this central problem by specialization and efficiently solve these with LMI-techniques. In order to illustrate this we recalled the LMI-solutions as well as the applied techniques of two well-known special cases and gave a short outline of further possible configurations that are covered and will be discussed in the subsequent chapters.

Chapter 4
Robust gain-scheduled estimation

4.1 Introduction

The online estimation of non-measured variables in a dynamical system constitutes one of the most important problems in systems and control theory. When the mathematical model represents the physical system exactly, this problem is solved efficiently within the framework of $\mathscr{H}_2$- or $\mathscr{H}_\infty$-control theory and the solutions are expressed in terms of convex optimization problems (see, e.g. [43]). In addition, it has recently been observed that in the presence of modeling uncertainties the robust estimation problem also admits a convex solution [51].

Since then, a number of papers have appeared investigating different uncertainty structures in the plant and different methods of solution. As is typical of all theoretical studies in the literature, the results vary with the dependency of the plant on the uncertainties. For example, systems which depend affinely on uncertain parameters that take their values in a convex polytope were studied in [51] and [52]. The conservatism of the techniques presented in these papers was reduced by the use of parametric Lyapunov functions in [53] as well as in e.g. [15, 186, 14, 36, 87]. On the other hand, in case of a linear fractional dependency of the plant on the uncertain parameters, static and dynamic DG-multipliers were considered in [180] and [175], respectively. Moreover, for dynamical systems involving uncertainties whose properties are captured by IQCs as in [116], a full solution for the robust estimation problem was given in [165] (see also [190]).

In this chapter, we focus on dynamical systems with a general linear fractional dependency on uncertainties that fall into two categories. The first category covers genuine uncertainties whose characteristics are known, but which are not measured online, and, therefore, cannot be used for scheduling the estimator, and the second category consists of parametric variations that are measurable online, with a priori known bounds and arbitrarily fast rates-of-variation. The objective is to design estimators, which guarantee stability of the overall system interconnection for all uncertainties in the first category, and which are scheduled on the uncertainties in the second category. The estimator is, hence, only partially gain-scheduled. In addition, the goal is to minimize the induced $\mathscr{L}_2$-gain from the disturbance input to the estimated error output. We emphasize that this problem is considerably more general than the one considered in e.g. [20], which only allows for plants with an affine dependency on parametric uncertainties from a convex polytope. This represents a very realistic engineering scenario that has, to the best of our knowledge, and apart from our conference paper [194] and book chapter [200], not been addressed in the literature.

The main result is given in terms of convex conditions for the existence of such estimators, together with a constructive proof in case these conditions are satisfied. We rely on the gain-scheduling synthesis results of [121, 174, 152], which we presented in Section 3.7, and exploit the robust estimation and feedforward synthesis techniques of [165, 102]. The solution is based on the robustness analysis of the overall system interconnection using IQCs in accordance with the developed tools of Chapter 2. For each category of the uncertainties, we use the most representative IQC-multiplier in the literature; for the genuine uncertainties, we allow for any appropriate, possibly dynamic (i.e. frequency-dependent) IQC-multiplier and for the measured parametric uncertainties we use static (i.e. frequency-independent) full-block multipliers as considered in [152] and Section 2.6.2. As a technical contribution, it will be shown that it is possible to apply the crucial step of 'dualization' (see Lemma C.1 in Appendix C). This not only allows us to eliminate the realization matrices of the estimator, but also to render the problem tractable. As an extra benefit, this gives us the possibility to design both static as well as full-order dynamic estimators (i.e. estimators whose McMillan degree is equal to zero or equal to the McMillan degree of the open-loop plant plus the McMillan degree of the IQC-multipliers respectively). Lastly, completely analogous results can be derived for the design of robust, partially gain-scheduled feedforward controllers employing the techniques here, only in dual form.

This chapter is organized as follows. We will start by introducing the robust estimation problem with partial gain-scheduling in Section 4.2 and continue by embedding the problem in the IQC-framework in Section 4.3. Subsequently, we present basic results regarding the structure of the multipliers and the issue of nominal stability characterization in the state-space in Section 4.4. The main result of the chapter, namely Theorem 4.2, follows then in Section 4.5. The findings are highlighted through a numerical example in Section 4.6 and we conclude the chapter with a summary in Section 4.7. We note that some of the technical proofs are deferred to the Appendix.

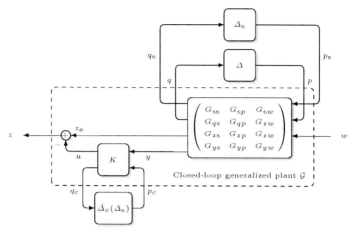

Fig. 4.1 The robust gain-scheduled estimation problem

4.2 The robust gain-scheduled estimation problem

Consider the uncertain linear parameter varying (LPV) system interconnection in Figure 4.1. Here the corresponding dynamical system is defined through the linear fractional representation

$$
\begin{pmatrix} q_s \\ q \\ z \\ y \end{pmatrix} = \underbrace{\begin{pmatrix} G_{ss} & G_{sp} & G_{sw} & 0 \\ G_{qs} & G_{qp} & G_{qw} & 0 \\ G_{zs} & G_{zp} & G_{zw} & -I_{n_z} \\ G_{ys} & G_{yp} & G_{yw} & 0 \end{pmatrix}}_{G} \begin{pmatrix} p_s \\ p \\ w \\ u \end{pmatrix}, \quad \begin{pmatrix} p_s \\ p \end{pmatrix} = \begin{pmatrix} \Delta_s q_s \\ \Delta(q) \end{pmatrix}, \qquad (4.1)
$$

where $p_s \to q_s$ represents the scheduling channel, $p \to q$ the uncertainty channel, $w \to z$ the performance channel and $u \to y$ the filtering channel. The open-loop plant $G \in \mathscr{RH}_\infty^{(n_{q_s}+n_q+n_z+n_y)\times(n_{p_s}+n_p+n_w+n_u)}$ is a real rational proper and stable transfer matrix, while the uncertainty and scheduling blocks Δ and Δ_s are contained in the sets $\boldsymbol{\Delta}$ and $\boldsymbol{\Delta}_s$ as defined in Section 2.4.1 and Section 3.3 respectively. We further assume that the system (4.1) is well-posed, which means that

$$
\begin{pmatrix} I & 0 \\ 0 & I \end{pmatrix} - \begin{pmatrix} G_{ss} & G_{sp} \\ G_{qs} & G_{qp} \end{pmatrix} \begin{pmatrix} \Delta_s & 0 \\ 0 & \Delta \end{pmatrix} \qquad (4.2)
$$

has a causal inverse for all $\Delta \in \boldsymbol{\Delta}$ and for all $\Delta_{\mathrm{s}} \in \boldsymbol{\Delta}_{\mathrm{s}}$. Note that the overall configuration is a special case of the general problem as formulated in Section 3.4, for $G_{su} := 0$, $G_{qu} := 0$, $G_{zu} := -I_{n_z}$ and $G_{yu} := 0$.

The main goal in robust gain-scheduled estimation is the synthesis of a stable filter

$$\begin{pmatrix} u \\ q_{\mathrm{c}} \end{pmatrix} = \underbrace{\begin{pmatrix} K_{11} & K_{12} \\ K_{21} & K_{22} \end{pmatrix}}_{K} \begin{pmatrix} y \\ p_{\mathrm{c}} \end{pmatrix}, \quad p_{\mathrm{c}} = \hat{\Delta}_{\mathrm{c}}(\Delta_{\mathrm{s}}) q_{\mathrm{c}} \tag{4.3}$$

which dynamically and causally processes the measurement y and the parameter vector $\delta : [0, \infty) \to \Lambda$, in order to provide an estimate u of the signal z_{e} in the sense that the induced $\mathscr{L}_2$-gain from w to $z = z_{\mathrm{e}} - u$ is rendered less than an a priori given $\gamma > 0$. Here $K \in \mathscr{R}\mathscr{H}_\infty^{(n_u + n_{q_{\mathrm{c}}}) \times (n_y + n_{p_{\mathrm{c}}})}$ is a to-be-designed stable LTI system and $\hat{\Delta}_{\mathrm{c}} : \mathbb{R}^{n_{p_{\mathrm{s}}} \times n_{q_{\mathrm{s}}}} \to \mathbb{R}^{n_{p_{\mathrm{c}}} \times n_{q_{\mathrm{c}}}}$ the to-be-constructed scheduling function in accordance with the definitions in Chapter 3.

4.3 Robust stability and performance analysis

Before we discuss the technical difficulties that need to be resolved, let us first embed the estimation problem in the IQC-framework. For this purpose, we reformulate the system interconnection in a standard fashion as

$$\begin{pmatrix} q_{\mathrm{e}} \\ q \\ z \end{pmatrix} = \underbrace{\begin{pmatrix} \mathcal{G}_{ee} & \mathcal{G}_{ep} & \mathcal{G}_{ew} \\ \mathcal{G}_{qe} & \mathcal{G}_{qp} & \mathcal{G}_{qw} \\ \mathcal{G}_{ze} & \mathcal{G}_{zp} & \mathcal{G}_{zw} \end{pmatrix}}_{\mathcal{G}} \begin{pmatrix} p_{\mathrm{e}} \\ p \\ w \end{pmatrix}, \quad \begin{pmatrix} p_{\mathrm{e}} \\ p \end{pmatrix} = \begin{pmatrix} \Delta_{\mathrm{e}}(\Delta_{\mathrm{s}}) q_{\mathrm{e}} \\ \Delta(q) \end{pmatrix}. \tag{4.4}$$

Here we recall from Section 3.3 that the extended scheduling block $\Delta_{\mathrm{e}}(\Delta_{\mathrm{s}}) = \Delta_{\mathrm{e}} \circ \Delta_{\mathrm{s}}$ is defined through the map (3.10), while $\mathcal{G}$ denotes the interconnection of G and the LTI part of the estimator K as given by

$$\underbrace{\begin{pmatrix} \mathcal{G}_{ee} & \mathcal{G}_{ep} & \mathcal{G}_{ew} \\ \mathcal{G}_{qe} & \mathcal{G}_{qp} & \mathcal{G}_{qw} \\ \mathcal{G}_{ze} & \mathcal{G}_{zp} & \mathcal{G}_{zw} \end{pmatrix}}_{\mathcal{G}} = \underbrace{\begin{pmatrix} G_{ss} & 0 & G_{sp} & G_{sw} & 0 & 0 \\ 0 & 0 & 0 & 0 & 0 & I_{n_{q_{\mathrm{c}}}} \\ G_{qs} & 0 & G_{qp} & G_{qw} & 0 & 0 \\ G_{zs} & 0 & G_{zp} & G_{zw} & -I_{n_z} & 0 \\ G_{ys} & 0 & G_{yp} & G_{yw} & 0 & 0 \\ 0 & I_{n_{p_{\mathrm{c}}}} & 0 & 0 & 0 & 0 \end{pmatrix}}_{G_{\mathrm{aug}}} \star \underbrace{\begin{pmatrix} K_{11} & K_{12} \\ K_{21} & K_{22} \end{pmatrix}}_{K}.$$

It is not hard to see that this can be more explicitly written as

$$
\begin{pmatrix} \mathcal{G}_{ee} & \mathcal{G}_{ep} & \mathcal{G}_{ew} \\ \mathcal{G}_{qe} & \mathcal{G}_{qp} & \mathcal{G}_{qw} \\ \mathcal{G}_{ze} & \mathcal{G}_{zp} & \mathcal{G}_{zw} \end{pmatrix} = \begin{pmatrix} G_{ss} & 0 & G_{sp} & G_{sw} \\ K_{21}G_{ys} & K_{22} & K_{21}G_{yp} & K_{21}G_{yw} \\ G_{qs} & 0 & G_{qp} & G_{qw} \\ G_{zs} - K_{11}G_{ys} & -K_{12} & G_{zp} - K_{11}G_{yp} & G_{zw} - K_{11}G_{yw} \end{pmatrix}.
$$

As a natural consequence, $\mathcal{G}$ is internally stable if both G and K are stable. Moreover, the system (4.4) is well-posed if the open-loop system (4.1) is well-posed (i.e. if (4.2) has a causal inverse for all $\Delta \in \boldsymbol{\Delta}$ and for all $\Delta_s \in \boldsymbol{\Delta}_s$) and if

$$
I - K_{22}\hat{\Delta}_c(\Delta_s) \tag{4.5}
$$

is well-posed for all $\Delta_s \in \boldsymbol{\Delta}_s$. Hence, since (4.2) is well-posed by assumption, we only require (4.5) to have a causal inverse for all $\Delta_s \in \boldsymbol{\Delta}_s$, which in turn just means that the estimator (4.3) must be well-posed, as will be guaranteed by the results in the sequel.

Let us further recall from Section 3.3 that $\Delta_e(\Delta_s)$ takes its values in the set of matrices $\hat{\boldsymbol{\Delta}}_e$ and that the LMI constraint (3.12) defines the set of extended full-block multipliers $\mathbf{P}_e$ and guarantees the IQC

$$
IQC(P_e, q_e, \tau \Delta_e(\Delta_s)q_e) \geq 0 \quad \forall q_e \in \mathscr{L}_2^{n_{qs}+n_{qc}}, \quad \forall \tau \in [0,1] \tag{4.6}
$$

to be satisfied for all $\Delta_s \in \boldsymbol{\Delta}_s$ and for all $P_e \in \mathbf{P}_e$. In a similar fashion, we follow the discussion in Section 3.2 and suppose that the properties of the uncertainty block Δ are captured with the IQC

$$
IQC(\Psi^* P \Psi, q, \Delta(q)) \geq 0 \quad \forall q \in \mathscr{L}_2^{n_q}, \tag{4.7}
$$

which holds for all $\Delta \in \boldsymbol{\Delta}$, for all $P \in \boldsymbol{P}$ and for $\Psi = (\Psi_1 \ \Psi_2) \in \mathscr{RH}_\infty^{\bullet \times (n_q + n_p)}$. If we also assume that the desired induced $\mathscr{L}_2$-gain performance objective is expressed through the IQC

$$
IQC(P_\gamma, z, w) \leq -\epsilon \|w\|_2^2, \tag{4.8}
$$

where

$$
P_\gamma = \mathrm{diag}(I_{n_z}, -\gamma^2 I_{n_w})
$$

then we arrive at the following analysis results.

Theorem 4.1. *Let the estimator (4.3) be given and assume that*

1. *both G and K are stable;*
2. *the open-loop system (4.1) is well-posed for all $\Delta \in \boldsymbol{\Delta}$ and for all $\Delta_s \in \boldsymbol{\Delta}_s$;*
3. *the IQC (4.6) is satisfied for all $\Delta_s \in \boldsymbol{\Delta}_s$ and for all $P_e \in \mathbf{P}_e$;*
4. *the IQC (4.7) is satisfied for all $\Delta \in \boldsymbol{\Delta}$ and for all $P \in \boldsymbol{P}$;*

5. *there exist $P \in \boldsymbol{P}$ and $P_{\mathrm{e}} \in \boldsymbol{P}_{\mathrm{e}}$ for which the following FDI holds:*

$$
\begin{pmatrix} \star \\ \star \\ -- \\ \star \\ \star \\ \star \end{pmatrix}^{*}
\left(\begin{array}{c|c|c} P_{\mathrm{e}} & 0 & 0 \\ \hline 0 & P & 0 \\ \hline 0 & 0 & P_{\gamma} \end{array} \right)
\left(\begin{array}{ccc} \mathcal{G}_{\mathrm{ee}} & \mathcal{G}_{\mathrm{ep}} & \mathcal{G}_{\mathrm{ew}} \\ I & 0 & 0 \\ \hline \Psi_{1}\mathcal{G}_{q\mathrm{e}} & \Psi_{1}\mathcal{G}_{qp}+\Psi_{2} & \Psi_{1}\mathcal{G}_{qw} \\ \hline \mathcal{G}_{z\mathrm{e}} & \mathcal{G}_{zp} & \mathcal{G}_{zw} \\ 0 & 0 & I \end{array} \right)
\prec 0 \text{ on } \mathbb{C}^{0}. \quad (4.9)
$$

Then, for all $\Delta \in \boldsymbol{\Delta}$ and for all $\Delta_{\mathrm{s}} \in \boldsymbol{\Delta}_{\mathrm{s}}$, the system (4.4) (and hence the estimator (4.3)) is well-posed and stable and the induced $\mathscr{L}_{2}$-gain from w to z is guaranteed to be less than $\gamma > 0$.

In order to turn Theorem 4.1 into computations, we recall (from Section 3.5) the realizations (3.19)-(3.23) for the systems K, G and Ψ respectively (as well as their properties) and observe that the right-hand outer-factor of (4.9) admits the realization

$$
\begin{pmatrix} \mathcal{G}_{\mathrm{ee}} & \mathcal{G}_{\mathrm{ep}} & \mathcal{G}_{\mathrm{ew}} \\ I & 0 & 0 \\ \hline \Psi_{1}\mathcal{G}_{q\mathrm{e}} & \Psi_{1}\mathcal{G}_{qp}+\Psi_{2} & \Psi_{1}\mathcal{G}_{qw} \\ \hline \mathcal{G}_{z\mathrm{e}} & \mathcal{G}_{zp} & \mathcal{G}_{zw} \\ 0 & 0 & I \end{pmatrix}
=
\left[\begin{array}{cc|ccc} A_{\Psi} & B_{\Psi 1}\mathcal{C}_{q} & B_{\Psi 1}\mathcal{D}_{q\mathrm{e}} & B_{\Psi 1}\mathcal{D}_{qp}+B_{\Psi 2} & B_{\Psi 1}\mathcal{D}_{qw} \\ 0 & \mathcal{A} & \mathcal{B}_{\mathrm{e}} & \mathcal{B}_{p} & \mathcal{B}_{w} \\ 0 & \mathcal{C}_{\mathrm{e}} & \mathcal{D}_{\mathrm{ee}} & \mathcal{D}_{\mathrm{ep}} & \mathcal{D}_{\mathrm{ew}} \\ 0 & 0 & I & 0 & 0 \\ \hline C_{\Psi} & D_{\Psi 1}\mathcal{C}_{q} & D_{\Psi 1}\mathcal{D}_{q\mathrm{e}} & D_{\Psi 1}\mathcal{D}_{qp}+D_{\Psi 2} & D_{\Psi 1}\mathcal{D}_{qw} \\ 0 & \mathcal{C}_{z} & \mathcal{D}_{z\mathrm{e}} & \mathcal{D}_{zp} & \mathcal{D}_{zw} \\ 0 & 0 & 0 & 0 & I \end{array} \right].
$$

Here the calligraphic matrices are given by

$$
\left(\begin{array}{c|c|cc} \mathcal{A} & \mathcal{B}_{\mathrm{e}} & \mathcal{B}_{\mathrm{p}} & \mathcal{B}_{w} \\ \hline \mathcal{C}_{\mathrm{e}} & \mathcal{D}_{\mathrm{ee}} & \mathcal{D}_{\mathrm{ep}} & \mathcal{D}_{\mathrm{ew}} \\ \hline \mathcal{C}_{q} & \mathcal{D}_{q\mathrm{e}} & \mathcal{D}_{qp} & \mathcal{D}_{qw} \\ \mathcal{C}_{z} & \mathcal{D}_{z\mathrm{e}} & \mathcal{D}_{zp} & \mathcal{D}_{zw} \end{array} \right)
=
\left(\begin{array}{cc|cc|cc} A & 0 & B_{\mathrm{s}} & 0 & B_{p} & B_{w} \\ 0 & 0 & 0 & 0 & 0 & 0 \\ \hline C_{\mathrm{s}} & 0 & D_{\mathrm{ss}} & 0 & D_{\mathrm{sp}} & D_{\mathrm{sw}} \\ 0 & 0 & 0 & 0 & 0 & 0 \\ \hline C_{q} & 0 & D_{q\mathrm{s}} & 0 & D_{qp} & D_{qw} \\ C_{z} & 0 & D_{z\mathrm{s}} & 0 & D_{zp} & D_{zw} \end{array} \right)
+
\left(\begin{array}{cc|c} 0 & 0 & 0 \\ I & 0 & 0 \\ \hline 0 & 0 & 0 \\ 0 & 0 & I \\ \hline 0 & 0 & 0 \\ 0 & -I & 0 \end{array} \right)
\times
$$

$$
\times
\begin{pmatrix} A_{K} & B_{K} \\ C_{K} & D_{K} \end{pmatrix}
\left(\begin{array}{cc|cc|cc} 0 & I & 0 & 0 & 0 & 0 \\ \hline C_{y} & 0 & D_{y\mathrm{s}} & 0 & D_{yp} & D_{yw} \\ 0 & 0 & I & 0 & 0 & 0 \end{array} \right). \quad (4.10)
$$

Since G and K are given stable transfer matrices, we can assume that both A and A_K are Hurwitz such that $\mathcal{A}$ is Hurwitz as well. Hence, we can infer by the KYP-lemma (i.e. Lemma 2.2 in Section 2.5.2) that the FDI (4.9) is equivalent to the existence of a symmetric matrix $\mathcal{X} \in \mathbb{S}^{n_\Psi + n_G + n_K}$ for which the following LMI is satisfied:

$$\begin{pmatrix} \star \\ \star \\ \star \\ \star \\ \star \\ \star \\ \star \\ \star \end{pmatrix}^{\mathsf{T}} \begin{pmatrix} 0 & \mathcal{X} & 0 & 0 & 0 \\ \mathcal{X} & 0 & 0 & 0 & 0 \\ 0 & 0 & P_e & 0 & 0 \\ 0 & 0 & 0 & P & 0 \\ 0 & 0 & 0 & 0 & P_\gamma \end{pmatrix} \begin{pmatrix} I & 0 & 0 & 0 \\ A_\Psi & B_{\Psi 1}\mathcal{C}_q & B_{\Psi 1}\mathcal{D}_{qe} & B_{\Psi 1}\mathcal{D}_{qp} + B_{\Psi 2} & B_{\Psi 1}\mathcal{D}_{qw} \\ 0 & \mathcal{A} & \mathcal{B}_e & \mathcal{B}_p & \mathcal{B}_w \\ 0 & \mathcal{C}_e & \mathcal{D}_{ee} & \mathcal{D}_{ep} & \mathcal{D}_{ew} \\ 0 & 0 & I & 0 & 0 \\ C_\Psi & D_{\Psi 1}\mathcal{C}_q & D_{\Psi 1}\mathcal{D}_{qe} & D_{\Psi 1}\mathcal{D}_{qp} + D_{\Psi 2} & D_{\Psi 1}\mathcal{D}_{qw} \\ 0 & \mathcal{C}_z & \mathcal{D}_{ze} & \mathcal{D}_{zp} & \mathcal{D}_{zw} \\ 0 & 0 & 0 & 0 & I \end{pmatrix} \prec 0.$$

(4.11)

Here $\mathcal{X}$ is again assumed to be partitioned as in (3.25) in Section 3.5.

Although the LMI (4.11) is identical to the one in Section 3.5 for general robust gain-scheduled controller synthesis, we stress that the underlying structure is specialized, as is visible in (4.10). The extra structure will be essential for our results in the sequel.

In the case of synthesis, the robust gain-scheduled estimation problem translates into the search for matrices A_K, B_K, C_K, D_K with A_K being Hurwitz as well as a scheduling function $\hat{\Delta}_c(\cdot)$ such that there exists some $\mathcal{X}$, $P \in \boldsymbol{P}$ and $P_e \in \mathbf{P}_e$ for which (4.11) is satisfied. Unfortunately, (4.11) is no longer affine in all variables, such that LMI solvers are unable to handle the synthesis problem. In the subsequent section we will formulate the technical results that are needed in order to overcome this issue and translate the latter synthesis problem into an LMI feasibility test.

4.4 From analysis to synthesis

As discussed in Section 3.6 and 3.7, a common procedure to resolve a synthesis problem is to eliminate the estimator or controller variables by applying the Elimination Lemma (See Lemma C.2 in Appendix C). However, there are three main issues appearing in the robust gain-scheduled estimator synthesis problem that need to be resolved.

1. Due to the generality of the IQC-multipliers, $\mathcal{X} \succ 0$ is no longer the appropriate condition in order to enforce $\mathcal{A}$ to be Hurwitz (see [165] and Remark 3.1 in Section 3.6).

2. In order to eliminate the estimator variables by applying Lemma C.2, and as we
   have seen in Section 3.6 and 3.7, it is required to formulate a second (dual) solvability condition by applying Lemma C.1 (See Appendix C). However, the outer-
   factors of the IQC-multiplier factorization $\Psi^* P \Psi$ are generally tall and, hence,
   cannot be inverted. Since the inverse of $\Psi^* P \Psi$ is essential in order to explicitly
   formulate the dual of matrix inequality (4.11), the (primal) matrix inequality
   (4.11) must be reformulated with a square factorization of $\Psi^* P \Psi$ as

$$\Psi^* P \Psi = \hat{\Psi}^* \hat{P} \hat{\Psi}$$

   with $\hat{P} := \mathrm{diag}(I, -I)$ and where $\hat{\Psi}$ is square and invertible (see also [103] and
   Remark 3.2 in Section 3.6).

3. In contrast to the standard nominal and nominal gain-scheduled controller synthesis problems from Section 3.6 and 3.7 respectively, it is not sufficient to only
   eliminate the realization matrices of K. The primal and dual solvability conditions
   typically induce non-convex constraints on the IQC-multipliers which are, in general, impossible to convexify. Indeed, it is well known that, after elimination, the
   resulting primal and dual matrix inequalities will depend on the IQC-multiplier
   $\Psi^* P \Psi$ and its inverse $(\Psi^* P \Psi)^{-1}$ respectively.

In the two subsections that follow, we will resolve the first two issues from above.
The third one will be addressed in Section 4.5 in the proof of Theorem 4.2.

4.4.1 A characterization of nominal stability

As we have seen in Section 3.6 and 3.7, it is possible to enforce $\mathcal{A}$ to be Hurwitz
and guarantee the corresponding system $\mathcal{G}$ to be internally stable if and only if the
following condition is satisfied:

$$\mathcal{X} \succ 0.$$

Unfortunately, this is neither true for the general robust gain-scheduling synthesis
problem of Section 3.4 nor for the considered configuration in this Chapter. Indeed,
observe that (4.11) (trivially) implies that the following inequality is satisfied:

$$\begin{pmatrix} I & 0 \\ 0 & I \\ \hline A_\Psi & B_{\Psi 1} C_q \\ 0 & \mathcal{A} \\ \hline C_\Psi & D_{\Psi 1} C_q \end{pmatrix}^\top \begin{pmatrix} 0 & 0 & X_{11} & X_{12} & 0 \\ 0 & 0 & \mathcal{X}_{12}^\top & \mathcal{X}_{22} & 0 \\ \hline X_{11} & \mathcal{X}_{12} & 0 & 0 & 0 \\ \mathcal{X}_{12}^\top & \mathcal{X}_{22} & 0 & 0 & 0 \\ \hline 0 & 0 & 0 & 0 & P \end{pmatrix} \begin{pmatrix} I & 0 \\ 0 & I \\ \hline A_\Psi & B_{\Psi 1} C_q \\ 0 & \mathcal{A} \\ \hline C_\Psi & D_{\Psi 1} C_q \end{pmatrix} \prec 0. \qquad (4.12)$$

This can be written as

$$\mathrm{He}\left[\begin{pmatrix} \mathcal{X}_{11} & \mathcal{X}_{12} \\ \mathcal{X}_{12}^\top & \mathcal{X}_{22} \end{pmatrix}\begin{pmatrix} A_\Psi & B_{\Psi 1}\mathcal{C}_q \\ 0 & \mathcal{A} \end{pmatrix}\right] + (\star)^\top P \begin{pmatrix} C_\Psi & D_{\Psi 1}\mathcal{C}_q \end{pmatrix} \prec 0.$$

Now observe that $P \in \boldsymbol{P}$ is generally an indefinite matrix. Hence, $\mathcal{A}$ being Hurwitz neither implies nor is implied by positive definiteness of the matrix $\mathcal{X}$. We conclude that $\mathcal{X} \succ 0$ is not an appropriate condition in order to enforce internal stability on the underlying system $\mathcal{G}$.

In order to overcome this issue it has been observed in [165] that we need to exploit the IQC (4.7). Indeed, since $0 \in \boldsymbol{\Delta}$ it is not hard to see that (4.7) implies that

$$\Psi_1^* P \Psi_1 \succcurlyeq 0 \text{ on } \mathbb{C}^0.$$

To avoid a perturbation argument, we assume that there exist at least one $P \in \boldsymbol{P}$ with

$$\Psi_1^* P \Psi_1 \succ 0 \text{ on } \mathbb{C}^0. \tag{4.13}$$

By the KYP-lemma we can then infer that (4.13) is equivalent to the existence of some symmetric matrix $X_\Psi \in \mathbb{S}^{n_\Psi}$ for which

$$\begin{pmatrix} I & 0 \\ A_\Psi & B_{\Psi 1} \\ \hline C_\Psi & D_{\Psi 1} \end{pmatrix}^\top \left(\begin{array}{cc|c} 0 & X_\Psi & 0 \\ X_\Psi & 0 & 0 \\ \hline 0 & 0 & P \end{array}\right) \begin{pmatrix} I & 0 \\ A_\Psi & B_{\Psi 1} \\ \hline C_\Psi & D_{\Psi 1} \end{pmatrix} \succ 0. \tag{4.14}$$

As one of the contributions of [165] it has been shown that internal stability of $\mathcal{G}$ can be characterized as follows.

Lemma 4.1. *A_Ψ and $\mathcal{A}$ are Hurwitz and the FDIs (4.9) as well as (4.13) hold if and only if there exist solutions $\mathcal{X}$ and X_Ψ of the LMIs (4.11) and (4.14) which are coupled as*

$$\mathcal{X} - \begin{pmatrix} X_\Psi & 0 \\ 0 & 0 \end{pmatrix} = \begin{pmatrix} \mathcal{X}_{11} - X_\Psi & \mathcal{X}_{12} \\ \mathcal{X}_{12}^\top & \mathcal{X}_{22} \end{pmatrix} \succ 0. \tag{4.15}$$

Proof. We proceed along the lines of [38]. Observe that (4.14) can be inflated to

$$\begin{pmatrix} I & 0 \\ 0 & I \\ \hline A_\Psi & B_{\Psi 1}\mathcal{C}_q \\ 0 & \mathcal{A} \\ \hline C_\Psi & D_{\Psi 1}\mathcal{C}_q \end{pmatrix}^\top \left(\begin{array}{cc|cc|c} 0 & 0 & -X_\Psi & 0 & 0 \\ 0 & 0 & 0 & 0 & 0 \\ \hline -X_\Psi & 0 & 0 & 0 & 0 \\ 0 & 0 & 0 & 0 & 0 \\ \hline 0 & 0 & 0 & 0 & -P \end{array}\right) \begin{pmatrix} I & 0 \\ 0 & I \\ \hline A_\Psi & B_{\Psi 1}\mathcal{C}_q \\ 0 & \mathcal{A} \\ \hline C_\Psi & D_{\Psi 1}\mathcal{C}_q \end{pmatrix} \preccurlyeq 0$$

Adding up (4.12) and the latter inequality then yields

$$\mathrm{He}\left[\begin{pmatrix} X_{11} - X_\Psi & X_{12} \\ X_{12}^\top & X_{22} \end{pmatrix}\begin{pmatrix} A_\Psi & B_{\Psi 1}\mathcal{C}_q \\ 0 & \mathcal{A} \end{pmatrix}\right] \prec 0.$$

Positive definiteness of (4.15) is, hence, equivalent to A_Ψ and $\mathcal{A}$ being Hurwitz. ∎

4.4.2 Reformulation of the analysis LMIs

In order to resolve the second issue we will rely on the results of [165] and as streamlined in [198] as well as [166, 168]. As a preparation, we need the following fact, which guarantees the existence of a factorization $\Psi^* P \Psi = \hat\Psi^* \hat P \hat\Psi$ with the desired properties as mentioned above.

Lemma 4.2. *Suppose that there exist matrices $P_e \in \boldsymbol{P}_e$ and $P \in \boldsymbol{P}$ for which the FDIs (4.9) and (4.13) are satisfied. Then*

$$\Psi^* P \Psi = \begin{pmatrix} \Psi_1^* P \Psi_1 & \Psi_1^* P \Psi_2 \\ \Psi_2^* P \Psi_1 & \Psi_2^* P \Psi_2 \end{pmatrix} = \begin{pmatrix} \Pi_{11} & \Pi_{12} \\ \Pi_{12}^* & \Pi_{22} \end{pmatrix} = \Pi$$

satisfies the inertia constraints

$$\Pi_{11} \succ 0 \text{ on } \mathbb{C}^0 \quad and \quad \Pi_{22} - \Pi_{12}^* \Pi_{11}^{-1} \Pi_{12} \prec 0 \text{ on } \mathbb{C}^0 \qquad (4.16)$$

Proof. Since the left-upper block Q_e and I of the multipliers P_e and P_γ respectively are positive definite, we infer that (4.9) implies

$$(\Psi_1 \mathcal{G}_{qp} + \Psi_2)^* P (\Psi_1 \mathcal{G}_{qp} + \Psi_2) \prec 0 \text{ on } \mathbb{C}^0.$$

This in turn can be written as

$$\begin{pmatrix} \mathcal{G}_{qp} \\ I \end{pmatrix}^* \begin{pmatrix} \Pi_{11} & \Pi_{12} \\ \Pi_{12}^* & \Pi_{22} \end{pmatrix}\begin{pmatrix} \mathcal{G}_{qp} \\ I \end{pmatrix} \prec 0 \text{ on } \mathbb{C}^0. \qquad (4.17)$$

Since (4.13) holds, Π can be factorized as

$$\begin{pmatrix} I & \Pi_{11}^{-1}\Pi_{12} \\ 0 & I \end{pmatrix}^* \begin{pmatrix} \Pi_{11} & 0 \\ 0 & \Pi_{22} - \Pi_{12}^* \Pi_{11}^{-1}\Pi_{12} \end{pmatrix}\begin{pmatrix} I & \Pi_{11}^{-1}\Pi_{12} \\ 0 & I \end{pmatrix}$$

Hence, (4.17) can be written as

$$\begin{pmatrix} \star \\ \star \end{pmatrix}^* \begin{pmatrix} \Pi_{11} & 0 \\ 0 & \Pi_{22} - \Pi_{12}^* \Pi_{11}^{-1}\Pi_{12} \end{pmatrix}\begin{pmatrix} \mathcal{G}_{qp} + \Pi_{11}^{-1}\Pi_{12} \\ I \end{pmatrix} \prec 0 \text{ on } \mathbb{C}^0,$$

which is nothing but

$$\left(\mathcal{G}_{qp} + \Pi_{11}^{-1}\Pi_{12}\right)^* \Pi_{11}\left(\mathcal{G}_{qp} + \Pi_{11}^{-1}\Pi_{12}\right) + \Pi_{22} - \Pi_{12}^*\Pi_{11}^{-1}\Pi_{12} \prec 0 \quad \text{on} \quad \mathbb{C}^0.$$

Again, since $\Pi_{11} \succ 0$ on $\mathbb{C}^0$ we conclude that indeed $\Pi_{22} - \Pi_{12}^*\Pi_{11}^{-1}\Pi_{12} \prec 0$ on $\mathbb{C}^0$.

∎

It is now possible to exploit Lemma 4.2 in order to guarantee the existence of the following factorization.

Lemma 4.3. *Let Ψ admit the stable and controllable realization (3.23) and suppose that $\Psi^*P\Psi$ satisfies the inertia constraints (4.16). Then there exist transfer matrices $\hat{\Psi}_j \in \mathscr{RH}_\infty^{\bullet\times\bullet}$, $j = 1,2,3$ with $\hat{\Psi}_2^{-1} \in \mathscr{RH}_\infty^{n_p\times n_p}$ as well as a symmetric matrix Z such that*

$$\Psi^*P\Psi = \hat{\Psi}^*\hat{P}\hat{\Psi} := \begin{pmatrix} \hat{\Psi}_1 & \hat{\Psi}_3 \\ 0 & \hat{\Psi}_2 \end{pmatrix}^* \begin{pmatrix} I_{n_q} & 0 \\ 0 & -I_{n_p} \end{pmatrix} \begin{pmatrix} \hat{\Psi}_1 & \hat{\Psi}_3 \\ 0 & \hat{\Psi}_2 \end{pmatrix} \tag{4.18}$$

with

$$\hat{\Psi} = \begin{pmatrix} \hat{\Psi}_1 & \hat{\Psi}_3 \\ 0 & \hat{\Psi}_2 \end{pmatrix} =$$

$$= \left[\begin{array}{c|cc} \hat{A}_\Psi & \hat{B}_{\Psi 1} & \hat{B}_{\Psi 2} \\ \hline \hat{C}_{\Psi 1} & \hat{D}_{\Psi 1} & \hat{D}_{\Psi 3} \\ \hat{C}_{\Psi 2} & 0 & \hat{D}_{\Psi 2} \end{array}\right] = \left[\begin{array}{cc|cc} A_1 & 0 & B_1 & 0 \\ 0 & \hat{A}_2 & 0 & \hat{B}_2 \\ \hline \hat{C}_1 & \hat{C}_3 & \hat{D}_1 & \hat{D}_3 \\ 0 & \hat{C}_2 & 0 & \hat{D}_2 \end{array}\right] = \left[\begin{array}{ccc|cc} A_1 & 0 & 0 & B_1 & 0 \\ 0 & A_{e1} & A_{e2} & 0 & B_e \\ 0 & 0 & A_2 & 0 & B_2 \\ \hline \hat{C}_1 & \hat{C}_{3,1} & \hat{C}_{3,2} & \hat{D}_1 & \hat{D}_3 \\ 0 & \hat{C}_{2,1} & \hat{C}_{2,2} & 0 & \hat{D}_2 \end{array}\right] \tag{4.19}$$

and

$$\begin{pmatrix} I & 0 & 0 \\ \hline \hat{A}_\Psi & \hat{B}_{\Psi 1} & \hat{B}_{\Psi 2} \\ \hat{C}_{\Psi 1} & \hat{D}_{\Psi 1} & \hat{D}_{\Psi 3} \\ \hat{C}_{\Psi 2} & 0 & \hat{D}_{\Psi 2} \\ \hline C_\Psi T_1 & D_{\Psi 1} & D_{\Psi 2} \end{pmatrix}^\top \begin{pmatrix} 0 & Z & 0 & 0 \\ \hline Z & 0 & 0 & 0 \\ \hline 0 & 0 & \hat{P} & 0 \\ \hline 0 & 0 & 0 & -P \end{pmatrix} \begin{pmatrix} I & 0 & 0 \\ \hline \hat{A}_\Psi & \hat{B}_{\Psi 1} & \hat{B}_{\Psi 2} \\ \hat{C}_{\Psi 1} & \hat{D}_{\Psi 1} & \hat{D}_{\Psi 3} \\ \hat{C}_{\Psi 2} & 0 & \hat{D}_{\Psi 2} \\ \hline C_\Psi T_1 & D_{\Psi 1} & D_{\Psi 2} \end{pmatrix} = 0. \tag{4.20}$$

Here $\hat{A}_\Psi \in \mathbb{R}^{n_{\hat{\Psi}}\times n_{\hat{\Psi}}}$, $\mathrm{eig}(\hat{A}_\Psi) \subset \mathbb{C}^-$ and T_1 is structured as

$$T_1 = \begin{pmatrix} I & 0 & 0 \\ 0 & 0 & I \end{pmatrix} \tag{4.21}$$

in accordance with the block structure of (4.19) *and such that* $C_\Psi T_1 = \begin{pmatrix} C_1 & 0 & C_2 \end{pmatrix}$.

The proof, which is found in Appendix G.1, originates from [165] and has been simplified and streamlined in our paper [198]. As shown, the results rely on solving an indefinite algebraic Riccati equation, whose solution Z is guaranteed to exist.

The key benefit of Lemma 4.3 is not only that the 'new' factorization of $\Psi^* P \Psi$ has a square and invertible outer-factor, but also that the 'new' realization matrices of $\hat{\Psi}$ are expressed in terms of the 'old' realization matrices of Ψ as in (3.23). This allows us to replace the initial multiplier factorization $\Psi^* P \Psi$ appearing the LMI (4.11) by the new factorization $\hat{\Psi}^* \hat{P} \hat{\Psi}$ as will be discussed next.

Lemma 4.4. *Suppose that there exist matrices* $\mathcal{X}$, $P_e \in \boldsymbol{P}_e$, $P \in \boldsymbol{P}$ *and* P_γ *for which* (4.11) *holds. Then there exists a* Z *with* (4.20), *some* X_e *with* $\mathrm{He}(X_e T_2 \hat{A}_\Psi T_2^\top) = \mathrm{He}(X_e A_{e1}) \prec 0$ *as well as an* $\epsilon > 0$ *(which can be chosen arbitrarily small) such that*

$$
\begin{pmatrix} \star \\ -- \\ \star \\ \star \\ -- \\ \star \\ \star \\ -- \\ -- \\ \star \\ \star \end{pmatrix}^\top
\begin{pmatrix} 0 & \hat{\mathcal{X}} & 0 & 0 & 0 \\ \hat{\mathcal{X}} & 0 & 0 & 0 & 0 \\ 0 & 0 & P_e & 0 & 0 \\ 0 & 0 & 0 & \hat{P} & 0 \\ 0 & 0 & 0 & 0 & P_\gamma \end{pmatrix}
\begin{pmatrix} I & 0 & 0 & 0 \\ \hat{A}_\Psi & \hat{B}_{\Psi 1} \mathcal{C}_q & \hat{B}_{\Psi 1} \mathcal{D}_{qe} & \hat{B}_{\Psi 1} \mathcal{D}_{qp} + \hat{B}_{\Psi 2} & \hat{B}_{\Psi 1} \mathcal{D}_{qw} \\ 0 & \mathcal{A} & \mathcal{B}_e & \mathcal{B}_p & \mathcal{B}_w \\ 0 & \mathcal{C}_e & \mathcal{D}_{ee} & \mathcal{D}_{ep} & \mathcal{D}_{ew} \\ 0 & 0 & I & 0 & 0 \\ \hat{C}_{\Psi 1} & \hat{D}_{\Psi 1} \mathcal{C}_q & \hat{D}_{\Psi 1} \mathcal{D}_{qe} & \hat{D}_{\Psi 1} \mathcal{D}_{qp} + \hat{D}_{\Psi 3} & \hat{D}_{\Psi 1} \mathcal{D}_{qw} \\ \hat{C}_{\Psi 2} & 0 & 0 & \hat{D}_{\Psi 2} & 0 \\ 0 & \mathcal{C}_z & \mathcal{D}_{ze} & \mathcal{D}_{zp} & \mathcal{D}_{zw} \\ 0 & 0 & 0 & 0 & I \end{pmatrix} \prec 0.
$$
(4.22)

Here $T_2^\top = \begin{pmatrix} 0 & I & 0 \end{pmatrix}^\top$ *is a basis for the kernel of* T_1 *and* $\hat{\mathcal{X}}$ *is, analogously to* $\mathcal{X}$, *structured as*

$$
\hat{\mathcal{X}} = \left(\begin{array}{c|c} T_1^\top X_{11} T_1 + \epsilon T_2^\top X_e T_2 + Z & T_1^\top \mathcal{X}_{12} \\ \hline \mathcal{X}_{12}^\top T_1 & \mathcal{X}_{22} \end{array} \right) =
$$

$$
= \left(\begin{array}{ccc|c} X_{11,11} + Z_{11} & Z_{12} & X_{11,12} + Z_{13} & \mathcal{X}_{12,1} \\ Z_{12}^\top & \epsilon X_e + Z_{22} & Z_{23} & 0 \\ X_{11,12}^\top + Z_{13}^\top & Z_{23}^\top & X_{11,22} + Z_{33} & \mathcal{X}_{12,2} \\ \hline \mathcal{X}_{12,1}^\top & 0 & \mathcal{X}_{12,1}^\top & \mathcal{X}_{22} \end{array} \right) =
$$

$$
= \left(\begin{array}{cc|c} T_1^\top X_{11} T_1 + \epsilon T_2^\top X_e T_2 + Z & T_1^\top \mathcal{X}_{12} & T_1^\top U_1 \\ \mathcal{X}_{12}^\top T_1 & \mathcal{X}_{22} & U_2 \\ \hline U_1^\top T_1 & U_2^\top & \bar{X} \end{array} \right) = \left(\begin{array}{c|c} \hat{X} & \hat{U} \\ \hline \hat{U}^\top & \bar{X} \end{array} \right). \quad (4.23)
$$

Proof. The proof is identical to the one of Lemma 6.2 in Section 6.6.1 and is omitted in order to avoid redundancies.

∎

Since the outer-factors of the multipliers are now square and invertible, it is possible to eliminate $\hat{C}_{\Psi 2}$ and $\hat{D}_{\Psi 2}$ by transforming (4.22) by congruence into

$$
\begin{pmatrix}
I & 0 & 0 & 0 & 0 \\
0 & I & 0 & 0 & 0 \\
\hline
0 & 0 & I & 0 & 0 \\
-\hat{D}_{\Psi 2}^{-1}\hat{C}_{\Psi 2} & 0 & 0 & \hat{D}_{\Psi 2}^{-1} & 0 \\
0 & 0 & 0 & 0 & I
\end{pmatrix}^{\top}
\quad (4.22) \quad
\begin{pmatrix}
I & 0 & 0 & 0 & 0 \\
0 & I & 0 & 0 & 0 \\
\hline
0 & 0 & I & 0 & 0 \\
-\hat{D}_{\Psi 2}^{-1}\hat{C}_{\Psi 2} & 0 & 0 & \hat{D}_{\Psi 2}^{-1} & 0 \\
0 & 0 & 0 & 0 & I
\end{pmatrix} \prec 0. \quad (4.24)
$$

This yields the inequality

$$
\begin{pmatrix}
\star \\
\star \\
\hline
\star \\
\star \\
\hline
\star \\
\star \\
\hline
\star \\
\star
\end{pmatrix}^{\top}
\begin{pmatrix}
0 & \hat{\mathcal{X}} & 0 & 0 & 0 \\
\hat{\mathcal{X}} & 0 & 0 & 0 & 0 \\
\hline
0 & 0 & P_e & 0 & 0 \\
\hline
0 & 0 & 0 & \hat{P} & 0 \\
\hline
0 & 0 & 0 & 0 & P_\gamma
\end{pmatrix}
\begin{pmatrix}
I & 0 & 0 & 0 \\
\hat{\mathcal{A}} & \hat{\mathcal{B}}_e & \hat{\mathcal{B}}_p & \hat{\mathcal{B}}_w \\
\hat{\mathcal{C}}_e & \hat{\mathcal{D}}_{ee} & \hat{\mathcal{D}}_{ep} & \hat{\mathcal{D}}_{ew} \\
0 & I & 0 & 0 \\
\hline
\hat{\mathcal{C}}_q & \hat{\mathcal{D}}_{qe} & \hat{\mathcal{D}}_{qp} & \hat{\mathcal{D}}_{qw} \\
0 & 0 & I & 0 \\
\hline
\hat{\mathcal{C}}_z & \hat{\mathcal{D}}_{ze} & \hat{\mathcal{D}}_{zp} & \hat{\mathcal{D}}_{zw} \\
0 & 0 & 0 & I
\end{pmatrix} \prec 0, \quad (4.25)
$$

where the calligraphic matrices are given by

$$
\left[\begin{array}{c|ccc}
\hat{\mathcal{A}} & \hat{\mathcal{B}}_e & \hat{\mathcal{B}}_p & \hat{\mathcal{B}}_w \\
\hline
\hat{\mathcal{C}}_e & \hat{\mathcal{D}}_{ee} & \hat{\mathcal{D}}_{ep} & \hat{\mathcal{D}}_{ew} \\
\hat{\mathcal{C}}_q & \hat{\mathcal{D}}_{qe} & \hat{\mathcal{D}}_{qp} & \hat{\mathcal{D}}_{qw} \\
\hat{\mathcal{C}}_z & \hat{\mathcal{D}}_{ze} & \hat{\mathcal{D}}_{zp} & \hat{\mathcal{D}}_{zw}
\end{array}\right] =
$$

$$
= \left[\begin{array}{cc|ccc}
\hat{A}_\Psi - \hat{\mathcal{L}}_1\hat{D}_{\Psi 2}^{-1}\hat{C}_{\Psi 2} & \hat{\mathcal{B}}_{\Psi 1}\mathcal{C}_q & \hat{\mathcal{B}}_{\Psi 1}\mathcal{D}_{qe} & \hat{\mathcal{L}}_1\hat{D}_{\Psi 2}^{-1} & \hat{\mathcal{B}}_{\Psi 1}\mathcal{D}_{qw} \\
-\mathcal{B}_p\hat{D}_{\Psi 2}^{-1}\hat{C}_{\Psi 2} & \mathcal{A} & \mathcal{B}_e & \mathcal{B}_p\hat{D}_{\Psi 2}^{-1} & \mathcal{B}_w \\
\hline
-\mathcal{D}_{ep}\hat{D}_{\Psi 2}^{-1}\hat{C}_{\Psi 2} & \mathcal{C}_e & \mathcal{D}_{ee} & \mathcal{D}_{ep}\hat{D}_{\Psi 2}^{-1} & \mathcal{D}_{ew} \\
\hat{C}_{\Psi 1} - \hat{\mathcal{L}}_2\hat{D}_{\Psi 2}^{-1}\hat{C}_{\Psi 2} & \hat{D}_{\Psi 1}\mathcal{C}_q & \hat{D}_{\Psi 1}\mathcal{D}_{qe} & \hat{\mathcal{L}}_2\hat{D}_{\Psi 2}^{-1} & \hat{D}_{\Psi 1}\mathcal{D}_{qw} \\
-\mathcal{D}_{zp}\hat{D}_{\Psi 2}^{-1}\hat{C}_{\Psi 2} & \mathcal{C}_z & \mathcal{D}_{ze} & \mathcal{D}_{zp}\hat{D}_{\Psi 2}^{-1} & \mathcal{D}_{zw}
\end{array}\right] \quad (4.26)
$$

and hence

$$
\begin{pmatrix}
\hat{\mathcal{A}} & \hat{\mathcal{B}}_e & \hat{\mathcal{B}}_p & \hat{\mathcal{B}}_w \\
\hline
\hat{\mathcal{C}}_e & \hat{\mathcal{D}}_{ee} & \hat{\mathcal{D}}_{ep} & \hat{\mathcal{D}}_{ew} \\
\hline
\hat{\mathcal{C}}_q & \hat{\mathcal{D}}_{qe} & \hat{\mathcal{D}}_{qp} & \hat{\mathcal{D}}_{qw} \\
\hat{\mathcal{C}}_z & \hat{\mathcal{D}}_{ze} & \hat{\mathcal{D}}_{zp} & \hat{\mathcal{D}}_{zw}
\end{pmatrix}
=
\begin{pmatrix}
\hat{A}_\Psi - \hat{L}_1 \hat{D}_{\Psi 2}^{-1} \hat{C}_{\Psi 2} & \hat{B}_{\Psi 1} C_q & 0 & \hat{B}_{\Psi 1} D_{qs} & 0 & \hat{L}_1 \hat{D}_{\Psi 2}^{-1} & \hat{B}_{\Psi 1} D_{qw} \\
-B_p \hat{D}_{\Psi 2}^{-1} \hat{C}_{\Psi 2} & A & 0 & B_s & 0 & B_p \hat{D}_{\Psi 2}^{-1} & B_w \\
0 & 0 & 0 & 0 & 0 & 0 & 0 \\
\hline
-D_{sp} \hat{D}_{\Psi 2}^{-1} \hat{C}_{\Psi 2} & C_s & 0 & D_{ss} & 0 & D_{sp} \hat{D}_{\Psi 2}^{-1} & D_{sw} \\
0 & 0 & 0 & 0 & 0 & 0 & 0 \\
\hline
\hat{C}_{\Psi 1} - \hat{L}_2 \hat{D}_{\Psi 2}^{-1} \hat{C}_{\Psi 2} & \hat{D}_{\Psi 1} C_q & 0 & \hat{D}_{\Psi 1} D_{qs} & 0 & \hat{L}_2 \hat{D}_{\Psi 2}^{-1} & \hat{D}_{\Psi 1} D_{qw} \\
-D_{zp} \hat{D}_{\Psi 2}^{-1} \hat{C}_{\Psi 2} & C_z & 0 & D_{zs} & 0 & D_{zp} \hat{D}_{\Psi 2}^{-1} & D_{zw}
\end{pmatrix}
+
$$

$$
\begin{pmatrix}
0 & 0 & 0 \\
0 & 0 & 0 \\
I & 0 & 0 \\
\hline
0 & 0 & 0 \\
0 & 0 & I \\
\hline
0 & 0 & 0 \\
0 & -I & 0
\end{pmatrix}
\begin{pmatrix}
A_K & B_K \\
C_K & D_K
\end{pmatrix}
\begin{pmatrix}
0 & 0 & I & 0 & 0 & 0 & 0 \\
\hline
-D_{yp} \hat{D}_{\Psi 2}^{-1} \hat{C}_{\Psi 2} & C_y & 0 & D_{ys} & 0 & D_{yp} \hat{D}_{\Psi 2}^{-1} & D_{zw} \\
0 & 0 & 0 & 0 & I & 0 & 0
\end{pmatrix}.
$$

$$(4.27)$$

Here $\hat{\mathcal{L}}_1 := \hat{B}_{\Psi 1} \mathcal{D}_{qp} + \hat{B}_{\Psi 2}$, $\hat{\mathcal{L}}_2 := \hat{D}_{\Psi 1} \mathcal{D}_{qp} + \hat{D}_{\Psi 3}$, $\hat{L}_1 := \hat{B}_{\Psi 1} D_{qp} + \hat{B}_{\Psi 2}$ and $\hat{L}_2 := \hat{D}_{\Psi 1} D_{qp} + \hat{D}_{\Psi 3}$.

It is now essential to observe that we have appropriately reformulated the LMI (4.11) in order to apply the crucial step of dualization (see Lemma C.1 in Appendix C). This yields the dual matrix inequality

$$
\begin{pmatrix}
\star \\
\hline
\star \\
\star \\
\hline
\star \\
\hline
\star \\
\hline
\star \\
\star
\end{pmatrix}^{\top}
\begin{pmatrix}
0 & \hat{\mathcal{X}} & 0 & 0 & 0 \\
\hat{\mathcal{X}} & 0 & 0 & 0 & 0 \\
\hline
0 & 0 & P_e & 0 & 0 \\
\hline
0 & 0 & 0 & \hat{P} & 0 \\
\hline
0 & 0 & 0 & 0 & P_\gamma
\end{pmatrix}^{-1}
\begin{pmatrix}
-\hat{A}^{\top} & -\hat{\mathcal{C}}_e^{\top} & -\hat{\mathcal{C}}_q^{\top} & -\hat{\mathcal{C}}_z^{\top} \\
I & 0 & 0 & 0 \\
0 & I & 0 & 0 \\
\hline
-\hat{\mathcal{B}}_e^{\top} & -\hat{\mathcal{D}}_{ee}^{\top} & -\hat{\mathcal{D}}_{qe}^{\top} & -\hat{\mathcal{D}}_{ze}^{\top} \\
0 & 0 & I & 0 \\
\hline
-\hat{\mathcal{B}}_p^{\top} & -\hat{\mathcal{D}}_{ep}^{\top} & -\hat{\mathcal{D}}_{qp}^{\top} & -\hat{\mathcal{D}}_{zp}^{\top} \\
0 & 0 & 0 & I \\
-\hat{\mathcal{B}}_w^{\top} & -\hat{\mathcal{D}}_{ew}^{\top} & -\hat{\mathcal{D}}_{qw}^{\top} & -\hat{\mathcal{D}}_{zw}^{\top}
\end{pmatrix}
\succ 0.
\quad (4.28)
$$

In view of (4.27), the primal and dual matrix inequalities (4.25) and (4.28) can now be written in the form (C.1) and (C.2) respectively, in the unknown realization matrices of K. We conclude that (4.25) and (4.28) satisfy all the required properties in order to apply Lemma C.2 in Appendix C. This resolves the second issue that we discussed in the beginning of this section.

Remark 4.1. Since the left-upper blocks of P_e, $\hat{P}$ and P_γ are all positive definite, the matrix inequality (4.25) implies that $\mathrm{He}(\hat{\mathcal{X}}\hat{\mathcal{A}}) \prec 0$. Because $\hat{\mathcal{A}}$ is Hurwitz (see the proof of Lemma 6.3 in Section 6.6.1), we conclude that positive definiteness of $\hat{\mathcal{X}}$ is equivalent to $\mathcal{A}$ being Hurwitz. Hence, if (4.25) is feasible with $\hat{\mathcal{X}} \succ 0$, the system (4.26) is guaranteed to be stable.

4.5 A convex solution

In parallel with the nominal and nominal gain-scheduling synthesis problem in Section 3.6 and 3.7 respectively, we can now proceed by eliminating the realization matrices of K in the inequalities (4.25) and (4.28) by applying Lemma C.2 in Appendix C. This involves the basis matrices

$$T_y = \begin{pmatrix} I_{n_\psi} & 0 \\ 0 & T_{y1} \\ 0 & T_{y2} \\ 0 & T_{y3} \\ 0 & T_{y4} \end{pmatrix} \ \text{of} \ \ker\begin{pmatrix} 0 & C_y & D_{ys} & D_{yp} & D_{yw} \end{pmatrix} \ \text{and} \ T_u = I \ \text{of} \ \ker(0).$$

The latter basis matrix T_u is trivial due to the simplified structure of the left-factor that multiplies the realization matrices of K in (4.27). As we will see in the sequel this specialized structure will turn out to be essential for resolving the third issue described in the beginning of Section 4.4.

Analogously to Section 3.7, the resulting conditions will neither involve the complete multiplier P_e nor its inverse $\tilde{P}_e$. However, different is that they can be expressed only in terms of the set of primal scalings $\mathbf{P}_s$ as defined in Section 3.7. Once more we stress that this set is described by a finite number of LMIs directly in terms of the scheduling function $\Delta_s(\cdot)$ of the open-loop system and the generators of the parameter set Λ; in particular it does not involve the to-be-constructed estimator scheduling function $\Delta_c(\cdot)$. We have now introduce all the necessary ingredients in order to state the main result of this chapter.

Theorem 4.2. *The following two statements are equivalent:*

1. *There exists a scheduling function $\hat{\Delta}_c(\cdot)$ as well as matrices A_K, B_K, C_K, D_K, $P_e \in \mathbf{P}_e$, $P \in \mathbf{P}$, $\mathcal{X}$ and X_Ψ for which (4.11) and (4.14) hold with (4.15).*
2. *There exist matrices X, Y, X_Ψ, $P \in \mathbf{P}$ and $P_s, \tilde{P}_s \in \mathbf{P}_s$ for which the following LMIs are satisfied:*

$$Y - \begin{pmatrix} X_\Psi & 0 \\ 0 & 0 \end{pmatrix} \succ 0, \quad X - Y \succ 0. \tag{4.29}$$

$$\begin{pmatrix} I & 0 \\ A_\Psi & B_{\Psi 1} \\ C_\Psi & D_{\Psi 1} \end{pmatrix}^\top \begin{pmatrix} 0 & X_\Psi & 0 \\ X_\Psi & 0 & 0 \\ 0 & 0 & P \end{pmatrix} \begin{pmatrix} I & 0 \\ A_\Psi & B_{\Psi 1} \\ C_\Psi & D_{\Psi 1} \end{pmatrix} \succ 0. \tag{4.30}$$

$$T_y^\top \mathcal{O}^\top \begin{pmatrix} 0 & X & 0 & 0 & 0 & 0 \\ X & 0 & 0 & 0 & 0 & 0 \\ 0 & 0 & P_{\mathrm{s}} & 0 & 0 & 0 \\ 0 & 0 & 0 & P & 0 & 0 \\ 0 & 0 & 0 & 0 & I & 0 \\ 0 & 0 & 0 & 0 & 0 & -\gamma^2 I \end{pmatrix} \mathcal{O} T_y \prec 0. \tag{4.31}$$

$$\mathcal{O}^\top \begin{pmatrix} 0 & Y & 0 & 0 & 0 & 0 \\ Y & 0 & 0 & 0 & 0 & 0 \\ 0 & 0 & \tilde{P}_{\mathrm{s}} & 0 & 0 & 0 \\ 0 & 0 & 0 & P & 0 & 0 \\ 0 & 0 & 0 & 0 & 0 & 0 \\ 0 & 0 & 0 & 0 & 0 & -\gamma^2 I \end{pmatrix} \mathcal{O} \prec 0. \tag{4.32}$$

Here $Y = Y^\top$ has a block structure identical to that of X and

$$\mathcal{O} = \begin{pmatrix} I & 0 & 0 & 0 \\ A_\Psi & B_{\Psi 1} C_q & B_{\Psi 1} D_{qs} & B_{\Psi 1} D_{qp} + B_{\Psi 2} & B_{\Psi 1} D_{qw} \\ 0 & A & B_{\mathrm{s}} & B_p & B_w \\ 0 & C_{\mathrm{s}} & D_{\mathrm{ss}} & D_{sp} & D_{sw} \\ 0 & 0 & I & 0 & 0 \\ C_\Psi & D_{\Psi 1} C_q & D_{\Psi 1} D_{qs} & D_{\Psi 1} D_{qp} + D_{\Psi 2} & D_{\Psi 1} D_{qw} \\ 0 & C_z & D_{zs} & D_{zp} & D_{zw} \\ 0 & 0 & 0 & 0 & I \end{pmatrix}.$$

Once the LMIs (4.29)-(4.32) in the variables X, Y, X_Ψ, P_{s}, $\tilde{P}_{\mathrm{s}}$, P and γ^2 are feasible, one can construct an $\mathcal{X}$, an extended multiplier $P_{\mathrm{e}} \in \boldsymbol{P}_{\mathrm{e}}$, the LTI part of the estimator $K = (A_K, B_K, C_K, D_K)$ of McMillan degree $n_K = n_G + n_\Psi$ and, last but not least, a scheduling function $\hat{\Delta}_{\mathrm{c}}(\cdot)$ such that (4.11) and (4.14) hold with (4.15). Then the estimator $K \star \hat{\Delta}_{\mathrm{c}}(\Delta_{\mathrm{s}})$ is well-posed and guarantees the system interconnection of Figure 4.1 to be stable for all $\Delta \in \boldsymbol{\Delta}$ and $\Delta_{\mathrm{s}} \in \boldsymbol{\Delta}_{\mathrm{s}}$, while the induced $\mathscr{L}_2$-gain for w to z is rendered less than $\gamma > 0$.

In summary we have merged the problem of designing robust LTI estimators using general dynamic IQC-multipliers [165] with the problem of designing nominal gain-scheduled estimators using full-block multipliers [152], yielding a convex feasibility test for the existence of robust gain-scheduled estimators. We have resolved the difficulties of enforcing internal stability, eliminating the realization matrices of K and removing the remaining non-convex constraints as raised in Section 4.4. It is one of the main differences between the existing results on gain-scheduled controller synthesis and Theorem 4.2 that both P_{s} and $\tilde{P}_{\mathrm{s}}$ can be identified as 'primal' IQC-multipliers. As a natural consequence, it is interesting to observe that we recover a special case of the robust estimation problem of [165], if we void the scheduling channel of the controller and enforce $P_{\mathrm{s}} = \tilde{P}_{\mathrm{s}}$. Then Δ_{s} is not considered to be a scheduling block anymore, but a genuine uncertainty, and feasibility of the LMIs would yield a genuine robust estimator. Finally, as discussed in the introduction, we emphasize that it is not only possible to design full-order estimators of McMillan degree $n_K = n_G + n_\Psi$, but also static one of McMillan degree zero. This can be done by enforcing $X = Y$ and ignoring the inequality $X - Y \succ 0$.

Although it is very nice to see that the robust estimation problem of [165] can be generalized to gain-scheduling based on static full-block multipliers, it remains an open question how to solve the more general case of gain-scheduling based on general dynamic multipliers. First results in this direction have shown that it is possible to solve the nominal gain-scheduled controller synthesis problem based on dynamic D- and DG-multipliers [167, 157, 158, 161]. However, it remains a conjecture that the two approaches can be merged. We further remark that it is straightforward to derive a convex feasibility test for the synthesis of robust gain-scheduled feedforward controllers by working with the dual of FDI (4.9) [103]. In fact, it has been recently shown in [155] that the robust estimator and feedforward controller synthesis problems can be unified into the following considerably more general robust controller synthesis problem

$$\begin{pmatrix} z \\ y \end{pmatrix} = \begin{pmatrix} \Sigma_{11}(\Delta_1, \Delta_2) & \Sigma_{12}(\Delta_2) \\ \Sigma_{21}(\Delta_1) & \Sigma_{22} \end{pmatrix} \begin{pmatrix} w \\ u \end{pmatrix},$$

where only the control channel is not affected by uncertainties. In Chapter 5 we reveal under which conditions this problem can be generalized to gain-scheduling as well.

Proof. To show that 1. implies 2., suppose there exists a scheduling function $\hat{\Delta}_{\mathrm{c}}(\cdot)$ as well as matrices A_K, B_K, C_K, D_K, $P_{\mathrm{e}} \in \boldsymbol{P}_{\mathrm{e}}$, $P \in \boldsymbol{P}$, $\mathcal{X}$ and $\hat{X}$ for which (4.11) and (4.14) hold with (4.15). We follow again the procedure of eliminating the estimator variables by working with primal and dual matrix inequalities and applying the elimination Lemma C.2 (See Appendix C). However, as discussed in detail, Ψ is a tall transfer matrix and, hence, cannot be inverted. This prevents us from applying the dualization Lemma C.1 (See Appendix C).

To overcome this trouble we reformulate (4.11) as suggested in Section 4.4.2 by applying Lemma 4.4. This yields the inequality (4.22) and leads us to the primal and dual conditions (4.25) and (4.28) respectively, which have the right structure in order to apply Lemma C.2. For this purpose, let us recall the realization matrices (4.27). These motivate us to choose the basis matrices

$$
\hat{T}_{ye} =
\begin{pmatrix}
I_{n_{\hat{\psi}}} & 0 \\
0 & T_{y1} \\
0 & 0 \\
0 & T_{y2} \\
0 & 0 \\
\hat{C}_{\Psi 2} & \hat{D}_{\Psi 2}T_{y3} \\
0 & T_{y4}
\end{pmatrix}
\quad \text{and} \quad
\hat{T}_{ue} =
\begin{pmatrix}
I_{n_{\hat{\psi}}} & 0 & 0 & 0 \\
0 & I & 0 & 0 \\
0 & 0 & 0 & 0 \\
0 & 0 & I & 0 \\
0 & 0 & 0 & 0 \\
0 & 0 & 0 & I \\
0 & 0 & 0 & 0
\end{pmatrix}
$$

of the kernels of

$$
\begin{pmatrix}
0 & 0 & I & 0 & 0 & 0 & 0 \\
-D_{yp}\hat{D}_{\Psi 2}^{-1}\hat{C}_{\Psi 2} & C_y & 0 & D_{ys} & 0 & D_{yp}\hat{D}_{\Psi 2}^{-1} & D_{zw} \\
0 & 0 & 0 & 0 & I & 0 & 0
\end{pmatrix}
\quad \text{and} \quad
\begin{pmatrix}
0 & 0 & I & 0 & 0 & 0 & 0 \\
0 & 0 & 0 & 0 & 0 & 0 & -I \\
0 & 0 & 0 & 0 & I & 0 & 0
\end{pmatrix}
$$

respectively. Note that in view of (4.10) we then have that

$$
\begin{pmatrix}
0 & 0 & I & 0 & 0 & 0 & 0 \\
-D_{yp}\hat{D}_{\Psi 2}^{-1}\hat{C}_{\Psi 2} & C_y & 0 & D_{ys} & 0 & D_{yp}\hat{D}_{\Psi 2}^{-1} & D_{zw} \\
0 & 0 & 0 & 0 & I & 0 & 0
\end{pmatrix}
\begin{pmatrix}
I_{n_{\hat{\psi}}} & 0 \\
0 & T_{y1} \\
0 & 0 \\
0 & T_{y2} \\
0 & 0 \\
\hat{C}_{\Psi 2} & \hat{D}_{\Psi 2}T_{y3} \\
0 & T_{y4}
\end{pmatrix}
=
$$

$$
=
\begin{pmatrix}
0 & 0 & I & 0 & 0 & 0 & 0 \\
0 & C_y & 0 & D_{ys} & 0 & D_{yp} & D_{zw} \\
0 & 0 & 0 & 0 & I & 0 & 0
\end{pmatrix}
\begin{pmatrix}
I_{n_{\hat{\psi}}} & 0 \\
0 & T_{y1} \\
0 & 0 \\
0 & T_{y2} \\
0 & 0 \\
0 & T_{y3} \\
0 & T_{y4}
\end{pmatrix}
= 0
$$

as well as

$$\begin{pmatrix} 0 & C_y & D_{ys} & D_{yp} & D_{zw} \end{pmatrix} \underbrace{\begin{pmatrix} I_{n_{\dot{\psi}}} & 0 \\ 0 & T_{y1} \\ 0 & T_{y2} \\ 0 & T_{y3} \\ 0 & T_{y4} \end{pmatrix}}_{\hat{T}_y} = 0$$

Hence, (4.25) and (4.28) clearly imply

$$\hat{T}_{ye}^{\top}(4.25)\hat{T}_{ye} \prec 0 \quad \text{and} \quad \hat{T}_{ue}^{\top}(4.28)\hat{T}_{ue} \succ 0. \tag{4.33}$$

If we now partition

$$P_e = \begin{pmatrix} Q_e & S_e \\ S_e^{\top} & R_e \end{pmatrix} = \begin{pmatrix} Q_s & Q_{12} & S_s & S_{12} \\ Q_{12}^{\top} & Q_{22} & S_{21} & S_{22} \\ S_s^{\top} & S_{21}^{\top} & R_s & R_{12} \\ S_{12}^{\top} & S_{22}^{\top} & R_{12}^{\top} & R_{22} \end{pmatrix} \tag{4.34}$$

according to $n_{q_s} + n_{p_s} + n_{q_c} + n_{p_c}$ and define

$$P_s = \begin{pmatrix} Q_s & S_s \\ S_s^{\top} & R_s \end{pmatrix}$$

it is easily verified (by canceling the right block-rows and -columns) that the first inequality simplifies into

$$\hat{T}_y^{\top} \hat{O}^{\top} \begin{pmatrix} 0 & \hat{X} & 0 & 0 & 0 \\ \hat{X} & 0 & 0 & 0 & 0 \\ 0 & 0 & P_s & 0 & 0 \\ 0 & 0 & 0 & P & 0 \\ 0 & 0 & 0 & 0 & P_{\gamma} \end{pmatrix} \hat{O}\hat{T}_y \prec 0,$$

where

$$
\hat{\mathcal{O}} := \begin{pmatrix}
I & \vdots & 0 & 0 & 0 \\
\hline
\hat{A}_\Psi & \hat{B}_{\Psi 1} C_q & \hat{B}_{\Psi 1} D_{qs} & \hat{B}_{\Psi 1} D_{qp} + \hat{B}_{\Psi 2} & \hat{B}_{\Psi 1} D_{qw} \\
0 & A & B_s & B_p & B_w \\
\hline
0 & C_s & D_{ss} & D_{sp} & D_{sw} \\
0 & 0 & I & 0 & 0 \\
\hline
\hat{C}_{\Psi 1} & \hat{D}_{\Psi 1} C_q & \hat{D}_{\Psi 1} D_{qs} & \hat{D}_{\Psi 1} D_{qp} + \hat{D}_{\Psi 3} & \hat{D}_{\Psi 1} D_{qw} \\
\hat{C}_{\Psi 2} & 0 & 0 & \hat{D}_{\Psi 2} & 0 \\
\hline
0 & C_z & D_{zs} & D_{zp} & D_{zw} \\
0 & 0 & 0 & 0 & I
\end{pmatrix}
$$

and where $\hat{X}$ is the left-upper block of $\hat{\mathcal{X}}$ as in (4.23). Since $P_\mathrm{e} \in \boldsymbol{P}_\mathrm{e}$ and if recalling the definitions of the set of extended full-block multipliers (3.13) in Section 3.3 and the set of primal full-block multipliers $\boldsymbol{P}_\mathrm{s}$, we infer that $P_\mathrm{s} \in \boldsymbol{P}_\mathrm{s}$. It is now crucial to observe that the steps taken in the proof of Lemma 4.4 can be applied in the reverse direction. This allows us to infer that the latter inequality implies that the first solvability condition (4.31) is satisfied.

Before we continue with the formulation of the dual solvability condition, let us discuss the following observation first: Since $\hat{\mathcal{X}} \succ 0$ (See Remark 4.1), it obviously holds that

$$
\hat{\mathcal{X}}^{-1} = \begin{pmatrix} \left(\hat{X} - \hat{U} \bar{X}^{-1} \hat{U}^\top \right)^{-1} & \star \\ \star & \star \end{pmatrix} =: \begin{pmatrix} \hat{Y}^{-1} & \star \\ \star & \star \end{pmatrix},
$$

with $\star$ of being no interest. Now let us define the new variable

$$
Y = \begin{pmatrix} Y_{11} & Y_{12} \\ Y_{12}^\top & Y_{22} \end{pmatrix} := \begin{pmatrix} X_{11} & X_{12} \\ X_{12}^\top & X_{22} \end{pmatrix} - \begin{pmatrix} U_1 \\ U_2 \end{pmatrix} \bar{X}^{-1} \begin{pmatrix} U_1^\top & U_2^\top \end{pmatrix}
$$

and observe that

$$
\hat{Y} = \hat{X} - \hat{U} \bar{X}^{-1} \hat{U}^\top = \begin{pmatrix} T_1^\top Y_{11} T_1 + \epsilon T_2^\top X_e T_2 + Z & T_1^\top Y_{12} \\ Y_{12}^\top T_1 & Y_{22} \end{pmatrix}. \tag{4.35}
$$

Hence, clearly the structured partitioning of $\hat{X}$ as in (4.23) is preserved. This is highly relevant in the sequel. Further observe that $\hat{\mathcal{X}} \succ 0$ implies that $\hat{X} \succ 0$, $\bar{X} \succ 0$ and thus $\hat{Y} \succ 0$ are all positive definite.

Let us now evaluate the second inequality in (4.33) and observe that after canceling the right block-rows and -columns it simplifies into

$$
\hat{\mathcal{O}}^\top
\begin{pmatrix}
0 & \hat{Y} & 0 & 0 & 0 \\
\hat{Y} & 0 & 0 & 0 & 0 \\
0 & 0 & \tilde{P}_{\mathrm{s}} & 0 & 0 \\
0 & 0 & 0 & \hat{P} & 0 \\
0 & 0 & 0 & 0 & -\gamma^2 I
\end{pmatrix}^{-1}
\hat{\mathcal{O}} \succ 0, \tag{4.36}
$$

for

$$
\hat{\mathcal{O}} :=
\begin{pmatrix}
-\hat{A}_\Psi^\top + \hat{C}_{\Psi 2}^\top \hat{D}_{\Psi 2}^{-\top} \hat{L}_1^\top & \hat{C}_{\Psi 2}^\top \hat{D}_{\Psi 2}^{-\top} B_p^\top & \hat{C}_{\Psi 2}^\top \hat{D}_{\Psi 2}^{-\top} D_{sp}^\top & -\hat{C}_{\Psi 1}^\top + \hat{C}_{\Psi 2}^\top \hat{D}_{\Psi 2}^{-\top} \hat{L}_2^\top \\
\;\;-C_q^\top \hat{B}_{\Psi 1}^\top & -A^\top & -C_{\mathrm{s}}^\top & \;\;-C_q^\top \hat{D}_{\Psi 1}^\top \\
I & 0 & 0 & 0 \\
0 & I & 0 & 0 \\
0 & 0 & I & 0 \\
-D_{qs}^\top \hat{B}_{\Psi 1}^\top & -B_{\mathrm{s}}^\top & -D_{ss}^\top & -D_{qs}^\top \hat{D}_{\Psi 1}^\top \\
0 & 0 & 0 & I \\
-\hat{D}_{\Psi 2}^{-\top} \hat{L}_1^\top & -\hat{D}_{\Psi 2}^{-\top} B_p^\top & -\hat{D}_{\Psi 2}^{-\top} D_{sp}^\top & -\hat{D}_{\Psi 2}^{-\top} \hat{L}_2^\top \\
-D_{qw}^\top \hat{B}_{\Psi 1}^\top & -B_w^\top & -D_{sw}^\top & -D_{qw}^\top \hat{D}_{\Psi 1}^\top
\end{pmatrix}
$$

and

$$
\tilde{P}_{\mathrm{s}} :=
\begin{pmatrix} \tilde{Q}_{\mathrm{s}} & \tilde{S}_{\mathrm{s}} \\ \tilde{S}_{\mathrm{s}}^\top & \tilde{R}_{\mathrm{s}} \end{pmatrix}
=
\begin{pmatrix} Q_{\mathrm{s}} & S_{\mathrm{s}} \\ S_{\mathrm{s}}^\top & R_{\mathrm{s}} \end{pmatrix}
-
\begin{pmatrix} \star \\ \star \end{pmatrix}^\top
\begin{pmatrix} Q_{22} & S_{22} \\ S_{22}^\top & R_{22} \end{pmatrix}^{-1}
\begin{pmatrix} Q_{12}^\top & S_{21} \\ S_{12}^\top & R_{12}^\top \end{pmatrix}. \tag{4.37}
$$

Note that in view of Lemma C.1 (See Appendix C) and since $P_{\mathrm{c}} \in \boldsymbol{P}_{\mathrm{c}}$ we can also infer that $\tilde{P}_{\mathrm{s}} \in \boldsymbol{P}_{\mathrm{s}}$.

As the key observation, we now resolve the third and final issue that prevented us from obtaining a convex feasibility test for the synthesis of robust gain-scheduled estimators. Indeed, observe that (4.36) has the right structure in order to apply Lemma C.1 for the second time. This yields the equivalent condition

$$
\hat{\mathcal{O}}^\top
\begin{pmatrix}
0 & \hat{Y} & 0 & 0 & 0 \\
\hat{Y} & 0 & 0 & 0 & 0 \\
0 & 0 & \tilde{P}_{\mathrm{s}} & 0 & 0 \\
0 & 0 & 0 & \hat{P} & 0 \\
0 & 0 & 0 & 0 & -\gamma^2 I
\end{pmatrix}
\hat{\mathcal{O}} \prec 0, \tag{4.38}
$$

for

$$\hat{\mathcal{O}} := \begin{pmatrix} I & 0 & 0 & 0 \\ \hat{A}_\Psi - \hat{L}_1\hat{D}_{\Psi 2}^{-1}\hat{C}_{\Psi 2} & \hat{B}_{\Psi 1}C_q & \hat{B}_{\Psi 1}D_{qs} & \hat{L}_1\hat{D}_{\Psi 2}^{-1} & \hat{B}_{\Psi 1}D_{qw} \\ -B_p\hat{D}_{\Psi 2}^{-1}\hat{C}_{\Psi 2} & A & B_s & B_p\hat{D}_{\Psi 2}^{-1} & B_w \\ -D_{sp}\hat{D}_{\Psi 2}^{-1}\hat{C}_{\Psi 2} & C_s & D_{ss} & D_{sp}\hat{D}_{\Psi 2}^{-1} & D_{sw} \\ 0 & 0 & I & 0 & 0 \\ \hat{C}_{\Psi 1} - \hat{L}_2\hat{D}_{\Psi 2}^{-1}\hat{C}_{\Psi 2} & \hat{D}_{\Psi 1}C_q & \hat{D}_{\Psi 1}D_{qs} & \hat{L}_2\hat{D}_{\Psi 2}^{-1} & \hat{D}_{\Psi 1}D_{qw} \\ 0 & 0 & 0 & I & 0 \\ 0 & 0 & 0 & 0 & I \end{pmatrix}.$$

By now applying the congruence transformation

$$\begin{pmatrix} I & 0 & 0 & 0 & 0 \\ 0 & I & 0 & 0 & 0 \\ 0 & 0 & I & 0 & 0 \\ \hat{C}_{\Psi 2} & 0 & 0 & \hat{D}_{\Psi 2} & 0 \\ 0 & 0 & 0 & 0 & I \end{pmatrix}^\top \quad (4.38) \quad \begin{pmatrix} I & 0 & 0 & 0 & 0 \\ 0 & I & 0 & 0 & 0 \\ 0 & 0 & I & 0 & 0 \\ \hat{C}_{\Psi 2} & 0 & 0 & \hat{D}_{\Psi 2} & 0 \\ 0 & 0 & 0 & 0 & I \end{pmatrix} \prec 0,$$

we obtain the inequality

$$\hat{\mathcal{O}}^\top \begin{pmatrix} 0 & \hat{Y} & 0 & 0 & 0 \\ \hat{Y} & 0 & 0 & 0 & 0 \\ 0 & 0 & \tilde{P}_s & 0 & 0 \\ 0 & 0 & 0 & P & 0 \\ 0 & 0 & 0 & 0 & -\gamma^2 I \end{pmatrix} \hat{\mathcal{O}} \prec 0, \qquad (4.39)$$

for

$$\hat{\mathcal{O}} := \begin{pmatrix} I & 0 & 0 & 0 \\ \hat{A}_\Psi & \hat{B}_{\Psi 1}C_q & \hat{B}_{\Psi 1}D_{qs} & \hat{B}_{\Psi 1}D_{qp} + \hat{B}_{\Psi 2} & \hat{B}_{\Psi 1}D_{qw} \\ 0 & A & B_s & B_p & B_w \\ 0 & C_s & D_{ss} & D_{sp} & D_{sw} \\ 0 & 0 & I & 0 & 0 \\ \hat{C}_{\Psi 1} & \hat{D}_{\Psi 1}C_q & \hat{D}_{\Psi 1}D_{qs} & \hat{D}_{\Psi 1}D_{qp} + \hat{D}_{\Psi 3} & \hat{D}_{\Psi 1}D_{qw} \\ \hat{C}_{\Psi 2} & 0 & 0 & \hat{D}_{\Psi 2} & 0 \\ 0 & 0 & 0 & 0 & I \end{pmatrix}.$$

Finally, since the structure of $\hat{X}$ is preserved in $\hat{Y}$, it is again possible to apply the steps taken in the proof of Lemma 4.4 in the reverse direction, in order to infer that

(4.39) implies (4.32). In conclusion, the structure of the estimation problem allows us to take the additional step of 'back-dualization' by applying Lemma C.1 for the second time. This yielded the second solvability condition of Theorem 4.2, which is affine in all the matrix variables Y, $\tilde{P}_{\mathrm{s}}$, P and γ^2.

The first part of the proof is completed by showing that (4.29) holds true. To this end, let us recall from Lemma 4.3 that stability of $\mathcal{A}$ is equivalent to

$$
\begin{pmatrix} X - \begin{pmatrix} X_\Psi & 0 \\ 0 & 0 \end{pmatrix} & U \\ U^\top & \bar{X} \end{pmatrix} \succ 0
$$

and hence

$$
\left(X - U\bar{X}^{-1}U^\top \right) - \begin{pmatrix} X_\Psi & 0 \\ 0 & 0 \end{pmatrix} \succ 0 \quad \text{and} \quad \bar{X} \succ 0.
$$

Since $Y = X - U\bar{X}^{-1}U^\top$ we can infer (by a slight perturbation if necessary) that $\bar{X} \succ 0$ implies $X \succ Y$ and thus (4.29).

In order to show that 2. implies 1., we assume that there exist X, Y, X_Ψ, $P \in \boldsymbol{P}$ and $P_{\mathrm{s}}, \tilde{P}_{\mathrm{s}} \in \boldsymbol{P}_{\mathrm{s}}$ such that (4.29)-(4.32) are feasible.

Let us first consider the non-singular matrices X and $X - Y$ (slightly perturb if necessary). Then, by defining

$$
\mathcal{X} = \begin{pmatrix} X & X - Y \\ X - Y & X - Y \end{pmatrix},
$$

we can infer that (4.29) holds thanks to the Schur complement.

For the given scalings P_{s} and $\tilde{P}_{\mathrm{s}}$, in the second step of the proof we follow [151], in order to extend P_{s} and $\tilde{P}_{\mathrm{s}}$ to $P_{\mathrm{s}} \in \boldsymbol{P}_{\mathrm{s}}$ as in (4.34) and such that (4.37) is satisfied. Furthermore, in [151] it is also shown how to explicitly define a quadratic scheduling function $\hat{\Delta}_{\mathrm{c}} : \mathbb{R}^{n_{p_{\mathrm{s}}} \times n_{q_{\mathrm{s}}}} \to \mathbb{R}^{n_{p_{\mathrm{c}}} \times n_{q_{\mathrm{c}}}}$ for which P_{c} is contained in the corresponding set of extended multipliers $\boldsymbol{P}_{\mathrm{c}}$.

In the final step it remains to construct the realization matrices of K. For this purpose, we substitute the constructed matrices $\mathcal{X}$, P and P_{c} in (4.11) and introduce the abbreviations

$$\begin{pmatrix}
\tilde{A} & \tilde{B}_s & \tilde{B}_p & \tilde{B}_w & \tilde{B}_u \\
\tilde{C}_s & \tilde{D}_{ss} & \tilde{D}_{sp} & \tilde{D}_{sw} & \tilde{D}_{su} \\
\tilde{C}_q & \tilde{D}_{qs} & \tilde{D}_{qp} & \tilde{D}_{qw} & 0 \\
\tilde{C}_z & \tilde{D}_{zs} & \tilde{D}_{zp} & \tilde{D}_{zw} & \tilde{D}_{zu} \\
\tilde{C}_y & \tilde{D}_{ys} & \tilde{D}_{yp} & \tilde{D}_{yw} & 0
\end{pmatrix} =$$

$$= \begin{pmatrix}
A_\Psi & B_{\Psi 1}C_q & 0 & B_{\Psi 1}D_{qs} & 0 & B_{\Psi 1}D_{qp}+B_{\Psi 2} & B_{\Psi 1}D_{qw} & 0 & 0 & 0 \\
0 & A & 0 & B_s & 0 & B_p & B_w & 0 & 0 & 0 \\
0 & 0 & 0 & 0 & 0 & 0 & 0 & I & 0 & 0 \\
0 & C_s & 0 & D_{ss} & 0 & D_{sp} & D_{sw} & 0 & 0 & 0 \\
0 & 0 & 0 & 0 & 0 & 0 & 0 & 0 & 0 & I \\
C_\Psi & D_{\Psi 1}C_q & 0 & D_{\Psi 1}D_{qs} & 0 & D_{\Psi 1}D_{qp}+D_{\Psi 2} & D_{\Psi 1}D_{qw} & 0 & 0 & 0 \\
0 & C_z & 0 & D_{zs} & 0 & D_{zp} & D_{zw} & 0 & -I & 0 \\
0 & 0 & I & 0 & 0 & 0 & 0 & 0 & 0 & 0 \\
0 & C_y & 0 & D_{ys} & 0 & D_{yp} & D_{yw} & 0 & 0 & 0 \\
0 & 0 & 0 & 0 & I & 0 & 0 & 0 & 0 & 0
\end{pmatrix}$$

as well as

$$\mathcal{K} = \begin{pmatrix} A_K & B_K \\ C_K & D_K \end{pmatrix} \quad \text{and} \quad \mathcal{O}(\mathcal{K}) = \mathcal{K} \begin{pmatrix} \tilde{C}_y & \tilde{D}_{ys} & \tilde{D}_{yp} & \tilde{D}_{yw} \end{pmatrix},$$

in order to obtain

$$\begin{pmatrix} \star \\ \star \\ \star \\ \star \\ \star \\ \star \\ \star \end{pmatrix}^\top \begin{pmatrix}
0 & \mathcal{X} & 0 & 0 & 0 & 0 & 0 \\
\mathcal{X} & 0 & 0 & 0 & 0 & 0 & 0 \\
0 & 0 & Q_e & S_e & 0 & 0 & 0 \\
0 & 0 & S_e^\top & R_e & 0 & 0 & 0 \\
0 & 0 & 0 & 0 & P & 0 & 0 \\
0 & 0 & 0 & 0 & 0 & I & 0 \\
0 & 0 & 0 & 0 & 0 & 0 & -\gamma^2 I
\end{pmatrix} \left(\begin{pmatrix}
I & 0 & 0 & 0 \\
\tilde{A} & \tilde{B}_s & \tilde{B}_p & \tilde{B}_w \\
\tilde{C}_s & \tilde{D}_{ss} & \tilde{D}_{sp} & \tilde{D}_{sw} \\
0 & I & 0 & 0 \\
\tilde{C}_q & \tilde{D}_{qs} & \tilde{D}_{qp} & \tilde{D}_{qw} \\
\tilde{C}_z & \tilde{D}_{zs} & \tilde{D}_{zp} & \tilde{D}_{zw} \\
0 & 0 & 0 & I
\end{pmatrix} + \begin{pmatrix} 0 \\ \tilde{B}_u \\ \tilde{D}_{su} \\ 0 \\ 0 \\ \tilde{D}_{zu} \\ 0 \end{pmatrix} \mathcal{O}(\mathcal{K}) \right) \prec 0.$$

If we also introduce

$$\tilde{Q}_e := -Q_e^{-1}, \quad \tilde{S}_e := Q_e^{-1}S_e \quad \text{and} \quad \tilde{R}_e := S_e - S_e^\top Q_e^{-1}S_e,$$

it is not hard to verify that the latter inequality can be rearranged as

$$
\mathrm{He}\left[\left(\begin{array}{ccc|c}
\mathcal{X}\tilde{A} & \mathcal{X}\tilde{B}_{\mathrm{s}} & \mathcal{X}\tilde{B}_p & \mathcal{X}\tilde{B}_w \\
\hline
0 & \frac{1}{2}\tilde{R}_{\mathrm{e}} & 0 & 0 \\
0 & 0 & 0 & 0 \\
0 & 0 & 0 & -\frac{1}{2}\gamma^2 I
\end{array}\right)+\left(\begin{array}{c}
\mathcal{X}\tilde{B}_u \\
\hline
0 \\
0 \\
0
\end{array}\right)\mathcal{O}(\mathcal{K})\right]+
$$

$$
+\left(\begin{array}{c}
\star \\
\star \\
\star
\end{array}\right)^{\top}\left(\begin{array}{ccc}
P & 0 & 0 \\
0 & Q_{\mathrm{e}} & 0 \\
0 & 0 & -I
\end{array}\right)\left(\left(\begin{array}{c|ccc}
\tilde{C}_q & \tilde{D}_{qs} & \tilde{D}_{qp} & \tilde{D}_{qw} \\
\tilde{C}_{\mathrm{s}} & \tilde{D}_{\mathrm{ss}}+\tilde{S}_{\mathrm{e}} & \tilde{D}_{\mathrm{s}p} & \tilde{D}_{\mathrm{s}w} \\
\tilde{C}_z & \tilde{D}_{z\mathrm{s}} & \tilde{D}_{zp} & \tilde{D}_{zw}
\end{array}\right)+\left(\begin{array}{c}
0 \\
\tilde{D}_{\mathrm{s}u} \\
\tilde{D}_{zu}
\end{array}\right)\mathcal{O}(\mathcal{K})\right)\prec 0.
$$

Applying the Schur complement with respect to the Q_{e} and $-I$ block then finally yields the LMI

$$
\mathrm{He}\left[\left(\begin{array}{cccc|cc}
\mathcal{X}\tilde{A} & \mathcal{X}\tilde{B}_{\mathrm{s}} & \mathcal{X}\tilde{B}_p & \mathcal{X}\tilde{B}_w & 0 & 0 \\
\hline
0 & \frac{1}{2}\tilde{R}_{\mathrm{e}} & 0 & 0 & 0 & 0 \\
0 & 0 & 0 & 0 & 0 & 0 \\
0 & 0 & 0 & -\frac{1}{2}\gamma^2 I & 0 & 0 \\
\hline
\tilde{C}_{\mathrm{s}} & \tilde{D}_{\mathrm{ss}}+\tilde{S}_{\mathrm{e}} & \tilde{D}_{\mathrm{s}p} & \tilde{D}_{\mathrm{s}w} & \frac{1}{2}Q_{\mathrm{e}} & 0 \\
\tilde{C}_z & \tilde{D}_{z\mathrm{s}} & \tilde{D}_{zp} & \tilde{D}_{zw} & 0 & -\frac{1}{2}I
\end{array}\right)+\left(\begin{array}{c}
\mathcal{X}\tilde{B}_u \\
\hline
0 \\
0 \\
0 \\
\hline
\tilde{D}_{\mathrm{s}u} \\
\tilde{D}_{zu}
\end{array}\right)\left(\mathcal{O}(\mathcal{K})\ 0\ 0\right)\right]+
$$

$$
+(\star)^{\top}P\left(\tilde{C}_q\ \tilde{D}_{qs}\ \tilde{D}_{qp}\ \tilde{D}_{qw}\ 0\ 0\right)\prec 0,
$$

which is affine in $\mathcal{K}$. The realization matrices of K can hence be obtained by solving a genuine LMI problem. Alternatively, one can as well follow the standard algebraic procedure described in [48]. We emphasize that both approaches might be numerically delicate. For further details we refer the reader e.g. to [48, 49]. ∎

4.6 Illustrations

In order to illustrate the results, let us recall the numerical example that we presented (almost literally) in [200]. To this end, consider the uncertain LPV system

$$
\left(\begin{array}{c}
\dot{x}(t) \\
\hline
z_{\mathrm{e}}(t) \\
y(t)
\end{array}\right)=\left(\begin{array}{cc|cc}
0 & -1+0.95\delta & -2 & 0 \\
1 & -0.5+0.25\delta_{\mathrm{s}}(t) & 1 & 0 \\
\hline
1 & \delta_{\mathrm{s}}(t) & 0 & 0 \\
1 & 0 & 0 & 0.01
\end{array}\right)\left(\begin{array}{c}
x(t) \\
\hline
w(t)
\end{array}\right), \tag{4.40}
$$

where $x(t)$ is the state, $\delta_\mathrm{s}(t) = \sin\frac{1}{10}t$ an online measurable scheduling parameter, $\delta \in [-1,1]$ a time-invariant parametric uncertainty and $w = \mathrm{col}(w_1, w_2)$. Then
the operators $\Delta_\mathrm{s}, \Delta : \mathcal{L}_2 \to \mathcal{L}_2$ are defined by $p_\mathrm{s}(t) = (\Delta_\mathrm{s} q_\mathrm{s})(t) = \delta_\mathrm{s}(t) q_\mathrm{s}(t)$ and
$p = (\Delta q)(t) = \delta q(t)$ respectively.

Completely analogously to Section 4.2, the goal is to design a robust gain-
scheduled estimator that dynamically processes the measurement y and the schedul-
ing parameter δ_s, in order to provide an estimate u of the signal z_e while the induced
$\mathcal{L}_2$-gain from w to $z = z_\mathrm{e} - u$ is rendered less than γ. For the following four scenarios
we determined estimators by applying Theorem 4.2:

1. A nominal LTI estimator K_nom, which renders the induced $\mathcal{L}_2$-gain from w to z
   less than $\gamma = 0.015$ if δ and δ_s are assumed to be zero. Note that this scenario
   can alternatively be handled with standard $\mathcal{H}_\infty$-synthesis techniques such as in
   [41, 48, 72].

2. A robust LTI estimator K_rob, which renders the induced $\mathcal{L}_2$-gain from w to z
   less than $\gamma = 29.19$ for all $\delta \in [-1,1]$, $\delta_\mathrm{s}(t) \in [-1,1]$ and $\dot{\delta}_\mathrm{s}(t) \in \frac{1}{10}[-1,1]$. Here
   we employed the IQC-multiplier 2.7 from Section 2.6.4 with basis-length $\nu = 2$
   and pole-location $\rho = -1$. This allowed us to bound the rate-of-variation of the
   scheduling variable δ_s. Without bounding the rate-of-variation of the scheduling
   variable (*i.e.* set $\nu = 0$) the synthesis LMIs are not feasible. We emphasize that
   this scenario can also be handled with the synthesis results of [165].

3. A nominal gain-scheduled estimator $K_\mathrm{nomgs}(\Delta_\mathrm{s})$, which renders the induced $\mathcal{L}_2$-
   gain from w to z less than $\gamma = 0.026$ for all $\delta_\mathrm{s}(t) \in [-1,1]$ and if δ is assumed
   to be zero. Here we employed the static full-block multipliers for the scheduling
   parameter δ_s as presented in Section 4.3 and 2.6.2.2. Note that we alternatively
   could have considered the synthesis results of [152] and as presented in Section 3.7.
   We further remark that gain-scheduling synthesis based on general dynamic IQC-
   multipliers is still an open problem. Bounding the rate-of-variation, like we did
   for the robust estimator design is, hence, not yet possible in this framework.

4. A robust gain-scheduled estimator $K_\mathrm{robgs}(\Delta_\mathrm{s})$ which renders the induced $\mathcal{L}_2$-
   gain from w to z less than $\gamma = 3.67$ for all $\delta \in [-1,1]$ and $\delta_\mathrm{s}(t) \in [-1,1]$. Here we
   employed the dynamic DG-multiplier 2.5 from Section 2.6.3.1, with basis-length
   $\nu = 2$ and pole-location $\rho = -1$ for the LTI parametric uncertainty δ and static
   full-block multipliers for the scheduling parameter δ_s as presented in Section 4.3
   and 2.6.2.2. Allowing δ to vary arbitrarily fast (*i.e.* set $\nu = 0$) leads to infeasibility
   of the synthesis LMIs.

Remark 4.2. It is interesting to see that considering static IQC-multipliers in this
example leads to infeasibility of the synthesis problem. This very nicely illustrates
the limitations of static IQC-multipliers, since allowing for dynamics in the multipliers
leads to feasible solutions as well as a considerable improvement of performance.

In accordance with Figure 4.1, let us now define the four plants $z := \Sigma_\mathrm{x}(\Delta, \Delta_\mathrm{s}) w$,
$\mathrm{x} \in \boldsymbol{x}$, where $\boldsymbol{x} = \{\mathrm{nom}, \mathrm{rob}, \mathrm{nomgs}, \mathrm{robgs}\}$, corresponding to the four designed

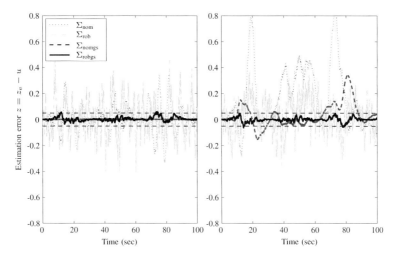

Fig. 4.2 Left: estimation error for $\delta = 0$. Right: estimation error for $\delta = 1$.

estimators K_x, and discuss the synthesis results. Figure 4.2 shows the estimation error for $\delta = 0$ (left) and $\delta = 1$ (right) for a random wave disturbance input w_1 (*i.e.* a sinusoid with a random uniformly distributed frequency) and a random uniformly distributed noise input w_2 with an amplitude of 1 and 0.0001 respectively.

It is very satisfactory to see that the gain-scheduled estimators K_{nomgs} and K_{robgs} outperform the LTI estimators K_{nom} and K_{rob} and that the LTI estimator K_{rob} outperforms the LTI estimator K_{nom} (in the sense that the estimation error is rendered small). This reveals that gain-scheduled estimation can be preferable over robust estimation in practice. Nevertheless, if we simulate the system with $\delta = 1$, the estimators K_{nom}, K_{rob} and K_{nomgs} show a drastic performance degradation. However, it is again very nice to see that the robust gain-scheduled estimator K_{robgs} keeps the estimation error small, despite the uncertainty in the system.

The results are consistent with the singular value plots from the disturbance input w to the estimation error output z in Figure 4.3, if fixing the parameter δ_s to various points in $[-1, 1]$. For example, if we compare, Figure 4.3.1.3 with Figure 4.3.2.3, it can be seen that the nominal gain-scheduled estimator K_{nomgs} is very sensitive to deviating values of the uncertainty δ from zero. Moreover, clearly, this is also the case if we consider the nominal and robust LTI estimator K_{nom} and K_{rob}, respectively, in Figure 4.3.1.1 - Figure 4.3.2.2, which are sensitive to variations of both δ_s and δ. On the other hand, it is very nice to observe that, by comparing Figure 4.3.1.4 with Figure 4.3.2.4, this is not the case for the robust gain-scheduled estimator K_{robgs}, which is insensitive to deviating values of the uncertainty δ_s from zero.

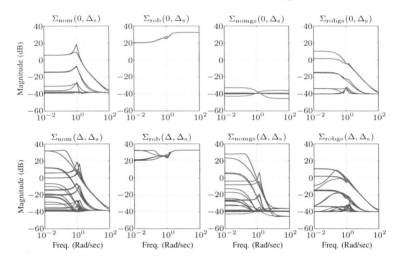

Fig. 4.3 Upper row: Singular value plots of $\Sigma_{\mathrm{x}}(\Delta, \Delta_{\mathrm{s}})$ evaluated at $\Delta = 0$ and different values of $\Delta_{\mathrm{s}} \in [-1, 1]$. Bottom row: Singular value plots of $\Sigma_{\mathrm{x}}(\Delta, \Delta_{\mathrm{s}})$ evaluated at different values of $\Delta_{\mathrm{s}} \in [-1, 1]$ and $\Delta \in [-1, 1]$.

4.7 Chapter summary

In this chapter, we considered the robust gain-scheduled estimation problem. For plants with a linear fractional dependency on both genuine uncertainties as well as scheduling parameters, we have shown how this problem can be embedded in the IQC-framework as a special case of the general robust gain-scheduling problem in Section 3.4. Through the use of some recent auxiliary results (i.e. Lemma 4.1, 4.3 and 4.4) we have been able to characterize the essential property of internal stability of the overall nominal system interconnection and appropriately reformulate the analysis LMI. The latter allowed us to take the crucial step of dualization, which enabled us the eliminate the realization matrices of the estimator and render the problem tractable. Indeed, as a contribution and as the main result of this chapter, we have shown how to obtain LMI conditions for the existence of robust partially gain-scheduled estimators, which guarantee a given bound on the induced $\mathscr{L}_2$-gain from the disturbance input to the estimated error output. Finally, we demonstrated the potential of results through an illustrative numerical example from our recent paper [200].

Chapter 5
Another general synthesis framework

5.1 Introduction

As discussed in Chapter 3, robust synthesis is much harder than the design of nominal and nominal gain-scheduled controllers since no convex solution for the associated optimization problem is known. Fortunately, and as we have already seen in Chapter 3 and 4, it is often possible for specialized problems to exploit the additional structure in order to transform the original non-convex problem into a convex one. For example, it is well-known that this is the case for the synthesis of robust estimators [51, 180, 165] or robust feedforward controllers [54, 103] for the configurations in Figure 5.1. For given uncertain systems $\Sigma_1(\Delta_1)$ or $\Sigma_2(\Delta_2)$ and performance weights W_z or W_w, the problem is to design an estimator K_e or a feedforward controller K_{ff} which minimizes

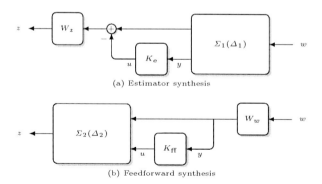

(a) Estimator synthesis

(b) Feedforward synthesis

Fig. 5.1 Interconnections for estimator and feedforward controller synthesis

the induced $\mathscr{L}_2$-gain of the channel $w \to z$. In fact, it has recently been established in
[155, 170] that synthesis problems which can be subsumed to the general configuration
shown in Figure 5.2, where only the control channel is not affected by uncertainties,
can be convexified. This framework not only covers robust estimator and feedforward

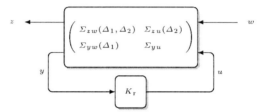

Fig. 5.2 Interconnection for the class of robust synthesis problems without control-channel
uncertainties

controller synthesis, but it also includes e.g. generalized l_2-synthesis, with extensions
to dynamically weighted signal constraints [32, 98, 33], as well as a rich class of
questions related to multi-objective and structured control [149]. Further it captures
generalizations of open-loop design problems as studied by [18] and those involving
uncertain input- or output-performance weights [38] as shown in Figure 5.3.

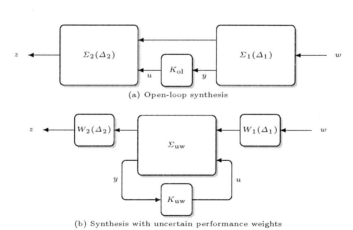

(a) Open-loop synthesis

(b) Synthesis with uncertain performance weights

Fig. 5.3 Interconnections for open-loop synthesis and feedback design with uncertain
weights

Despite the generality of the setup in Figure 5.2, and although special cases have been considered in e.g. Chapter 4 and [20], it does not cover the class of gain-scheduling problems addressed in Section 3.3. This motivates the first goal of this chapter, which is to reveal how a combination of the gain-scheduled controller synthesis approaches in [121, 174, 150] and the robust controller synthesis techniques of [155, 170] can be handled in the unified framework as schematically shown in Figure 5.4. Here we assume that the system is described by a standard linear fractional representation and that the properties of the uncertainties and scheduling parameters are captured by suitable families of IQCs. In accordance with Chapter 2, the results of this chapter offer the full flexibility to handle diagonally structured uncertainty blocks Δ_1, Δ_2 that might include time-invariant or time-varying rate-bounded parametric uncertainties as well as linear dynamic or infinite dimensional uncertainties and static nonlinearities. We develop LMIs whose feasibility ensures the existence of gain-scheduled controllers that guarantee robust stability and performance for the corresponding uncertain closed-loop system.

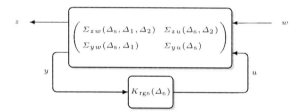

Fig. 5.4 Interconnection for the class of robust gain-scheduling synthesis problems without control-channel uncertainties

This is achieved by subsuming the resulting optimization problem to a generic gain-scheduling feasibility problem that involves both parametric as well as dynamic decision variables. As our second and main contribution, we show how this feasibility problem can be convexified, while avoiding the rather sophisticated technical step of appropriately factorizing the IQC-multipliers as was required in Chapter 4. Moreover, we argue that this setup is surprisingly flexible. It not only enables to handle the synthesis configuration in Figure 5.4, but also other problems that have been individually covered or not at all considered in the literature.

Let us also emphasize some key technical benefits of the proposed approach. Robust controller design with IQCs requires to handle unstable performance weights in nominal $\mathcal{H}_\infty$-synthesis, which turns out to be non-trivial [118, 69]. Similar issues arise in regulation problems [90, 108, 119, 142, 117] and in gain-scheduling control with dynamic multipliers [157]. Even for nominal synthesis, the present chapter extends the recent results in [157] to unstable weights both at the input and the output of the performance channel. Furthermore, if designing robust controllers for plants with a general linear fractional dependence on the uncertainties, one typically has to rely on an iteration between nominal synthesis and robustness analysis [143, 40, 11, 28, 146, 144, 198]. After each analysis step, the involved scalings are factorized to obtain stable weights for a new augmented plant in order to design a robustified controller for continuing the iteration. These intermediate factorization steps often cause numerical ill-conditioning in computations. We will show in this chapter how to completely avoid any such factorizations which leads to a significant simplification of the existing μ- and IQC-synthesis tools. Let us finally stress that the results in this chapter have already appeared in [197] and that large portions of the text overlap.

Remark 5.1. Although the previous and the present chapter both address general versions of the robust gain-scheduled estimation problem, we explicitly stress that the results in this chapter do not cover the estimation problem of Chapter 4 in its full generality. The results in Chapter 4 are, hence, justified as an separate contribution. In the sequel, we will discuss the differences between the approaches in more detail.

This chapter is organized as follows. In Section 5.2 we present a generic feasibility problem and Section 5.3 contains our main result; the construction of a semi-definite program for solving the generic feasibility problem. Subsequently, in Section 5.4, it is demonstrated how the synthesis problem for the setup in Figure 5.4 can be subsumed under our main result; and in Section 5.5 we provide a list of further concrete design problems that are covered by our framework. We conclude the chapter with two numerical examples as well as a summary in Sections 5.6 and 5.7 respectively.

5.2 A generic feasibility problem

In order to motivate the formulation of our generic design framework, let us start by briefly recalling the nominal gain-scheduled controller synthesis configuration from Section 3.3. As discussed, this problem translates into the search for K that internally stabilizes the plant G_{aug} and a scheduling function $\hat{\Delta}_c(\cdot)$ such that there exists some $P_e \in \boldsymbol{P}_e$ for which the following FDI holds true:

$$
\begin{pmatrix} \mathcal{G}_{ee} & \mathcal{G}_{ew} \\ \hline I & 0 \\ \hline \mathcal{G}_{ze} & \mathcal{G}_{zw} \\ 0 & I \end{pmatrix}^{*} \begin{pmatrix} P_e & 0 \\ 0 & P_\gamma \end{pmatrix} \begin{pmatrix} \mathcal{G}_{ee} & \mathcal{G}_{ew} \\ \hline I & 0 \\ \hline \mathcal{G}_{ze} & \mathcal{G}_{zw} \\ 0 & I \end{pmatrix} \prec 0.
\tag{5.1}
$$

On the basis of this formulation and as shown in Section 3.7, it well-known how to convexify the corresponding synthesis problem [121, 174, 150].

Now observe that (5.1) can be rendered affine in γ by applying the Schur-complement. This yields the equivalent FDI

$$
\begin{pmatrix} \mathcal{H}_{11} & \mathcal{H}_{12} \\ I & 0 \\ \hline \mathcal{H}_{21} & \mathcal{H}_{22} \\ 0 & I \end{pmatrix}^{*}
\begin{pmatrix} P_e & 0 & 0 \\ \hline 0 & 0 & I \\ 0 & I & P_r \end{pmatrix}
\begin{pmatrix} \mathcal{H}_{11} & \mathcal{H}_{12} \\ I & 0 \\ \hline \mathcal{H}_{21} & \mathcal{H}_{22} \\ 0 & I \end{pmatrix} \prec 0,
\tag{5.2}
$$

where

$$
P_r := -\gamma I_{n_w + n_z} \quad \text{and} \quad
\begin{pmatrix} \mathcal{H}_{11} & \mathcal{H}_{12} \\ \mathcal{H}_{21} & \mathcal{H}_{22} \end{pmatrix} :=
\begin{pmatrix} \mathcal{G}_{ee} & \mathcal{G}_{ew} & 0 \\ \hline 0 & 0 & 0 \\ \mathcal{G}_{ze} & \mathcal{G}_{zw} & 0 \end{pmatrix}.
\tag{5.3}
$$

A suitable generalization of the FDI (5.2) forms the basis of the generic framework considered in this chapter. In particular, it consists of the following ingredients:

1. a dynamical system $\Gamma(\Delta_s)$, which is assumed to admit the linear fractional representation

$$
\underbrace{\begin{pmatrix} \Gamma_{11}(\Delta_s) & \Gamma_{12}(\Delta_s) \\ \Gamma_{21}(\Delta_s) & \Gamma_{22}(\Delta_s) \end{pmatrix}}_{\Gamma(\Delta_s)} = \Delta_s \star
\underbrace{\begin{pmatrix} H_{11} & H_{12} & H_{13} \\ \hline H_{21} & H_{22} & H_{23} \\ H_{31} & H_{32} & H_{33} \end{pmatrix}}_{H}
$$

that is well-posed for all $\Delta_s \in \boldsymbol{\Delta}_s$. Here the scheduling block $\Delta_s \in \boldsymbol{\Delta}_s$ is defined as previously, while H is a transfer matrix with the realization

$$
H = \begin{pmatrix} H_{11} & H_{12} & H_{13} \\ H_{21} & H_{22} & H_{23} \\ H_{31} & H_{32} & H_{33} \end{pmatrix} = \left[\begin{array}{c|ccc} A & B_1 & B_2 & B_3 \\ \hline C_1 & D_{11} & D_{12} & D_{13} \\ C_2 & D_{21} & D_{22} & D_{23} \\ C_3 & D_{31} & D_{32} & 0 \end{array} \right],
$$

where $A \in \mathbb{R}^{n_H \times n_H}$, (A, B_3) is stabilizable, (A, C_3) is detectable and, without loss of generality, $H_{33}(\infty) = D_{33} = 0$.
2. a gain-scheduled controller $K_{gs}(\Delta_s) = K \star \hat{\Delta}_c(\Delta_s)$.
3. possibly unstable weights $T_z, T_w \in \mathscr{RL}_\infty^{\bullet \times n_i}$ with the realizations (A_z, B_z, C_z, D_z) and (A_w, B_w, C_w, D_w) respectively. Here $A_z \in \mathbb{R}^{n_{T_z} \times n_{T_z}}$ and $A_w \in \mathbb{R}^{n_{T_w} \times n_{T_w}}$ have no eigenvalues on $\mathbb{C}^0$.
4. a family of transfer matrices $T_r^* P_r T_r$, where P_r is some LMIable set of real symmetric matrices and $T_r \in \mathscr{RL}_\infty^{\bullet \times n_i}$ is a fixed and possibly unstable transfer matrix with the realization (A_r, B_r, C_r, D_r), $A_r \in \mathbb{R}^{n_{T_r} \times n_{T_r}}$, $\text{eig}(A_r) \cap \mathbb{C}^0 = \emptyset$.

As before, the closed-loop system $\Gamma(\Delta_s) \star K_{gs}(\Delta_s)$ is described by the linear fractional representation

$$\Delta_c(\Delta_s) \star \underbrace{\begin{pmatrix} \mathcal{H}_{11} & \mathcal{H}_{12} \\ \mathcal{H}_{12} & \mathcal{H}_{22} \end{pmatrix}}_{\mathcal{H}},$$

with the extended scheduling block $\Delta_c(\Delta_s)$ and where $\mathcal{H}$ denotes the interconnection of H with the LTI part of the controller K as in

$$\underbrace{\left(\begin{array}{c|c} \mathcal{H}_{11} & \mathcal{H}_{12} \\ \hline \mathcal{H}_{12} & \mathcal{H}_{22} \end{array} \right)}_{\mathcal{H}} = \underbrace{\left(\begin{array}{cc|cc|c} H_{11} & 0 & H_{12} & H_{13} & 0 \\ 0 & 0 & 0 & 0 & I_{n_{q_c}} \\ \hline H_{21} & 0 & H_{22} & H_{23} & 0 \\ H_{31} & 0 & H_{32} & H_{33} & 0 \\ \hline 0 & I_{n_{p_c}} & 0 & 0 & 0 \end{array} \right)}_{H_{\text{aug}}} \star \underbrace{\begin{pmatrix} K_{11} & K_{12} \\ K_{21} & K_{22} \end{pmatrix}}_{K}.$$

We are now ready to formulate the central problem considered in this chapter.

Problem 5.1. Check the existence of an LTI system K that internally stabilizes H_{aug} and a scheduling function $\hat{\Delta}_c(\cdot)$ such that there exist $P_r \in \boldsymbol{P}_r$ and $P_e \in \boldsymbol{P}_e$ for which the following FDI holds true:

$$\begin{pmatrix} \mathcal{H}_{11} & \mathcal{H}_{12}T_w \\ \hline I & 0 \\ \hline T_z^* \mathcal{H}_{21} & T_z^* \mathcal{H}_{22}T_w \\ 0 & I \end{pmatrix}^* \begin{pmatrix} P_e & 0 & 0 \\ \hline 0 & 0 & I \\ \hline 0 & I & T_r^* P_r T_r \end{pmatrix} \begin{pmatrix} \mathcal{H}_{11} & \mathcal{H}_{12}T_w \\ \hline I & 0 \\ \hline T_z^* \mathcal{H}_{21} & T_z^* \mathcal{H}_{22}T_w \\ 0 & I \end{pmatrix} \prec 0. \quad (5.4)$$

If the conditions in Problem 5.1 are satisfied then $K_{gs}(\Delta_s) = K \star \hat{\Delta}_c(\Delta_s)$ stabilizes $\Gamma(\Delta_s)$ for all $\Delta_s \in \boldsymbol{\Delta}_s$. Let us further observe that the FDI (5.4) directly specializes into (5.2) for $T_r := I$, $T_z := I$, $T_w := I$ and P_r, $\mathcal{H}$ as in (5.3). Moreover, if the weights T_z and T_w are stable and are merged with the plant $\mathcal{H}$ (as is typically done in $\mathscr{H}_\infty$-synthesis), then Problem 5.1 is in turn a special case of the general robust gain-scheduling synthesis problem from Section 3.4. Although this is not straightforward to see, we emphasize that the 2×2 lower-block of the middle matrix of (5.4) is nothing but an IQC-multiplier of the form $\Psi^* P \Psi$ with the following specialized structure:

$$\begin{pmatrix} 0 & I \\ I & T_r^* P_r T_r \end{pmatrix} = \underbrace{\begin{pmatrix} I & 0 \\ 0 & T_r \\ 0 & I \end{pmatrix}^*}_{\Psi^*} \underbrace{\begin{pmatrix} 0 & 0 & I \\ 0 & P_r & 0 \\ I & 0 & 0 \end{pmatrix}}_{P} \underbrace{\begin{pmatrix} I & 0 \\ 0 & T_r \\ 0 & I \end{pmatrix}}_{\Psi}.$$

It will come forward in the sequel that this particular IQC-multiplier evolves from performance specifications and the inclusion of IQC-multipliers in order to capture the properties of uncertainties in robust synthesis. On the other hand, and as a major difference with the general synthesis from Section 3.4, we do not merge the weights T_z and T_w with the plant $\mathcal{H}$ and assume that they are, more generally, unstable transfer matrices in $\mathscr{RL}_\infty$. We will argue that the structure of the problem, which is in itself non-convex, is particulary well-suited for obtaining a finite dimensional convex feasibility test in an insightful fashion. Moreover, in Sections 5.4 and 5.5 we will extensively illustrate that Problem 5.1 encompasses a rich class of concrete design scenarios and also motivate the benefit of using unstable weights.

5.3 A convex solution

5.3.1 Analysis

This section serves to show how to translate the analysis FDI (5.4) into an LMI formulation that allows us to approach the central result of this chapter, the construction of synthesis LMIs which characterize the existence of a controller that solves Problem 5.1. In parallel to (5.4), and as we have seen in Chapter 3 and 4, this requires to consider the dual FDI as obtained with Lemma C.1 in Appendix C. Fortunately, we do not need to factorize the IQC-multipliers as was required in Chapter 4. Instead, the dual FDI of (5.4) is straightforwardly obtained in the standard fashion as

$$\begin{pmatrix} I & 0 \\ \hline -\mathcal{H}_{11}^* & -\mathcal{H}_{21}^* T_z \\ 0 & I \\ -T_w^* \mathcal{H}_{12}^* & -T_w^* \mathcal{H}_{22}^* T_z \end{pmatrix}^* \begin{pmatrix} \tilde{P}_e & 0 & 0 \\ \hline 0 & -T_r^* P_r T_r & I \\ 0 & I & 0 \end{pmatrix} \begin{pmatrix} I & 0 \\ \hline -\mathcal{H}_{11}^* & -\mathcal{H}_{21}^* T_z \\ 0 & I \\ -T_w^* \mathcal{H}_{12}^* & -T_w^* \mathcal{H}_{22}^* T_z \end{pmatrix} \succ 0.$$

(5.5)

Here $\tilde{P}_e := P_e^{-1}$ takes its values from the class of dual extended full-block multipliers $\tilde{P}_e$ as defined in Section 3.7. Let us emphasize the crucial fact that the particular structure of the primal FDI (5.4) causes the dual FDI (5.5) to depend affinely on P_r as well. We arrive at the following dual formulation of the conditions in Problem 5.1.

Lemma 5.1. *Suppose that K internally stabilizes H_{aug} and let $\hat{\Delta}_c(\cdot)$ be given. Then there exist $P_e \in \boldsymbol{P}_e$, $P_r \in \boldsymbol{P}_r$ satisfying (5.4) if and only if there are $\tilde{P}_e \in \tilde{\boldsymbol{P}}_e$, $P_r \in \boldsymbol{P}_r$ with (5.5).*

Now we translate the FDIs (5.4) and (5.5) into LMIs. First, simple manipulations reveal that the weight T_z^* in the right outer-factor of the primal FDI (5.4) can be shifted to the left outer-factor, which yields

$$
\begin{pmatrix} \mathcal{H}_{11} & \mathcal{H}_{12}T_w \\ I & 0 \\ \hline \mathcal{H}_{21} & \mathcal{H}_{22}T_w \\ 0 & T_r \\ 0 & T_z \end{pmatrix}^{*}
\begin{pmatrix} P_{\mathrm{e}} & 0 & 0 & 0 \\ \hline 0 & 0 & 0 & I \\ 0 & 0 & P_r & 0 \\ 0 & I & 0 & 0 \end{pmatrix}
\begin{pmatrix} \mathcal{H}_{11} & \mathcal{H}_{12}T_w \\ I & 0 \\ \hline \mathcal{H}_{21} & \mathcal{H}_{22}T_w \\ 0 & T_r \\ 0 & T_z \end{pmatrix} \prec 0. \tag{5.6}
$$

In the same fashion, we can shift the weight T_w^{*} from the right outer-factor of the dual FDI (5.5) to the left outer-factor as follows:

$$
\begin{pmatrix} I & 0 \\ -\mathcal{H}_{11}^{*} & -\mathcal{H}_{21}^{*}T_z \\ \hline 0 & T_w \\ 0 & T_r \\ -\mathcal{H}_{12}^{*} & -\mathcal{H}_{22}^{*}T_z \end{pmatrix}^{*}
\begin{pmatrix} \tilde{P}_{\mathrm{e}} & 0 & 0 & 0 \\ \hline 0 & 0 & 0 & I \\ 0 & 0 & -P_r & 0 \\ 0 & I & 0 & 0 \end{pmatrix}
\begin{pmatrix} I & 0 \\ -\mathcal{H}_{11}^{*} & -\mathcal{H}_{21}^{*}T_z \\ \hline 0 & T_w \\ 0 & T_r \\ -\mathcal{H}_{12}^{*} & -\mathcal{H}_{22}^{*}T_z \end{pmatrix} \succ 0. \tag{5.7}
$$

Note that (5.6) and (5.7) are no longer related by duality through Lemma C.1 (see Appendix C). However, as will become clearer in the sequel, the latter steps enable us to algebraically characterize the fact that K stabilizes H_{aug} even though the weights T_w, T_r and T_z might be unstable.

In order to turn the FDIs (5.6) and (5.7) into LMIs by making use of the KYP-Lemma (see Section 2.5.2), let us define

$$
\begin{pmatrix} T_w \\ T_r \\ T_z \end{pmatrix} =
\left[\begin{array}{ccc|c} A_w & 0 & 0 & B_w \\ 0 & A_r & 0 & B_r \\ 0 & 0 & A_z & B_z \\ \hline C_w & 0 & 0 & D_w \\ 0 & C_r & 0 & D_r \\ 0 & 0 & C_z & D_z \end{array} \right]
=: \left[\begin{array}{c|c} A_{\mathrm{t}} & B_{\mathrm{t}} \\ \hline C_{\mathrm{t}1} & D_{\mathrm{t}1} \\ C_{\mathrm{t}2} & D_{\mathrm{t}2} \\ C_{\mathrm{t}3} & D_{\mathrm{t}3} \end{array} \right]
$$

with A_{t} of dimension $n_{\mathrm{t}} = n_{T_w} + n_{T_r} + n_{T_z}$, and introduce the realization

$$
\mathcal{H} = H_{\mathrm{aug}} \star K = \begin{pmatrix} \mathcal{H}_{11} & \mathcal{H}_{12} \\ \mathcal{H}_{21} & \mathcal{H}_{22} \end{pmatrix} = \left[\begin{array}{c|cc} \mathcal{A} & \mathcal{B}_1 & \mathcal{B}_2 \\ \hline \mathcal{C}_1 & \mathcal{D}_{11} & \mathcal{D}_{12} \\ \mathcal{C}_2 & \mathcal{D}_{21} & \mathcal{D}_{22} \end{array} \right].
$$

Here the calligraphic matrices are, as well-known, given by

$$
\begin{pmatrix} \mathcal{A} & \mathcal{B}_1 & \mathcal{B}_2 \\ \mathcal{C}_1 & \mathcal{D}_{11} & \mathcal{D}_{12} \\ \mathcal{C}_2 & \mathcal{D}_{21} & \mathcal{D}_{22} \end{pmatrix} = \left(\begin{array}{cc|cc|c} A & 0 & B_1 & 0 & B_2 \\ 0 & 0 & 0 & 0 & 0 \\ \hline C_1 & 0 & D_{11} & 0 & D_{12} \\ 0 & 0 & 0 & 0 & 0 \\ \hline C_2 & 0 & D_{21} & 0 & D_{22} \end{array}\right) + \left(\begin{array}{cc|c} 0 & B_3 & 0 \\ I & 0 & 0 \\ \hline 0 & D_{13} & 0 \\ 0 & 0 & I \\ \hline 0 & D_{23} & 0 \end{array}\right) \times
$$

$$
\times \begin{pmatrix} A_K & B_K \\ C_K & D_K \end{pmatrix} \left(\begin{array}{cc|cc|c} 0 & I & 0 & 0 & 0 \\ \hline C_3 & 0 & D_{31} & 0 & D_{32} \\ 0 & 0 & 0 & I & 0 \end{array}\right). \tag{5.8}
$$

As a consequence, the outer-factor of the primal FDI (5.6) admits the realization

$$
\begin{pmatrix} \mathcal{H}_{11} & \mathcal{H}_{12}T_w \\ I & 0 \\ \hline \mathcal{H}_{21} & \mathcal{H}_{22}T_w \\ 0 & T_r \\ 0 & T_z \end{pmatrix} = \left[\begin{array}{cc|cc} A_t & 0 & 0 & B_t \\ \mathcal{B}_2 C_{t1} & \mathcal{A} & \mathcal{B}_1 & \mathcal{B}_2 D_{t1} \\ \mathcal{D}_{12}C_{t1} & \mathcal{C}_1 & \mathcal{D}_{11} & \mathcal{D}_{12}D_{t1} \\ \hline 0 & 0 & I & 0 \\ \mathcal{D}_{22}C_{t1} & \mathcal{C}_2 & \mathcal{D}_{21} & \mathcal{D}_{22}D_{t1} \\ C_{t2} & 0 & 0 & D_{t2} \\ C_{t3} & 0 & 0 & D_{t3} \end{array}\right].
$$

Recall that K internally stabilizes H_{aug} if and only if $\mathcal{A}$ is Hurwitz. For stabilizing controllers, the KYP-lemma implies that (5.6) is equivalent to the existence of some symmetric matrix $\mathcal{X}$ with

$$
\begin{pmatrix} \star \\ \star \\ \star \\ \star \\ \star \\ \star \\ \star \\ \star \\ \star \end{pmatrix}^{\top} \left(\begin{array}{cc|cc|c|c|c} 0 & 0 & X_{11} & \mathcal{X}_{12} & 0 & 0 & 0 & 0 \\ 0 & 0 & \mathcal{X}_{12}^{\top} & \mathcal{X}_{22} & 0 & 0 & 0 & 0 \\ \hline X_{11} & \mathcal{X}_{12} & 0 & 0 & 0 & 0 & 0 & 0 \\ \mathcal{X}_{12}^{\top} & \mathcal{X}_{22} & 0 & 0 & 0 & 0 & 0 & 0 \\ \hline 0 & 0 & 0 & 0 & P_e & 0 & 0 & 0 \\ \hline 0 & 0 & 0 & 0 & 0 & 0 & 0 & I \\ 0 & 0 & 0 & 0 & 0 & 0 & P_r & 0 \\ 0 & 0 & 0 & 0 & 0 & I & 0 & 0 \end{array}\right) \left(\begin{array}{cc|cc} I & 0 & 0 & 0 \\ 0 & I & 0 & 0 \\ \hline A_t & 0 & 0 & B_t \\ \mathcal{B}_2 C_{t1} & \mathcal{A} & \mathcal{B}_1 & \mathcal{B}_2 D_{t1} \\ \hline \mathcal{D}_{12}C_{t1} & \mathcal{C}_1 & \mathcal{D}_{11} & \mathcal{D}_{12}D_{t1} \\ \hline 0 & 0 & I & 0 \\ \mathcal{D}_{22}C_{t1} & \mathcal{C}_2 & \mathcal{D}_{21} & \mathcal{D}_{22}D_{t1} \\ C_{t2} & 0 & 0 & D_{t2} \\ C_{t3} & 0 & 0 & D_{t3} \end{array}\right) \prec 0.
$$

For the ease of subsequent manipulations, this can be more elegantly and compactly expressed as

$$
\text{He}\left[\begin{pmatrix} X_{11} & \mathcal{X}_{12} & C_{\text{t}3}^\top \\ \mathcal{X}_{12}^\top & \mathcal{X}_{22} & 0 \\ \hline 0 & 0 & 0 \\ 0 & 0 & D_{\text{t}3}^\top \end{pmatrix}\begin{pmatrix} A_{\text{t}} & 0 & 0 & B_{\text{t}} \\ \mathcal{B}_2 C_{\text{t}1} & \mathcal{A} & \mathcal{B}_1 & \mathcal{B}_2 D_{\text{t}1} \\ \hline \mathcal{D}_{22}C_{\text{t}1} & \mathcal{C}_2 & \mathcal{D}_{21} & \mathcal{D}_{22}D_{\text{t}1} \end{pmatrix}\right] +
$$

$$
\begin{pmatrix} \star \\ \star \\ \hline \star \end{pmatrix}^\top \begin{pmatrix} P_{\text{e}} & 0 \\ 0 & P_r \end{pmatrix}\begin{pmatrix} \mathcal{D}_{12}C_{\text{t}1} & \mathcal{C}_1 & \mathcal{D}_{11} & \mathcal{D}_{12}D_{\text{t}1} \\ 0 & 0 & I & 0 \\ \hline C_{\text{t}2} & 0 & 0 & D_{\text{t}2} \end{pmatrix} \prec 0. \quad (5.9)
$$

Analogously, (5.7) is equivalent to the existence of some symmetric matrix $\mathcal{Y}$ with

$$
\text{He}\left[\begin{pmatrix} Y_{11} & \mathcal{Y}_{12} & C_{\text{t}1}^\top \\ \mathcal{Y}_{12}^\top & \mathcal{Y}_{22} & 0 \\ \hline 0 & 0 & 0 \\ 0 & 0 & D_{\text{t}1}^\top \end{pmatrix}\begin{pmatrix} A_{\text{t}} & 0 & 0 & B_{\text{t}} \\ -\mathcal{C}_2^\top C_{\text{t}3} & -\mathcal{A}^\top & -\mathcal{C}_1^\top & -\mathcal{C}_2^\top D_{\text{t}3} \\ \hline -\mathcal{D}_{22}^\top C_{\text{t}3} & -\mathcal{B}_2^\top & -\mathcal{D}_{12}^\top & -\mathcal{D}_{22}^\top D_{\text{t}3} \end{pmatrix}\right] +
$$

$$
\begin{pmatrix} \star \\ \star \\ \hline \star \end{pmatrix}^\top \begin{pmatrix} \tilde{P}_{\text{e}} & 0 \\ 0 & -P_r \end{pmatrix}\begin{pmatrix} 0 & 0 & I & 0 \\ -\mathcal{D}_{21}^\top C_{\text{t}3} & -\mathcal{B}_1^\top & -\mathcal{D}_{11}^\top & -\mathcal{D}_{21}^\top D_{\text{t}3} \\ \hline C_{\text{t}2} & 0 & 0 & D_{\text{t}2} \end{pmatrix} \succ 0. \quad (5.10)
$$

Whenever required, we partition $\mathcal{Z} \in \{\mathcal{X}, \mathcal{Y}\}$ as

$$
\mathcal{Z} = \left(\begin{array}{c|c} Z_{11} & \mathcal{Z}_{12} \\ \hline \mathcal{Z}_{12}^\top & \mathcal{Z}_{22} \end{array}\right) = \left(\begin{array}{c|cc} Z_{11} & Z_{12} & Z_{13} \\ \hline Z_{12}^\top & Z_{22} & Z_{23} \\ Z_{13}^\top & Z_{23}^\top & Z_{33} \end{array}\right) = \left(\begin{array}{c|c} Z & \star \\ \hline \star & \star \end{array}\right), \quad (5.11)
$$

where Z_{11}, Z_{22} and Z_{33} have compatible dimensions with A_{t}, A and A_K, respectively; as indicated by the partition lines, this implies the same for $\mathcal{Z}_{22}$ and $\mathcal{A}$ as well as for Z and $\text{diag}(A_{\text{t}}, A)$.

It is now possible to formulate a key analysis result, which comprises an elementary characterization for K to internally stabilize H_{aug} together with a link between the solutions of the primal and dual analysis LMIs (5.9) and (5.10) respectively. This involves the involution

$$
\mathcal{I}\begin{pmatrix} T_{11} & T_{12} \\ T_{12}^\top & T_{22} \end{pmatrix} := \begin{pmatrix} T_{12}T_{22}^{-1}T_{12}^\top - T_{11} & -T_{12}T_{22}^{-1} \\ -T_{22}^{-1}T_{12}^\top & T_{22}^{-1} \end{pmatrix}
$$

if T_{22} is nonsingular; here T_{11} and T_{22} have the same dimensions as A_{t} and $\mathcal{A}$ respectively.

Lemma 5.2.

1. *For any solution* $\mathcal{X} = \mathcal{X}^\top$, $P_r \in \boldsymbol{P}_r$, $P_e \in \boldsymbol{P}_e$ *of the primal LMI* (5.9) *it holds that*

   a. $\mathcal{A}$ *is Hurwitz if and only if* $\mathcal{X}_{22} \succ 0$;
   b. *the dual LMI* (5.10) *is satisfied with* $\mathcal{Y} := \mathcal{I}(\mathcal{X})$, $P_r \in \boldsymbol{P}_r$, $\tilde{P}_e := P_e^{-1} \in \tilde{\boldsymbol{P}}_e$.

2. *For any solution* $\mathcal{Y} = \mathcal{Y}^\top$, $P_r \in \boldsymbol{P}_r$, $\tilde{P}_e \in \tilde{\boldsymbol{P}}_e$ *of the dual LMI* (5.10) *it holds that*

   a. $\mathcal{A}$ *is Hurwitz if and only if* $\mathcal{Y}_{22} \succ 0$;
   b. *the primal LMI* (5.9) *is satisfied with* $\mathcal{X} := \mathcal{I}(\mathcal{Y})$, $P_r \in \boldsymbol{P}_r$, $P_e := \tilde{P}_e^{-1} \in \boldsymbol{P}_e$.

Proof. In order to prove the first statement, suppose there exist $\mathcal{X} = \mathcal{X}^\top$, $P_r \in \boldsymbol{P}_r$, $P_e \in \boldsymbol{P}_e$ for which (5.9) holds. Just by considering the sub-block in the $(2,2)$ position of (5.9) this implies $\mathrm{He}(\mathcal{X}_{22}\mathcal{A}) + \mathcal{C}_1^\top Q_e \mathcal{C}_1 \prec 0$ and hence $\mathrm{He}(\mathcal{X}_{22}\mathcal{A}) \prec 0$ since $Q_e \succ 0$. Therefore $\mathcal{A}$ is Hurwitz if and only if $\mathcal{X}_{22} \succ 0$. On the other hand, it shows that $\mathcal{X}_{22}$ is non-singular such that $\mathcal{Y} := \mathcal{I}(\mathcal{X})$ is well-defined. Further observe that $\mathcal{I} \circ \mathcal{I}(\mathcal{X}) = \mathcal{X}$ and that $\mathcal{X}$ can be written in terms of $\mathcal{Y}$ as follows:

$$\begin{pmatrix} X_{11} & \mathcal{X}_{12} \\ \mathcal{X}_{12}^\top & \mathcal{X}_{22} \end{pmatrix} = \hat{\mathcal{Y}}^{-1} \begin{pmatrix} -Y_{11} & 0 \\ -\mathcal{Y}_{12}^\top & I \end{pmatrix}, \quad \hat{\mathcal{Y}} = \begin{pmatrix} I & \mathcal{Y}_{12} \\ 0 & \mathcal{Y}_{22} \end{pmatrix} \tag{5.12}$$

Next to $\hat{\mathcal{Y}}$ we introduce the abbreviation

$$\begin{pmatrix} \hat{\mathcal{A}} & \hat{\mathcal{B}}_1 & \hat{\mathcal{B}}_2 \\ \hat{\mathcal{C}}_1 & \hat{\mathcal{D}}_{11} & \hat{\mathcal{D}}_{12} \\ \hat{\mathcal{C}}_2 & \hat{\mathcal{D}}_{21} & \hat{\mathcal{D}}_{22} \end{pmatrix} = \left(\begin{array}{c|c|c} \frac{1}{2}C_{t2}^\top P_r C_{t2} - Y_{11}A_t & 0 & 0 & -Y_{11}B_t \\ \hline -\mathcal{Y}_{12}^\top A_t & 0 & 0 & -\mathcal{Y}_{12}^\top B_t \\ \hline 0 & 0 & 0 & 0 \\ \hline D_{t2}^\top P_r C_{t2} & 0 & 0 & \frac{1}{2}D_{t2}^\top P_r D_{t2} \end{array} \right) +$$

$$+ \begin{pmatrix} 0 & 0 & C_{t3}^\top \\ \hline I & 0 & 0 \\ \hline 0 & I & 0 \\ \hline 0 & 0 & D_{t3}^\top \end{pmatrix} \begin{pmatrix} \mathcal{A} & \mathcal{B}_1 & \mathcal{B}_2 \\ \mathcal{C}_1 & \mathcal{D}_{11} & \mathcal{D}_{12} \\ \mathcal{C}_2 & \mathcal{D}_{21} & \mathcal{D}_{22} \end{pmatrix} \left(\begin{array}{c|c|c} 0 & I & 0 & 0 \\ 0 & 0 & I & 0 \\ C_{t1} & 0 & 0 & D_{t1} \end{array} \right). \tag{5.13}$$

By exploiting (5.12), it is not hard to see that (5.9) can be written as

$$\mathrm{He}\begin{pmatrix} \hat{\mathcal{Y}}^{-1}\hat{\mathcal{A}} & \hat{\mathcal{Y}}^{-1}\hat{\mathcal{B}}_1 & \hat{\mathcal{Y}}^{-1}\hat{\mathcal{B}}_2 \\ 0 & 0 & 0 \\ \hat{\mathcal{C}}_2 & \hat{\mathcal{D}}_{21} & \hat{\mathcal{D}}_{22} \end{pmatrix} + \begin{pmatrix} \star \\ \star \\ \star \end{pmatrix}^\top P_e \begin{pmatrix} \hat{\mathcal{C}}_1 & \hat{\mathcal{D}}_{11} & \hat{\mathcal{D}}_{12} \\ 0 & I & 0 \end{pmatrix} \prec 0$$

which can be obviously expressed as

$$
\begin{pmatrix} \star \\ \star \\ \text{--} \\ \star \\ \star \\ \text{--} \\ \star \\ \star \end{pmatrix}^{\top}
\begin{pmatrix}
0 & \hat{y}^{-1} & \vline & 0 & \vline & 0 & 0 \\
\hat{y}^{-\top} & 0 & \vline & 0 & \vline & 0 & 0 \\
\hline
0 & 0 & \vline & P_{\mathrm{e}} & \vline & 0 & 0 \\
\hline
0 & 0 & \vline & 0 & \vline & 0 & I \\
0 & 0 & \vline & 0 & \vline & I & 0
\end{pmatrix}
\begin{pmatrix}
\hat{\mathcal{A}} & \hat{\mathcal{B}}_1 & \hat{\mathcal{B}}_2 \\
I & 0 & 0 \\
\hline
\hat{\mathcal{C}}_1 & \hat{\mathcal{D}}_{11} & \hat{\mathcal{D}}_{12} \\
0 & I & 0 \\
\hline
\hat{\mathcal{C}}_2 & \hat{\mathcal{D}}_{21} & \hat{\mathcal{D}}_{22} \\
0 & 0 & I
\end{pmatrix}
\prec 0. \tag{5.14}
$$

Although elementary, these re-arrangements make it possible to apply the dualization Lemma C.1. Therefore, (5.14) is equivalent to

$$
\begin{pmatrix} \star \\ \text{--} \\ \star \\ \star \\ \text{--} \\ \star \\ \star \end{pmatrix}^{\top}
\begin{pmatrix}
0 & \hat{y}^{\top} & \vline & 0 & \vline & 0 & 0 \\
\hat{y} & 0 & \vline & 0 & \vline & 0 & 0 \\
\hline
0 & 0 & \vline & P_{\mathrm{e}}^{-1} & \vline & 0 & 0 \\
\hline
0 & 0 & \vline & 0 & \vline & 0 & I \\
0 & 0 & \vline & 0 & \vline & I & 0
\end{pmatrix}
\begin{pmatrix}
I & 0 & 0 \\
-\hat{\mathcal{A}}^{\top} & -\hat{\mathcal{C}}_1^{\top} & -\hat{\mathcal{C}}_2^{\top} \\
\hline
0 & I & 0 \\
-\hat{\mathcal{B}}_1^{\top} & -\hat{\mathcal{D}}_{11}^{\top} & -\hat{\mathcal{D}}_{21}^{\top} \\
\hline
0 & 0 & I \\
-\hat{\mathcal{B}}_2^{\top} & -\hat{\mathcal{D}}_{12}^{\top} & -\hat{\mathcal{D}}_{22}^{\top}
\end{pmatrix}
\succ 0, \tag{5.15}
$$

which is in turn nothing but

$$
\mathrm{He}
\begin{pmatrix}
-\hat{y}\hat{\mathcal{A}}^{\top} & -\hat{y}\hat{\mathcal{C}}_1^{\top} & -\hat{y}\hat{\mathcal{C}}_2^{\top} \\
0 & 0 & 0 \\
-\hat{\mathcal{B}}_2^{\top} & -\hat{\mathcal{D}}_{12}^{\top} & -\hat{\mathcal{D}}_{22}^{\top}
\end{pmatrix}
+
\begin{pmatrix} \star \\ \star \end{pmatrix}^{\top}
P_{\mathrm{e}}^{-1}
\begin{pmatrix}
0 & I & 0 \\
-\hat{\mathcal{B}}_1^{\top} & -\hat{\mathcal{D}}_{11}^{\top} & -\hat{\mathcal{D}}_{21}^{\top}
\end{pmatrix}
\succ 0.
$$

By direct computation this is seen to be identical to (5.10) with $\mathcal{Y} = \mathcal{I}(\mathcal{X})$, $P_r \in \boldsymbol{P}_r$, $\tilde{P}_{\mathrm{e}} = P_{\mathrm{e}}^{-1} \in \tilde{\boldsymbol{P}}_{\mathrm{e}}$. Statement (2) follows from identical arguments. ∎

We emphasize that $\mathcal{X}_{22} \succ 0$ (or dually $\mathcal{Y}_{22} \succ 0$) characterizes the stability of only $\mathcal{A}$, which is the state-matrix of the controlled but unweighted system $\mathcal{H} = H_{\mathrm{aug}} \star K$ as defined in (5.8). This is the key point to allow for possibly unstable weights (i.e. unstable matrices A_{t}) and should be contrasted with standard approaches in which the entire matrix $\mathcal{X}$ (or dually $\mathcal{Y}$) is required to be positive definite in order to guarantee internal stability. Lemma 5.2, which generalizes [157, Lemma 1], turns out to be very useful for synthesis as seen next.

5.3.2 Synthesis

Controller synthesis requires to design K and the scheduling function $\hat{\Delta}_{\mathrm{c}}(\cdot)$ in order to render (5.9) satisfied for some $\mathcal{X}_{22} \succ 0$, $P_r \in \boldsymbol{P}_r$ and $P_{\mathrm{e}} \in \boldsymbol{P}_{\mathrm{e}}$. As formulated, this problem is non-convex. Analogously to Chapter 3 and 4, we follow the existing approaches for gain-scheduled controller synthesis in the LFT-framework and eliminate the realization matrices of K in the analysis inequality (5.9) by applying Lemma C.2 in Appendix C. In our context this involves the basis matrices

$$T_{\mathrm{p}} \;\; \text{of} \;\; \ker\left(\begin{array}{cccc} D_{32}C_{\mathrm{t}1} & C_3 & D_{31} & D_{32}D_{\mathrm{t}1} \end{array} \right),$$

$$T_{\mathrm{d}} \;\; \text{of} \;\; \ker\left(\begin{array}{cccc} D_{23}^{\top}C_{\mathrm{t}3} & B_3^{\top} & D_{13}^{\top} & D_{23}^{\top}D_{\mathrm{t}3} \end{array} \right),$$

whose rows are, if required, partitioned accordingly as

$$T_{\mathrm{p}} = \mathrm{col}(T_{\mathrm{p}1},T_{\mathrm{p}2},T_{\mathrm{p}3},T_{\mathrm{p}4}) \quad \text{and} \quad T_{\mathrm{d}} = \mathrm{col}(T_{\mathrm{d}1},T_{\mathrm{d}2},T_{\mathrm{d}3},T_{\mathrm{d}4}).$$

It is crucial to observe that, after elimination, the resulting conditions do not involve the complete multiplier P_{e} or its inverse $\tilde{P}_{\mathrm{e}}$ any more, but that they again can be expressed in terms of the set of primal and dual scalings $\boldsymbol{P}_{\mathrm{s}}$ and $\tilde{\boldsymbol{P}}_{\mathrm{s}}$ respectively as defined in Section 3.7. We are now ready to state the main result of this chapter.

Theorem 5.1. *The following statements are equivalent:*

1. There exists a controller scheduling function $\Delta_{\mathrm{c}}(\cdot)$ and matrices A_K, B_K, C_K, D_K, $\mathcal{X} = \mathcal{X}^{\top}$, $P_r \in \boldsymbol{P}_r$, $P_{\mathrm{e}} \in \boldsymbol{P}_{\mathrm{e}}$ for which (5.9) holds with $\mathcal{X}_{22} \succ 0$.
2. There exist X, Y, $P_r \in \boldsymbol{P}_r$, $P_{\mathrm{s}} \in \boldsymbol{P}_{\mathrm{s}}$, $\tilde{P}_{\mathrm{s}} \in \tilde{\boldsymbol{P}}_{\mathrm{s}}$ for which the following LMIs hold:

$$T_{\mathrm{p}}^{\top} M_{\mathrm{p}} T_{\mathrm{p}} \prec 0, \tag{5.16a}$$

$$T_{\mathrm{d}}^{\top} M_{\mathrm{d}} T_{\mathrm{d}} \succ 0, \tag{5.16b}$$

$$M_{\mathrm{c}} \succcurlyeq 0. \tag{5.16c}$$

Here M_{p}, M_{d} and M_{c} are respectively given by

$$M_{\mathrm{p}} = \mathrm{He}\left[\begin{pmatrix} X_{11} & X_{12} & C_{\mathrm{t}3}^{\top} \\ X_{12}^{\top} & X_{22} & 0 \\ \hline 0 & 0 & 0 \\ 0 & 0 & D_{\mathrm{t}3}^{\top} \end{pmatrix} \begin{pmatrix} A_{\mathrm{t}} & 0 & 0 & B_{\mathrm{t}} \\ B_2C_{\mathrm{t}1} & A & B_1 & B_2D_{\mathrm{t}1} \\ \hline D_{22}C_{\mathrm{t}1} & C_2 & D_{21} & D_{22}D_{\mathrm{t}1} \end{pmatrix} \right]$$
$$+ \begin{pmatrix} \star \\ \star \\ \hline \star \end{pmatrix}^{\top} \begin{pmatrix} P_{\mathrm{s}} & 0 \\ 0 & P_r \end{pmatrix} \begin{pmatrix} D_{12}C_{\mathrm{t}1} & C_1 & D_{11} & D_{12}D_{\mathrm{t}1} \\ 0 & 0 & I & 0 \\ \hline C_{\mathrm{t}2} & 0 & 0 & D_{\mathrm{t}2} \end{pmatrix},$$

$$M_{\mathrm{d}} = \mathrm{He}\left[\left(\begin{array}{ccc} Y_{11} & Y_{12} & C_{\mathrm{t}1}^{\top} \\ Y_{12}^{\top} & Y_{22} & 0 \\ \hline 0 & 0 & 0 \\ 0 & 0 & D_{\mathrm{t}1}^{\top} \end{array}\right)\left(\begin{array}{cccc} A_{\mathrm{t}} & 0 & 0 & B_{\mathrm{t}} \\ -C_2^{\top}C_{\mathrm{t}3} & -A^{\top} & -C_1^{\top} & -C_2^{\top}D_{\mathrm{t}3} \\ \hline -D_{22}^{\top}C_{\mathrm{t}3} & -B_2^{\top} & -D_{12}^{\top} & -D_{22}^{\top}D_{\mathrm{t}3} \end{array}\right)\right]$$

$$+\left(\begin{array}{c} \star \\ \star \\ \hline \star \end{array}\right)^{\top}\left(\begin{array}{cc} \tilde{P}_{\mathrm{s}} & 0 \\ 0 & -P_r \end{array}\right)\left(\begin{array}{cccc} 0 & 0 & I & 0 \\ -D_{21}^{\top}C_{\mathrm{t}3} & -B_1^{\top} & -D_{11}^{\top} & -D_{21}^{\top}D_{\mathrm{t}3} \\ \hline C_{\mathrm{t}2} & 0 & 0 & D_{\mathrm{t}2} \end{array}\right)$$

and

$$M_{\mathrm{c}} = \left(\begin{array}{ccc} Y_{11}+X_{11} & Y_{12} & -X_{12} \\ Y_{12}^{\top} & Y_{22} & I \\ -X_{12}^{\top} & I & X_{22} \end{array}\right).$$

Once the LMIs (5.16) in the variables X, Y, P_r, P_{s} and $\tilde{P}_{\mathrm{s}}$ are feasible, it is sketched in the proof how to construct some K of McMillan degree $n_H + n_{\mathrm{t}}$ and a scheduling function $\hat{\Delta}_{\mathrm{c}}(\cdot)$ such that (5.9) is feasible with $\mathcal{X}_{22} \succ 0$. Then K internally stabilizes H_{aug} and (5.4) is satisfied, which implies that the gain-scheduled controller achieves the desired goals as formulated in Problem 5.1.

In summary we have shown how Problem 5.1 can be handled by solving a standard semi-definite program. The crux of the proof relies on Lemma 5.2 and enables us to eliminate the realization matrices of K. This leads to the primal and dual solvability conditions (5.16a) and (5.16b) respectively, with the solutions being coupled by (5.16c). Observe that (5.16a) and (5.16b) share their structure with the analysis inequalities (5.9) and (5.10). Further note that the coupling condition (5.16c) also emerged in [98, 195, 155, 170, 157] for specialized versions of Theorem 5.1. As a new ingredient and in contrast to [98, 195], X_{11} and $-Y_{11}$ need not be positive definite. This reflects the requirement that K internally stabilizes H_{aug} without incorporating any stability properties of the weights T_w, T_r, T_z. For $T_z := I$, $T_w := I$ and $T_r := I$, we have $X := X_{22}$, $Y := Y_{22}$ and (5.16c) reduces to the standard inequality

$$\left(\begin{array}{cc} Y_{22} & I \\ I & X_{22} \end{array}\right) \succcurlyeq 0$$

as well-known from the literature; see [170, 157] for more detailed remarks. In the same vein, the results seamlessly specialize e.g. to the nominal- and nominal gain-scheduling synthesis results of [41, 48, 72, 121, 174, 150] and as discussed in Section 4.6 and 4.7 respectively. The proposed setting offers the possibility to formulate the synthesis result in a rather insightful fashion and to cover a large multitude of concrete design problems as will be exposed in the sequel.

Remark 5.2. If the scheduling block Δ_s is composed of diagonally repeated scheduling parameters, one can replace $\boldsymbol{P}_e$ in Problem 5.1 by other multiplier classes which are computationally less demanding but possibly more conservative. This encompasses the D- and DG-scalings, as considered in [121] and [174], and in accordance with the IQC-multipliers of Section 2.6.1 and 2.6.2.1 respectively. Also note that $\boldsymbol{P}_e$ consists of static (frequency independent) multipliers. For full-block dynamic (frequency dependent) IQC-multipliers the gain-scheduling synthesis problem is still open, while results for dynamic D and D/G-scalings have been recently obtained in [167, 157, 158, 161].

Proof. In order to show that (1) implies (2), suppose that there exist a scheduling function $\hat{\Delta}_c(\cdot)$ and A_K, B_K, C_K, D_K, $\mathcal{X} = \mathcal{X}^\top$, $P_r \in \boldsymbol{P}_r$, $P_e \in \boldsymbol{P}_e$ for which (5.9) holds with $\mathcal{X}_{22} \succ 0$. Then, by Lemma 5.2, $\mathcal{A}$ is Hurwitz and we can infer that the dual matrix inequality (5.10) is satisfied with $\mathcal{Y} = \mathcal{I}(\mathcal{X})$, $\mathcal{Y}_{22} \succ 0$, $P_r \in \boldsymbol{P}_r$, $\tilde{P}_e = P_e^{-1} \in \tilde{\boldsymbol{P}}_e$.

Let us now recall from the proof of Lemma 5.2 that the inequalities (5.9) and (5.10) are identical to (5.14) and (5.15) respectively. In view of (5.8), we can hence eliminate from these inequalities the controller realization matrices A_K, B_K, C_K, D_K by applying Lemma C.2 in Appendix C. For this purpose we note that

$$T_{\text{pe}} = \text{col}(T_{\text{p}1}, T_{\text{p}2}, 0_{n_K \times \bullet}, T_{\text{p}3}, 0_{p_c \times \bullet}, T_{\text{p}4})$$

and

$$T_{\text{de}} = \text{col}(T_{\text{d}1}, T_{\text{d}2}, 0_{n_K \times \bullet}, T_{\text{d}3}, 0_{q_c \times \bullet}, T_{\text{d}4})$$

are basis-matrices of the kernels of

$$\begin{pmatrix} 0 & I & 0 & 0 & 0 \\ C_3 & 0 & D_{31} & 0 & D_{32} \\ 0 & 0 & 0 & I & 0 \end{pmatrix} \begin{pmatrix} 0 & I & 0 & 0 \\ 0 & 0 & I & 0 \\ C_{t1} & 0 & 0 & D_{t1} \end{pmatrix} \quad \text{and} \quad \begin{pmatrix} 0 & I & 0 & 0 & 0 \\ B_3^\top & 0 & D_{13}^\top & 0 & D_{23}^\top \\ 0 & 0 & 0 & I & 0 \end{pmatrix} \begin{pmatrix} 0 & I & 0 & 0 \\ 0 & 0 & I & 0 \\ C_{t3} & 0 & 0 & D_{t3} \end{pmatrix}$$

respectively. Then (5.14) and (5.15) clearly imply

$$T_{\text{pe}}^\top (5.14) T_{\text{pe}} \prec 0 \quad \text{and} \quad T_{\text{de}}^\top (5.15) T_{\text{de}} \succ 0. \tag{5.17}$$

If we partition

$$P_e = \begin{pmatrix} Q_e & S_e \\ S_e^\top & R_e \end{pmatrix} = \begin{pmatrix} Q_s & \star & S_s & \star \\ \star & \star & \star & \star \\ S_s^\top & \star & R_s & \star \\ \star & \star & \star & \star \end{pmatrix} \quad \text{and} \quad \tilde{P}_e = \begin{pmatrix} \tilde{Q}_e & \tilde{S}_e \\ \tilde{S}_e^\top & \tilde{R}_e \end{pmatrix} = \begin{pmatrix} \tilde{Q}_s & \star & \tilde{S}_s & \star \\ \star & \star & \star & \star \\ \tilde{S}_s^\top & \star & \tilde{R}_s & \star \\ \star & \star & \star & \star \end{pmatrix} \tag{5.18}$$

according to $n_{q_s} + n_{p_s} + n_{q_c} + n_{p_c}$ and notice the structure of the first summand on the right in (5.8), these two inequalities simplify into (5.16a) and (5.16b) for

$$P_s := \begin{pmatrix} Q_s & S_s \\ S_s^\top & R_s \end{pmatrix} \quad \text{and} \quad \tilde{P}_s := \begin{pmatrix} \tilde{Q}_s & \tilde{S}_s \\ \tilde{S}_s^\top & \tilde{R}_s \end{pmatrix}. \tag{5.19}$$

As in Chapter 3 and 4, we infer again that $P_\mathrm{c} \in \boldsymbol{P}_\mathrm{c}$ and $\tilde{P}_\mathrm{c} \in \tilde{\boldsymbol{P}}_\mathrm{c}$ imply that $P_\mathrm{s} \in \boldsymbol{P}_\mathrm{s}$ and $\tilde{P}_\mathrm{s} \in \tilde{\boldsymbol{P}}_\mathrm{s}$ respectively.

The first part of the proof is finished by showing (5.16c). For this purpose we make use of the fine-partitions of the matrices $\mathcal{X}$ and $\mathcal{Y}$ in (5.11). Since $\mathcal{X}_{22} \succ 0$ we can infer that $X_{22} \succ 0$, $\tilde{X}_{33} := X_{33} - X_{23}^\top X_{22}^{-1} X_{23} \succ 0$ and

$$\mathcal{X}_{22}^{-1} = \mathrm{diag}(-X_{22}^{-1}, 0) + (\star)^\top \tilde{X}_{33}^{-1} \left(-X_{23}^\top X_{22}^{-1} \quad I \right) \succ 0.$$

Therefore $\mathcal{X}_{22}^{-1} \succcurlyeq \mathrm{diag}(-X_{22}^{-1}, 0)$ and hence

$$(\star)^\top \mathcal{X}_{22}^{-1} \left(-\mathcal{X}_{12}^\top \quad I \right) \succcurlyeq (\star)^\top X_{22}^{-1} \left(-X_{12}^\top \quad I \quad 0 \right). \tag{5.20}$$

Due to $\mathcal{Y} = \mathcal{I}(\mathcal{X})$ we have

$$\begin{pmatrix} Y_{11} + X_{11} & \mathcal{Y}_{12} \\ \mathcal{Y}_{12}^\top & \mathcal{Y}_{22} \end{pmatrix} = \begin{pmatrix} -\mathcal{X}_{12} \\ I \end{pmatrix} \mathcal{X}_{22}^{-1} \left(-\mathcal{X}_{12}^\top \quad I \right).$$

Hence, by exploiting (5.20) we conclude

$$\begin{pmatrix} Y_{11} + X_{11} & Y_{12} \\ Y_{12}^\top & Y_{22} \end{pmatrix} \succcurlyeq \begin{pmatrix} -X_{12} \\ I \end{pmatrix} X_{22}^{-1} \left(-X_{12}^\top \quad I \right),$$

which leads to (5.16c) by taking the Schur-complement.

For proving (2) $\Rightarrow$ (1), suppose that there exist matrices X, Y, $P_r \in \boldsymbol{P}_r$, $P_\mathrm{s} \in \boldsymbol{P}_\mathrm{s}$ and $\tilde{P}_\mathrm{s} \in \tilde{\boldsymbol{P}}_\mathrm{s}$ for which (5.16a) and (5.16b) hold, with (5.16c) being strict (possibly after a perturbation). After eliminating the off-diagonal terms in the second block row/column of the left-hand side by congruence, we can proceed with

$$\begin{pmatrix} X_{11} + Y_{11} - Y_{12} Y_{22}^{-1} Y_{12}^\top & -X_{12} - Y_{12} Y_{22}^{-1} \\ -X_{12}^\top - Y_{22}^{-1} Y_{12}^\top & X_{22} - Y_{22}^{-1} \end{pmatrix} \succ 0. \tag{5.21}$$

For the given X and Y of dimension $n_\mathrm{t} + n_H$, the first step of the proof consists in finding matrices X_{13}, X_{23} and X_{33} with column dimension n_K such that

$$\begin{pmatrix} Y_{11} + X_{11} & Y_{12} & \star \\ Y_{12}^\top & Y_{22} & \star \\ \star & \star & \star \end{pmatrix} = \begin{pmatrix} -X_{12} & -X_{13} \\ I & 0 \\ 0 & I \end{pmatrix} \begin{pmatrix} X_{22} & X_{23} \\ X_{23}^\top & X_{33} \end{pmatrix}^{-1} \begin{pmatrix} \star \\ \star \\ \star \end{pmatrix}^\top \tag{5.22}$$

and

$$\mathcal{X}_{22} = \begin{pmatrix} X_{22} & X_{23} \\ X_{23}^\top & X_{33} \end{pmatrix} \succ 0. \tag{5.23}$$

One such choice with $n_K := n_H + n_l$ is

$$
\begin{pmatrix} X_{13} \\ \hline X_{23} \\ \hline X_{33} \end{pmatrix} := \left(\begin{array}{c|c} X_{12} + Y_{12} Y_{22}^{-1} & I \\ \hline X_{22} - Y_{22}^{-1} & 0 \\ \hline X_{22} - Y_{22}^{-1} & 0 \\ \hline 0 & X_{\mathrm{e}}^{-1} \end{array} \right)
$$

where

$$
X_{\mathrm{e}} := X_{11} + Y_{11} - Y_{12} Y_{22}^{-1} Y_{12}^{\top} - \left(X_{12} + Y_{12} Y_{22}^{-1} \right) \left(X_{22} - Y_{22}^{-1} \right)^{-1} (\star)^{\top};
$$

in view of (5.21), we stress that all inverses are well-defined and that X_{e} is positive definite (Schur). We infer

$$
\mathcal{X}_{22} = \left(\begin{array}{c|cc} X_{22} & X_{22} - Y_{22}^{-1} & 0 \\ \hline X_{22} - Y_{22}^{-1} & X_{22} - Y_{22}^{-1} & 0 \\ 0 & 0 & X_{\mathrm{e}}^{-1} \end{array} \right) \succ 0
$$

which is (5.23). (5.22) is verified by direct computation.

For the given scalings (5.19), in the second step of the proof we follow again [150] in order to extend P_{s} and $\tilde{P}_{\mathrm{s}}$ to P_{e} and $\tilde{P}_{\mathrm{e}}^{-1}$ as in (5.18) such that $P_{\mathrm{e}} = \tilde{P}_{\mathrm{e}}^{-1}$. This also yields a quadratic scheduling function $\hat{\Delta}_{\mathrm{c}} : \mathbb{R}^{n_{p_{\mathrm{s}}} \times n_{q_{\mathrm{s}}}} \to \mathbb{R}^{n_{p_{\mathrm{c}}} \times n_{q_{\mathrm{c}}}}$ for which P_{e} is contained in the corresponding set of extended multipliers $\boldsymbol{P}_{\mathrm{e}}$.

In the final step it remains to construct the LTI part K of the gain-scheduled controller. If we define $\hat{\mathcal{Y}}$ through (5.12), we infer as in the first part of the proof that (5.16a), (5.16b) are identical to (5.17), respectively. We can hence apply Lemma C.2 (Appendix C) to (5.14) to ensure the existence of A_K, B_K, C_K and D_K which render (5.14) satisfied. Completely analogously to the proof of Theorem 4.2 in Section 4.5 these matrices can be obtained by solving a genuine LMI, after taking the Schur-complement. Alternatively, one can as well follow the standard algebraic procedure described in [48]. For further details we refer the reader e.g. to [48, 49]. ∎

5.4 Robust gain-scheduling control for systems without control-channel uncertainties

Let us now discuss how to handle the synthesis problem for the interconnection in Figure 5.4. Here $w \to z$ is the performance channel, $u \to y$ is the control channel and $\Delta_{\mathrm{s}} \in \boldsymbol{\Delta}_{\mathrm{s}}$ is the scheduling block. In addition, the uncertainty blocks $\Delta_{\mathrm{u}} : \mathscr{L}_2^{n_{q_{\mathrm{u}}}} \to \mathscr{L}_2^{n_{p_{\mathrm{u}}}}$

are causal and bounded operators which belong to given sets $\boldsymbol{\Delta}_\mathrm{u}$, $\mathrm{u} \in \{1,2\}$, which satisfy Assumption 2.1. Note that this encompasses the possibility $\Delta_1 = \Delta_2 = \Delta$ such that Σ_{zw}, Σ_{zu} and Σ_{yw} all are affected by the very same uncertainty. Furthermore, we suppose that $\Sigma(\Delta_\mathrm{s}, \Delta_1, \Delta_2)$ admits the linear fractional representation

$$
\begin{pmatrix} \Delta_\mathrm{s} & 0 & 0 \\ 0 & \Delta_1 & 0 \\ 0 & 0 & \Delta_2 \end{pmatrix} \star \underbrace{\left(\begin{array}{ccc|cc} G_{\mathrm{ss}} & G_{\mathrm{s}1} & 0 & G_{\mathrm{s}w} & G_{\mathrm{s}u} \\ 0 & G_{11} & 0 & G_{1w} & 0 \\ G_{2\mathrm{s}} & G_{21} & G_{22} & G_{2w} & G_{2u} \\ \hline G_{z\mathrm{s}} & G_{z1} & G_{z2} & G_{zw} & G_{zu} \\ G_{y\mathrm{s}} & G_{y1} & 0 & G_{yw} & G_{yu} \end{array} \right)}_{G} \tag{5.24}
$$

with some (structured) transfer matrix G, which is well-posed for all $\Delta_\mathrm{s} \in \boldsymbol{\Delta}_\mathrm{s}$ and for all $\Delta_\mathrm{u} \in \boldsymbol{\Delta}_\mathrm{u}$, $\mathrm{u} \in \{1,2\}$.

Problem 5.2. Given the transfer matrix G as in (5.24) with stable G_{11}, G_{1w}, G_{22}, G_{z2} and the uncertainty sets $\boldsymbol{\Delta}_\mathrm{s}$, $\boldsymbol{\Delta}_1$, $\boldsymbol{\Delta}_2$, design a robust gain-scheduled controller $K_{\mathrm{rgs}}(\Delta_\mathrm{s}) = K \star \hat{\Delta}_\mathrm{c}(\Delta_\mathrm{s})$ such that, for all $\Delta_\mathrm{s} \in \boldsymbol{\Delta}_\mathrm{s}$ and for all $\Delta_1 \in \boldsymbol{\Delta}_1$, $\Delta_2 \in \boldsymbol{\Delta}_2$, the closed-loop system $\Sigma(\Delta_\mathrm{s}, \Delta_1, \Delta_2) \star K_{\mathrm{rgs}}(\Delta_\mathrm{s})$ is stabilized and its induced $\mathscr{L}_2$-gain is smaller than some a priori given $\gamma > 0$.

Just for notational simplicity we confine the discussion to worst-case $\mathscr{L}_2$-gain performance to keep the exposition simple. The extension to more general quadratic performance costs is straightforward [155, 170].

Remark 5.3. For linear uncertainties Δ_1 and Δ_2, (5.24) structurally boils down to the system $\Sigma(\Delta_\mathrm{s}, \Delta_1, \Delta_2)$ being a trilinear function

$$
\begin{pmatrix} G_{zw} & G_{zu} \\ G_{yw} & G_{yu} \end{pmatrix} + \begin{pmatrix} G_{z\mathrm{s}} & G_{z1} & G_{z2} \\ G_{y\mathrm{s}} & G_{y1} & 0 \end{pmatrix} \tilde{G}_{22} \times
$$

$$
\times \begin{pmatrix} I & 0 & 0 \\ 0 & I & 0 \\ G_{2\mathrm{s}} & G_{21} & I \end{pmatrix} \tilde{G}_{\mathrm{ss}} \begin{pmatrix} I & G_{\mathrm{s}1} & 0 \\ 0 & I & 0 \\ 0 & 0 & I \end{pmatrix} \tilde{G}_{11} \begin{pmatrix} G_{\mathrm{s}w} & G_{\mathrm{s}u} \\ G_{1w} & 0 \\ G_{2w} & G_{2u} \end{pmatrix}
$$

in the blocks

$$
\begin{aligned} \tilde{G}_{\mathrm{ss}} &= \mathrm{diag}\left(\Delta_\mathrm{s}(I - G_{\mathrm{ss}}\Delta_\mathrm{s})^{-1}, I, I \right), \\ \tilde{G}_{11} &= \mathrm{diag}\left(I, \Delta_1(I - G_{11}\Delta_1)^{-1}, I \right), \\ \tilde{G}_{22} &= \mathrm{diag}\left(I, I, \Delta_2(I - G_{22}\Delta_2)^{-1} \right). \end{aligned}
$$

Although seemingly restrictive, we stress that this description encompasses and generalizes other structures considered in the literature. For example, by canceling the scheduling-channel (i.e. the first block row/column of G), we retrieve the configuration for robust controller synthesis without control channel uncertainties as studied in [155, 170] and by excluding the uncertainty channels (i.e. the second and third block rows/columns of G), we obtain the problem class of nominal gain-scheduled controller synthesis as considered in [121, 174, 150] and Section 3.3. We refer to Section 5.5 for an illustrative overview of configurations that have and have not been covered so far, which is a further account of the flexibility of our framework.

Remark 5.4. Let us also come back to Remark 5.1 and observe the crucial difference between Problem 5.2 and the one studied in Chapter 4 for robust gain-scheduled estimation. If adapting the plant (4.4) in Section 4.3 to the notation of this chapter, we obtain the linear fractional representation

$$
\begin{pmatrix} \Delta_{\mathrm{s}} & 0 \\ 0 & \Delta_1 \end{pmatrix} \star
\left(\begin{array}{cc|cc}
G_{\mathrm{ss}} & G_{\mathrm{s}1} & G_{\mathrm{s}w} & 0 \\
G_{1\mathrm{s}} & G_{11} & G_{1w} & 0 \\ \hline
G_{z\mathrm{s}} & G_{z1} & G_{zw} & -I_{n_z} \\
G_{y\mathrm{s}} & G_{y1} & G_{yw} & 0
\end{array} \right).
$$

We conclude that the estimation problem as considered in Chapter 4 allows for a fully general linear fractional dependency on the blocks Δ_{s} and Δ_1. This is certainly not the case for Problem 5.2 since we require $G_{1\mathrm{s}}$ to be zero.

Due to the structure of (5.24) and with $\Delta_{\mathrm{e}}(\cdot)$, we can describe the closed-loop system $\Sigma(\Delta_{\mathrm{s}}, \Delta_1, \Delta_2) \star K_{\mathrm{rgs}}(\Delta_{\mathrm{s}})$ as

$$
\begin{pmatrix} \Delta_{\mathrm{e}}(\Delta_{\mathrm{s}}) & 0 & 0 \\ 0 & \Delta_1 & 0 \\ 0 & 0 & \Delta_2 \end{pmatrix} \star
\underbrace{\left(\begin{array}{ccc|c}
\mathcal{G}_{\mathrm{ee}} & \mathcal{G}_{\mathrm{e}1} & 0 & \mathcal{G}_{\mathrm{e}w} \\
0 & G_{11} & 0 & G_{1w} \\
\mathcal{G}_{2\mathrm{e}} & \mathcal{G}_{21} & G_{22} & \mathcal{G}_{2w} \\ \hline
\mathcal{G}_{z\mathrm{e}} & \mathcal{G}_{z1} & G_{z2} & \mathcal{G}_{zw}
\end{array} \right)}_{\mathcal{G}}. \tag{5.25}
$$

Here the calligraphic matrices denote those blocks that explicitly depend on K and are given by

$$
\begin{pmatrix} \mathcal{G}_{\mathrm{ee}} & \mathcal{G}_{\mathrm{e}1} & \mathcal{G}_{\mathrm{e}w} \\ \mathcal{G}_{2\mathrm{e}} & \mathcal{G}_{21} & \mathcal{G}_{2w} \\ \mathcal{G}_{z\mathrm{e}} & \mathcal{G}_{z1} & \mathcal{G}_{zw} \end{pmatrix} =
\underbrace{\left(\begin{array}{cc|ccc|cc}
G_{\mathrm{ss}} & 0 & G_{\mathrm{s}1} & G_{\mathrm{s}w} & G_{\mathrm{s}u} & 0 \\
0 & 0 & 0 & 0 & 0 & I_{n_{q_c}} \\ \hline
G_{2\mathrm{s}} & 0 & G_{21} & G_{2w} & G_{2u} & 0 \\
G_{z\mathrm{s}} & 0 & G_{z1} & G_{zw} & G_{zu} & 0 \\ \hline
G_{y\mathrm{s}} & 0 & G_{y1} & G_{yw} & G_{yu} & 0 \\
0 & I_{n_{p_c}} & 0 & 0 & 0 & 0
\end{array} \right)}_{G_{\mathrm{aug}}} \star
\underbrace{\begin{pmatrix} K_{11} & K_{12} \\ K_{21} & K_{22} \end{pmatrix}}_{K}.
$$

The particular structure of $\mathcal{G}$ and the way it interacts with the LTI part of the controller K as well as with the scheduling and uncertainty blocks Δ_{s}, Δ_1 and Δ_2 will be of crucial importance in the sequel. Note that the limitations placed on the class of plants in Problem 5.2 stem from the fact that the control channel is not affected by uncertainties. Consequently, the second block row and the third block column of $\mathcal{G}$ are independent of K and the existence of an internally stabilizing K requires G_{11}, G_{1w}, G_{22} and G_{z2} to be stable.

To continue, let $\boldsymbol{\Pi}_{\mathrm{u}}$ denote classes of valid IQC-multipliers for the uncertainty blocks Δ_{u}, $\mathrm{u} \in \{1,2\}$. This means that any $\Pi_{\mathrm{u}}^* = \Pi_{\mathrm{u}} \in \boldsymbol{\Pi}_{\mathrm{u}} \subseteq \mathscr{RL}_{\infty}^{\bullet \times \bullet}$ satisfies the IQC

$$IQC\left(\Pi_{\mathrm{u}}, q_{\mathrm{u}}, \Delta_{\mathrm{u}}(q_{\mathrm{u}})\right) \geq 0 \quad \forall q_{\mathrm{u}} \in \mathscr{L}_2^{n_{q_{\mathrm{u}}}} \tag{5.26}$$

and for all $\Delta_{\mathrm{u}} \in \boldsymbol{\Delta}_{\mathrm{u}}$, $\mathrm{u} \in \{1,2\}$. Furthermore, if we also define $\boldsymbol{P}_{\mathrm{e}}$ and P_{γ} as previously, then it is possible to state the following result.

Lemma 5.3. *For a given well-posed controller $K_{\mathrm{rgs}}(\Delta_{\mathrm{s}}) = K \star \hat{\Delta}_{\mathrm{c}}(\Delta_{\mathrm{s}})$, assume that K internally stabilizes G_{aug} and that there exist $P_{\mathrm{e}} \in \boldsymbol{P}_{\mathrm{e}}$, $\Pi_1 \in \boldsymbol{\Pi}_1$, $\Pi_2 \in \boldsymbol{\Pi}_2$ for which the following FDI is satisfied:*

$$\begin{pmatrix} \star \\ \star \\ \star \\ \star \\ \star \\ \star \\ \star \end{pmatrix}^* \begin{pmatrix} P_{\mathrm{e}} & 0 & 0 & 0 \\ 0 & \Pi_1 & 0 & 0 \\ 0 & 0 & \Pi_2 & 0 \\ 0 & 0 & 0 & P_{\gamma} \end{pmatrix} \begin{pmatrix} \mathcal{G}_{\mathrm{ee}} & \mathcal{G}_{\mathrm{e}1} & 0 & \mathcal{G}_{\mathrm{e}w} \\ I & 0 & 0 & 0 \\ 0 & G_{11} & 0 & G_{1w} \\ 0 & I & 0 & 0 \\ \mathcal{G}_{2\mathrm{e}} & \mathcal{G}_{21} & G_{22} & \mathcal{G}_{2w} \\ 0 & 0 & I & 0 \\ \mathcal{G}_{z\mathrm{e}} & \mathcal{G}_{z1} & G_{z2} & \mathcal{G}_{zw} \\ 0 & 0 & 0 & I \end{pmatrix} \prec 0. \tag{5.27}$$

Then, for all $\Delta_{\mathrm{s}} \in \boldsymbol{\Delta}_{\mathrm{s}}$, $\Delta_1 \in \boldsymbol{\Delta}_1$ and $\Delta_2 \in \boldsymbol{\Delta}_2$, the closed-loop system (5.25) is well-posed and stable, and the induced $\mathscr{L}_2$-gain of $w \to z$ is less than $\gamma > 0$.

Proof. Let us actually prove well-posedness of (5.25) for all $\Delta_{\mathrm{s}} \in \boldsymbol{\Delta}_{\mathrm{s}}$, $\Delta_1 \in \boldsymbol{\Delta}_1$, $\Delta_2 \in \boldsymbol{\Delta}_2$ and if replacing $\Delta_{\mathrm{e}}(\Delta_{\mathrm{s}})$ with $\tau \Delta_{\mathrm{e}}(\Delta_{\mathrm{s}})$ for $\tau \in [0,1]$. Then the statements about robust stability and performance in Lemma 5.3 follow directly from the IQC-theorem (See Section 2.3).

For $\Delta_{\mathrm{s}} \in \boldsymbol{\Delta}_{\mathrm{s}}$, $\Delta_1 \in \boldsymbol{\Delta}_1$, $\Delta_2 \in \boldsymbol{\Delta}_2$, well-posedness of (5.24) means that for every disturbance $\mathrm{col}(\mu_{\mathrm{s}}, \mu_1, \mu_2) \in \mathscr{L}_{2e}^{\bullet}$ there exists a unique response $\mathrm{col}(q_{\mathrm{s}}, q_1, q_2) \in \mathscr{L}_{2e}^{\bullet}$ with causal dependence on the disturbance satisfying

$$\begin{pmatrix} q_{\mathrm{s}} \\ q_1 \\ q_2 \end{pmatrix} = \begin{pmatrix} G_{\mathrm{ss}} & G_{\mathrm{s}1} & 0 \\ 0 & G_{11} & 0 \\ G_{2\mathrm{s}} & G_{21} & G_{22} \end{pmatrix} \begin{pmatrix} \Delta_{\mathrm{s}} q_{\mathrm{s}} \\ \Delta_1(q_1) \\ \Delta_2(q_2) \end{pmatrix} + \begin{pmatrix} \mu_{\mathrm{s}} \\ \mu_1 \\ \mu_2 \end{pmatrix}.$$

This equation can be expressed as

$$(I - G_{11}\Delta_1)(q_1) = \mu_1,$$

$$(I - G_{ss}\Delta_s)q_s - G_{s1}\Delta_1(q_1) = \mu_s,$$

$$(I - G_{22}\Delta_2)(q_2) - G_{2s}\Delta_s q_e - G_{21}\Delta_1(q_1) = \mu_2.$$

We can sequentially conclude that (5.24) is well-posed iff $I - G_{11}\Delta_1$, $I - G_{ss}\Delta_s$ and $I - G_{22}\Delta_2$ have causal inverses. In compete analogy, (5.25) is well-posed iff $I - \mathcal{G}_{ee}\Delta_e(\Delta_s)$, $I - G_{11}\Delta_1$ and $I - G_{22}\Delta_2$ have causal inverses.

If fixing $\Delta_s \in \boldsymbol{\Delta}_s$, $\Delta_1 \in \boldsymbol{\Delta}_1$ and $\Delta_2 \in \boldsymbol{\Delta}_2$ and if the FDI (5.27) holds, let us now prove well-posedness of (5.25) for $\Delta_e(\Delta_s)$ replaced by $\tau\Delta_e(\Delta_s)$ and any $\tau \in [0, 1]$. Since (5.24) was assumed to be well-posed, we conclude that $I - G_{11}\Delta_1$ and $I - G_{22}\Delta_2$ have causal inverses. To show the same for $I - \tau\mathcal{G}_{ee}\Delta_e(\Delta_s)$, it suffices to prove that the time-varying matrix $I - \tau\mathcal{G}_{ee}(\infty)\Delta_e(\Delta_s)$ is invertible. This is indeed guaranteed, since the FDI (5.27) and $P_e \in \boldsymbol{P}_e$ imply the validity of

$$\begin{pmatrix} \mathcal{G}_{ee}(\infty) \\ I \end{pmatrix}^* P_e \begin{pmatrix} \mathcal{G}_{ee}(\infty) \\ I \end{pmatrix} \prec 0 \quad \text{and} \quad \begin{pmatrix} I \\ \tau\Delta_e(\Delta_s) \end{pmatrix}^\top P_e \begin{pmatrix} I \\ \tau\Delta_e(\Delta_s) \end{pmatrix} \succ 0.$$

This completes the proof. ∎

Now note that $0 \in \boldsymbol{\Delta}_2$ and the FDI (5.27) imply

$$\begin{pmatrix} G_{22} \\ I \end{pmatrix}^* \Pi_2 \begin{pmatrix} G_{22} \\ I \end{pmatrix} \prec 0 \quad \text{and} \quad \begin{pmatrix} I \\ 0 \end{pmatrix}^* \Pi_2 \begin{pmatrix} I \\ 0 \end{pmatrix} \succcurlyeq 0.$$

Hence, we can conclude that the sum of $\mathrm{im}(\mathrm{col}(G_{22}(i\omega), I))$ and $\mathrm{im}(\mathrm{col}(I, 0))$ is direct and equals $\mathbb{C}^{(n_{q2}+n_{p2})\times(n_{q2}+n_{p2})}$ for all $\omega \in [0, \infty]$. If Π_2 is invertible, these are equivalent to

$$\begin{pmatrix} I \\ -G_{22}^* \end{pmatrix}^* \Pi_2^{-1} \begin{pmatrix} I \\ -G_{22}^* \end{pmatrix} \succ 0 \quad \text{and} \quad \begin{pmatrix} 0 \\ I \end{pmatrix}^* \Pi_2^{-1} \begin{pmatrix} 0 \\ I \end{pmatrix} \prec 0.$$

by Lemma C.1 (see Appendix C). This motivates the assumptions on Π_2 in the following reformulation of (5.27) [155, 170].

Lemma 5.4. *Let $\Pi_2 \in \boldsymbol{\Pi}_2$ be nonsingular and the right-lower block of Π_2^{-1} (in the partition $\mathrm{col}(I, -G_{22}^*)$) be negative semi-definite. Then (5.27) is equivalent to*

$$
\text{He}
\begin{pmatrix}
0 & | & 0 & 0 & 0 & 0 \\
\hline
0 & | & 0 & 0 & 0 & 0 \\
0 & | & 0 & -\frac{\gamma}{2}I & 0 & 0 \\
\mathcal{G}_{2e} & | & \mathcal{G}_{21} & \mathcal{G}_{2w} & 0 & 0 \\
\mathcal{G}_{ze} & | & \mathcal{G}_{z1} & \mathcal{G}_{zw} & 0 & -\frac{\gamma}{2}I
\end{pmatrix}
+
\begin{pmatrix} \star \\ \star \end{pmatrix}^{*}
P_e
\begin{pmatrix}
\mathcal{G}_{ee} & | & \mathcal{G}_{e1} & \mathcal{G}_{ew} & 0 & 0 \\
\hline
I & | & 0 & 0 & 0 & 0
\end{pmatrix}
+
$$

$$
+
\begin{pmatrix} \star \\ \hline \star \\ \star \end{pmatrix}^{*}
\begin{pmatrix} \Pi_1 & 0 \\ 0 & -\Pi_2^{-1} \end{pmatrix}
\begin{pmatrix}
0 & | & G_{11} & G_{1w} & 0 & 0 \\
0 & | & I & 0 & 0 & 0 \\
\hline
0 & | & 0 & 0 & I & 0 \\
0 & | & 0 & 0 & -G_{22}^{*} & -G_{z2}^{*}
\end{pmatrix}
\prec 0. \quad (5.28)
$$

Proof. Consider the partition

$$
\Pi_2 = \begin{pmatrix} Q_2 & S_2 \\ S_2^{*} & R_2 \end{pmatrix}
$$

according to the structure of $\mathrm{col}(q_2, \Delta_2(q_2))$. Since Δ_2 satisfies Assumption 2.1 it contains 0. Therefore (5.26) implies that $Q_2 \succcurlyeq 0$. Let us first assume that $Q_2 \succ 0$ and recall that $\Pi_2 \in \boldsymbol{\Pi}_2$ is nonsingular. Then, we can consider the factorization

$$
\Pi_2^{-1} = \begin{pmatrix} I & -Q_2^{-1}S_2 \\ 0 & I \end{pmatrix}
\begin{pmatrix} Q_2^{-1} & 0 \\ 0 & \tilde{R}_2 \end{pmatrix}
\begin{pmatrix} I & -Q_2^{-1}S_2 \\ 0 & I \end{pmatrix}^{*},
$$

where $\tilde{R}_2 = (R_2 - S_2^{*}Q_2^{-1}S_2)^{-1}$, and rearrange (5.27) as

$$
\begin{pmatrix} \star \\ \star \\ \hline \star \\ \star \end{pmatrix}^{*}
\begin{pmatrix} P_e & 0 \\ 0 & \Pi_1 \end{pmatrix}
\begin{pmatrix}
\mathcal{G}_{ee} & \mathcal{G}_{e1} & 0 & \mathcal{G}_{ew} \\
I & 0 & 0 & 0 \\
\hline
0 & G_{11} & 0 & G_{1w} \\
0 & I & 0 & 0
\end{pmatrix}
+
$$

$$
\begin{pmatrix} \star \\ \star \\ \hline \star \\ \star \end{pmatrix}^{*}
\begin{pmatrix}
\tilde{R}_2^{-1} & 0 & | & 0 & 0 \\
0 & -\gamma I & | & 0 & 0 \\
\hline
0 & 0 & | & Q_2 & 0 \\
0 & 0 & | & 0 & \frac{1}{\gamma}I
\end{pmatrix}
\begin{pmatrix}
0 & 0 & I & 0 \\
0 & 0 & 0 & I \\
\hline
\mathcal{G}_{2e} & \mathcal{G}_{21} & \tilde{G}_{22} & \mathcal{G}_{2w} \\
\mathcal{G}_{ze} & \mathcal{G}_{z1} & G_{z2} & \mathcal{G}_{zw}
\end{pmatrix}
\prec 0.
$$

Here $\tilde{G}_{22} = G_{22} + Q_2^{-1}S_2$. Subsequently, it is possible to apply the Schur complement with respect to the Q_2 and $\gamma^{-1}I$ blocks. This yields

$$
\begin{pmatrix}
0 & 0 & 0 & 0 & \vline & \mathcal{G}_{2\mathrm{e}}^* & \mathcal{G}_{z\mathrm{e}}^* \\
0 & 0 & 0 & 0 & \vline & \mathcal{G}_{21}^* & \mathcal{G}_{z1}^* \\
0 & 0 & \tilde{R}_2^{-1} & 0 & \vline & \tilde{G}_{22}^* & G_{z2}^* \\
0 & 0 & 0 & -\gamma I & \vline & \mathcal{G}_{2w}^* & \mathcal{G}_{zw}^* \\
\hdashline
\mathcal{G}_{2\mathrm{e}} & \mathcal{G}_{21} & \tilde{G}_{22} & \mathcal{G}_{2w} & \vline & -Q_2^{-1} & 0 \\
\mathcal{G}_{z\mathrm{e}} & \mathcal{G}_{z1} & G_{z2} & \mathcal{G}_{zw} & \vline & 0 & -\gamma I
\end{pmatrix} +
$$

$$
+ \begin{pmatrix} \star \\ \star \\ \hdashline \star \\ \star \end{pmatrix}^*
\begin{pmatrix} P_{\mathrm{e}} & 0 \\ 0 & \Pi_1 \end{pmatrix}
\begin{pmatrix}
\mathcal{G}_{\mathrm{ee}} & \mathcal{G}_{\mathrm{e}1} & 0 & \mathcal{G}_{\mathrm{ew}} & \vline & 0 & 0 \\
I & 0 & 0 & 0 & \vline & 0 & 0 \\
\hdashline
0 & G_{11} & 0 & G_{1w} & \vline & 0 & 0 \\
0 & I & 0 & 0 & \vline & 0 & 0
\end{pmatrix} \prec 0. \quad (5.29)
$$

Now observe that all off-diagonal blocks in the third block row/column of (5.29) can be eliminated, by applying the congruence transformation

$$
\begin{pmatrix}
I & 0 & 0 & 0 & \vline & 0 & 0 \\
0 & I & 0 & 0 & \vline & 0 & 0 \\
0 & 0 & I & 0 & \vline & -\tilde{R}_2\tilde{G}_{22}^* & -\tilde{R}_2 G_{z2}^* \\
0 & 0 & 0 & I & \vline & 0 & 0 \\
\hdashline
0 & 0 & 0 & 0 & \vline & I & 0 \\
0 & 0 & 0 & 0 & \vline & 0 & I
\end{pmatrix}^*
(5.29)
\begin{pmatrix}
I & 0 & 0 & 0 & \vline & 0 & 0 \\
0 & I & 0 & 0 & \vline & 0 & 0 \\
0 & 0 & I & 0 & \vline & -\tilde{R}_2\tilde{G}_{22}^* & -\tilde{R}_2 G_{z2}^* \\
0 & 0 & 0 & I & \vline & 0 & 0 \\
\hdashline
0 & 0 & 0 & 0 & \vline & I & 0 \\
0 & 0 & 0 & 0 & \vline & 0 & I
\end{pmatrix},
$$

which yields

$$
\mathrm{He}
\begin{pmatrix}
0 & 0 & 0 & 0 & \vline & 0 & 0 \\
0 & 0 & 0 & 0 & \vline & 0 & 0 \\
0 & 0 & \frac{1}{2}\tilde{R}_2^{-1} & 0 & \vline & 0 & 0 \\
0 & 0 & 0 & -\frac{\gamma}{2}I & \vline & 0 & 0 \\
\hdashline
\mathcal{G}_{2\mathrm{e}} & \mathcal{G}_{21} & 0 & \mathcal{G}_{2w} & \vline & 0 & 0 \\
\mathcal{G}_{z\mathrm{e}} & \mathcal{G}_{z1} & 0 & \mathcal{G}_{zw} & \vline & 0 & -\frac{\gamma}{2}I
\end{pmatrix}
+ \begin{pmatrix} \star \\ \star \end{pmatrix}^* P_{\mathrm{e}}
\begin{pmatrix}
\mathcal{G}_{\mathrm{ee}} & \mathcal{G}_{\mathrm{e}1} & 0 & \mathcal{G}_{\mathrm{ew}} & \vline & 0 & 0 \\
\hdashline
I & 0 & 0 & 0 & \vline & 0 & 0
\end{pmatrix} +
$$

$$
+ \begin{pmatrix} \star \\ \star \\ \hdashline \star \\ \star \end{pmatrix}^*
\begin{pmatrix} \Pi_1 & 0 \\ 0 & -\Pi_2^{-1} \end{pmatrix}
\begin{pmatrix}
0 & G_{11} & 0 & G_{1w} & \vline & 0 & 0 \\
0 & I & 0 & 0 & \vline & 0 & 0 \\
\hdashline
0 & 0 & 0 & 0 & \vline & I & 0 \\
0 & 0 & 0 & 0 & \vline & -G_{22}^* & -G_{z2}^*
\end{pmatrix} \prec 0.
$$

Dropping the third block row/column finally yields (5.28).

In case Q_2 is nonnegative, one can replace the right-hand side of (5.27) by $-\delta I$ for some sufficiently small $\delta > 0$ and repeat all previous arguments by considering $Q_2 + \epsilon I$ for some $\epsilon > 0$. Then, if we denote the corresponding multiplier by Π_2^ϵ, it is not hard to see that $(\Pi_2^\epsilon)^{-1}$ converges to Π_2^{-1} as $\epsilon \to 0$ just because Π_2 is nonsingular.

The converse statement is proved by reversing the arguments and starting with the assumption that $\tilde{R}_2 \prec 0$. Subsequently, one can apply a perturbation argument similar to the one that we just provided.

$\blacksquare$

This so-called partial dualization leads to an additive separation of the FDI into a part that is affected by the gain-scheduling IQC-multiplier P_e and by K, while the other part only depends on the uncertainty IQC-multipliers Π_1 and Π_2^{-1} and is, in particular, independent from K. This is the essential feature that enables us to subsume the design question considered in this section under Problem 5.1.

To do so we assume that all elements in $\boldsymbol{\Pi}_2$ are invertible and that IQC-multipliers Π_1 and Π_2^{-1} are parameterized as in

$$\boldsymbol{\Pi}_1 = \{\Psi^* P_1 \Psi : P_1 \in \boldsymbol{P}_1\} \quad \text{and} \quad \boldsymbol{\Pi}_2^{-1} = \{\Phi P_2 \Phi^* : P_2 \in \boldsymbol{P}_2\} \tag{5.30}$$

with LMIable sets $\boldsymbol{P}_1$ and $\boldsymbol{P}_2$ of real symmetric matrices and with some fixed transfer-matrices $\Psi, \Phi \in \mathscr{RL}_\infty^{\bullet \times \bullet}$. The latter are assumed to be partitioned as $\Psi = (\Psi_1 \ \Psi_2)$ and $\Phi^* = (\Phi_1^* \ -\Phi_2^*)$ compatibly with the structure of $\mathrm{col}(G_{11}, I)$ and $\mathrm{col}(I, -G_{22}^*)$ respectively. In view of the remark before Lemma 5.4, it causes no loss of generality to assume that $\Phi_2 P_2 \Phi_2^* \preceq 0$. If not satisfied, this condition can be enforced by using the KYP Lemma with a minimal realization of Φ_2^* and adjoining an extra LMI constraint in the description of $\boldsymbol{P}_2$. Again we emphasize that Ψ or Φ are not required to be stable.

Remark 5.5. As we have seen in Chapter 2 parameterizations of the form $\boldsymbol{\Pi}_1 = \Psi^* \boldsymbol{P}_1 \Psi$ exist for many IQC-multiplier classes. On the other hand, parameterizations of the form $\boldsymbol{\Pi}_2^{-1} = \Phi \boldsymbol{P}_2 \Phi^*$ with $\Phi_2 \boldsymbol{P}_2 \Phi_2^* \preceq 0$ have been suggested for various linear uncertainty blocks as well as for the convex conic hull of finitely many multipliers. For some more details on this issue we refer the reader to [103].

With (5.30), the FDI (5.28) is identical to (5.4) for

$$T_z := I, \quad T_w := I, \quad \mathcal{H} = \begin{pmatrix} \mathcal{H}_{11} & \mathcal{H}_{12} \\ \mathcal{H}_{21} & \mathcal{H}_{22} \end{pmatrix} := \begin{pmatrix} \mathcal{G}_{ee} & \mathcal{G}_{e1} & \mathcal{G}_{ew} & 0 & 0 \\ 0 & 0 & 0 & 0 & 0 \\ 0 & 0 & 0 & 0 & 0 \\ \mathcal{G}_{2e} & \mathcal{G}_{21} & \mathcal{G}_{2w} & 0 & 0 \\ \mathcal{G}_{ze} & \mathcal{G}_{z1} & \mathcal{G}_{zw} & 0 & 0 \end{pmatrix}$$

and

$$
P_r := \begin{pmatrix} -\gamma I & 0 & 0 & 0 \\ 0 & -\gamma I & 0 & 0 \\ 0 & 0 & P_1 & 0 \\ 0 & 0 & 0 & -P_2 \end{pmatrix}, \quad T_r := \begin{pmatrix} 0 & I & 0 & 0 \\ 0 & 0 & 0 & I \\ \Psi_1 G_{11} + \Psi_2 & \Psi_1 G_{1w} & 0 & 0 \\ 0 & 0 & \Phi_2^* G_{22}^* + \Phi_1^* & \Phi_2^* G_{z2}^* \end{pmatrix}.
$$

Hence, with $n_a := n_{p1} \times n_{q2}$, $n_b := n_w \times n_z$ and

$$
\begin{pmatrix} H_{11} & H_{12} & H_{13} \\ H_{21} & H_{22} & H_{23} \\ H_{31} & H_{32} & H_{33} \end{pmatrix} := \left(\begin{array}{ccccc|c} G_{ss} & G_{s1} & G_{sw} & 0 & 0 & G_{su} \\ \hline 0 & 0 & 0 & 0_{n_a} & 0 & 0 \\ 0 & 0 & 0 & 0 & 0_{n_b} & 0 \\ G_{2s} & G_{21} & G_{2w} & 0 & 0 & G_{2u} \\ G_{zs} & G_{z1} & G_{zw} & 0 & 0 & G_{zu} \\ \hline G_{ys} & G_{y1} & G_{yw} & 0 & 0 & G_{yu} \end{array}\right),
$$

we infer $H_{\mathrm{aug}} \star K = \mathcal{H}$ and have thus fitted the design question of this section into the setting of Problem 5.1. To summarize, $K_{\mathrm{rgs}}(\Delta_s) = K \star \hat{\Delta}_c(\Delta_s)$ solves Problem 5.2 if K stabilizes H_{aug} and if there exist $P_r \in \boldsymbol{P}_r := \mathrm{diag}(-\gamma I, -\gamma I, \boldsymbol{P}_1, -\boldsymbol{P}_2)$ and $P_e \in \boldsymbol{P}_e$ for which the FDI (5.4) is satisfied; controller synthesis can hence be performed on the basis of Theorem 5.1 and is reduced to the solution of a convex optimization problem.

5.5 Concrete applications

Apart form all the configurations in the introduction and in Sections 5.4, let us explore a few more that are covered by our results. The list is by no means complete, but is rather meant as an illustration of the framework's flexibility. Throughout this section, we assume as before that Δ_s and Δ_u belong to the sets $\boldsymbol{\Delta}_s$ and $\boldsymbol{\Delta}_u$, and that the linear fractional representations under consideration are well-posed for all $\Delta_s \in \boldsymbol{\Delta}_s$ and for all $\Delta_u \in \boldsymbol{\Delta}_u$, $u \in \{1,2\}$.

5.5.1 Generalized $\mathscr{L}_2$-synthesis

Generalized $\mathscr{L}_2$-synthesis has been introduced in [32] in order to handle nominal performance synthesis with independently norm-bounded disturbances or robust synthesis against unstructured but element-by-element bounded uncertainties. The extension to dynamically weighted signal constraints with a multitude of interesting motivations has been proposed in [33] and has also been discussed and further generalized in Section 2.7.5. Without any technical complications, the corresponding synthesis problem is covered by our framework as follows.

Let $z = (F \star K)w$ denote the closed-loop system for some stabilizing controller K and some plant F. Define $\mathscr{E}$, $\mathscr{W}$ to be the set of those signals $w \in \mathscr{L}_2^{n_w}$, $e \in \mathscr{L}_2^{n_z}$ which respectively satisfy

$$L_w \left(\int_{-\infty}^{\infty} [\Psi_w(i\omega)\hat{w}(i\omega)][\Psi_w(i\omega)\hat{w}(i\omega)]^* d\omega \right) + L_{w0} \preccurlyeq 0,$$

$$L_e \left(\int_{-\infty}^{\infty} [\Psi_e(i\omega)\hat{e}(i\omega)][\Psi_e(i\omega)\hat{e}(i\omega)]^* d\omega \right) + L_{e0} \preccurlyeq 0.$$

Here L_w, L_e are linear maps that take symmetric matrices into symmetric matrices, $L_{w0} = L_{w0}^\top$, $L_{e0} = L_{e0}^\top$ are real matrices, $\Psi_w, \Psi_e \in \mathscr{RL}_\infty$ are dynamic filters and $\hat{}$ denotes taking the Fourier-transform. The performance constraint in [33] is then expressed as

$$\sup_{w \in \mathscr{W}, e \in \mathscr{E}} \int_{-\infty}^{\infty} \hat{e}(i\omega)^* (F \star K)(i\omega)\hat{w}(i\omega) d\omega < 1. \tag{5.31}$$

As shown in Section 2.7.5 it is elementary to check that (5.31) is achieved if there exist real symmetric matrices W and E with

$$W \preccurlyeq 0, \quad E \preccurlyeq 0, \quad \text{trace}(W L_{w0}) + \text{trace}(E L_{e0}) < 2$$

that satisfy, with the adjoint maps L_w^*, L_e^*, the FDI

$$\text{He} \begin{pmatrix} 0 & 0 \\ F \star K & 0 \end{pmatrix} + \begin{pmatrix} \Psi_w^* L_w^*(W)\Psi_w & 0 \\ 0 & \Psi_e^* L_e^*(E)\Psi_e \end{pmatrix} \prec 0.$$

This fits into Problem 5.1 by substituting P_r with $\text{diag}(L_w^*(W), L_e^*(E))$ and by taking

$$T_z := I, \quad T_w := I, \quad T_r := \begin{pmatrix} \Psi_w & 0 \\ 0 & \Psi_e \end{pmatrix} \quad \text{and} \quad \mathcal{H}_{22} := \begin{pmatrix} 0 & 0 \\ F \star K & 0 \end{pmatrix}.$$

Both nominal and gain-scheduled controllers can hence be designed on the basis of Theorem 5.1. As discussed in Section 2.7.5 our setting even permits the inclusion of dynamically weighted LMI-constraints on the cross-correlation

$$\int_{-\infty}^{\infty} \hat{w}(i\omega)\hat{e}(i\omega)^* d\omega.$$

As argued in [33], the specifications in this section allow to model arbitrarily tight approximations of the $\mathscr{H}_2$-cost. Since they can be easily incorporated in the setting of Section 5.4, we actually cover the design of robust gain-scheduled $\mathscr{H}_2$-controllers as well. In contrast to [33], there is no need to introduce any factorization of the filters

Ψ_w, Ψ_e, which was also the motivation for [98] to address a particular version of the considered problem.

5.5.2 Multi-objective and structured controller synthesis

Similarly, the results in [149] generalize to gain-scheduling and allow to address a rich class of questions related to multi-objective and structured controller synthesis. Given the stable transfer matrices Ψ_0, Ψ_1, Ψ_2 and the plant $\Sigma(\Delta_s)$, the goal is to design a gain-scheduled controller $K_{gs}(\Delta_s)$ such that, for all $\Delta_s \in \boldsymbol{\Delta}_s$,

$$\|\Psi_0 + \Psi_1 P_{12}\Psi_2 + \Sigma(\Delta_s) \star K_{gs}(\Delta_s)\|_{i2} < \gamma. \tag{5.32}$$

Here $P_{12} \in \boldsymbol{P}_{12}$ is some real matrix in the LMIable set $\boldsymbol{P}_{12}$. If $\Sigma(\Delta_s) \star K_{gs}(\Delta_s)$ is described as in (3.9) in Section 3.3, then (5.32) can be translated into the FDI

$$\mathrm{He}\begin{pmatrix} 0 & 0 & 0 \\ 0 & -\frac{\gamma}{2}I & 0 \\ \mathcal{G}_{ze} & \mathcal{G}_{zw} + \Psi_0 + \Psi_1 P_{12}\Psi_2 & -\frac{\gamma}{2}I \end{pmatrix} + \begin{pmatrix} \star \\ \star \\ \star \end{pmatrix}^* P_e \begin{pmatrix} \mathcal{G}_{ee} & \mathcal{G}_{ew} & 0 \\ I & 0 & 0 \end{pmatrix} \prec 0.$$

In fact, this inequality is identical to (5.4) for

$$T_w := I, \quad T_z := I, \quad \begin{pmatrix} \mathcal{H}_{11} & \mathcal{H}_{12} \\ \mathcal{H}_{21} & \mathcal{H}_{22} \end{pmatrix} := \begin{pmatrix} \mathcal{G}_{ee} & \mathcal{G}_{ew} & 0 \\ 0 & 0 & 0 \\ \mathcal{G}_{ze} & \mathcal{G}_{zw} & 0 \end{pmatrix},$$

and

$$P_r := \begin{pmatrix} -\gamma I & 0 & 0 & 0 & 0 \\ 0 & -\gamma I & I & 0 & 0 \\ 0 & I & 0 & 0 & 0 \\ 0 & 0 & 0 & 0 & P_{12} \\ 0 & 0 & 0 & P_{12}^\top & 0 \end{pmatrix}, \quad T_r := \begin{pmatrix} I & 0 \\ 0 & I \\ \Psi_0 & 0 \\ 0 & \Psi_1^* \\ \Psi_2 & 0 \end{pmatrix}.$$

5.5.3 Robust gain-scheduled observer design

Let us also consider the generalization of a so-called augmented-gain observer (AGO) structure from [17] to gain-scheduling as shown in Figure 5.5; it is composed of an

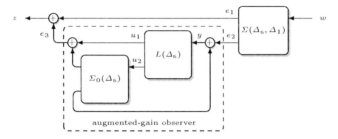

Fig. 5.5 Interconnection for the class of robust observer design problems

uncertain system $\Sigma(\Delta_s, \Delta_1)$, some model $\Sigma_0(\Delta_s)$ deduced from $\Sigma(\Delta_s, \Delta_1)$, and a to-be-designed dynamical component $L(\Delta_s)$. The goal is to synthesize $L(\Delta_s)$ such that w is suppressed at z as much as possible; the AGO can hence be interpreted as estimating the signal $-e_1$ based on measuring e_2 by a gain-scheduled robust estimator. This problem has been suggested in [17] as a one shot combination of an observer and an estimator structure, which allows to design robust observers for a specific class of unstable systems. We assume that $\Sigma(\Delta_s, \Delta_1)$ admits the linear fractional representation

$$\Sigma(\Delta_s, \Delta_1) = \begin{pmatrix} \Delta_s & 0 \\ 0 & \Delta_1 \end{pmatrix} \star \underbrace{\begin{pmatrix} F_{11} & F_{12} & F_{13} \\ 0 & F_{22} & F_{23} \\ \hline F_{31} & F_{32} & F_{33} \\ F_{41} & F_{42} & F_{43} \end{pmatrix}}_{F}, \tag{5.33}$$

where F is a transfer matrix with the realization

$$F = \begin{pmatrix} F_{11} & F_{12} & F_{13} \\ 0 & F_{22} & F_{23} \\ \hline F_{31} & F_{32} & F_{33} \\ F_{41} & F_{42} & F_{43} \end{pmatrix} = \left[\begin{array}{cc|ccc} A_{11} & 0 & B_{11} & B_{12} & B_{13} \\ 0 & A_{22} & 0 & B_{22} & B_{23} \\ \hline C_{11} & 0 & D_{11} & D_{12} & D_{13} \\ 0 & C_{22} & 0 & D_{22} & D_{23} \\ \hline C_{31} & 0 & D_{31} & D_{32} & D_{33} \\ C_{41} & 0 & D_{41} & D_{42} & D_{43} \end{array} \right].$$

As opposed to the estimator synthesis techniques in [51, 180, 165] and Chapter 4, we only require A_{22} to be Hurwitz, which means that F_{22} and F_{23} must be stable.

In order to cover both estimation as well as observer structures, we choose $\Sigma_0(\Delta_s) = \Delta_s \star F_0$ with F_0 being defined through the realization

$$F_0 = \begin{pmatrix} F_{11} & F_{14} \\ F_{31} & F_{34} \\ F_{41} & F_{44} \end{pmatrix} = \left[\begin{array}{c|cc} A_{11} & B_{11} & I \\ \hline C_{11} & D_{11} & 0 \\ C_{31} & D_{31} & 0 \\ C_{41} & D_{41} & 0 \end{array} \right].$$

In this fashion, the estimator framework of [51, 180, 165] and the standard Kalman or Luenberger observer structure [89, 111] can be obtained from the AGO through specialization, either by removing the model $\Sigma_0(\Delta_s)$ from the AGO and or by restricting $L(\Delta_s)$ to be a static gain.

The precise goal is to design $L(\Delta_s) = K \star \hat{\Delta}_c(\Delta_s)$ such that, for all $\Delta_s \in \boldsymbol{\Delta}_s$ and $\Delta_1 \in \boldsymbol{\Delta}_1$, the interconnection in Figure 5.5 is stable and the induced $\mathscr{L}_2$-gain from the disturbance input w to the estimated error output z is rendered less than $\gamma > 0$.

The open-loop plant $\operatorname{col}(w, u_1, u_2) \to \operatorname{col}(z, y)$ for the configuration in Figure 5.5 is given by

$$\begin{pmatrix} \Delta_s & 0 \\ 0 & \Delta_1 \end{pmatrix} \star \begin{pmatrix} F_{11} & F_{12} & F_{13} & 0 & F_{14} \\ 0 & F_{22} & F_{23} & 0 & 0 \\ \text{-} & \text{-} & \text{-} & \text{-} & \text{-} \\ F_{31} & F_{32} & F_{33} & I & F_{34} \\ F_{41} & F_{42} & F_{43} & 0 & F_{44} \end{pmatrix}$$

and can be viewed as a special case of the linear fractional representation (5.24) with $\Delta := \operatorname{diag}(\Delta_s, \Delta_1)$ and

$$G = \begin{pmatrix} G_{ss} & G_{s1} & G_{sw} & G_{su} \\ 0 & G_{11} & G_{1w} & 0 \\ \text{-} & \text{-} & \text{-} & \text{-} \\ G_{zs} & G_{z1} & G_{zw} & G_{zu} \\ G_{ys} & G_{y1} & G_{yw} & G_{yu} \end{pmatrix} = \begin{pmatrix} F_{11} & F_{12} & F_{13} & 0 & F_{14} \\ 0 & F_{22} & F_{23} & 0 & 0 \\ \text{-} & \text{-} & \text{-} & \text{-} & \text{-} \\ F_{31} & F_{32} & F_{33} & I & F_{34} \\ F_{41} & F_{42} & F_{43} & 0 & F_{44} \end{pmatrix}.$$

This illustrates that the class of robust gain-scheduling $\mathscr{L}_2$-gain observation problems can be seen as a special case of the general problem considered in Section 5.4. We refer the reader to Section 5.6 for a numerical example.

5.5.4 Open-loop controller synthesis

Our results also allow us to generalize the open-loop design problem of Figure 5.3(a) to robust gain-scheduling synthesis based on the configuration in Figure 5.6. Here $\Sigma_1(\Delta_1, \Delta_s)$ and $\Sigma_2(\Delta_2, \Delta_s)$ admit the linear fractional representation

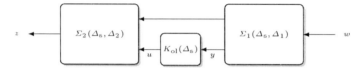

Fig. 5.6 Interconnection for the class of robust open-loop gain-scheduling design problems

$$
\Sigma_1(\Delta_1,\Delta_{\mathrm{s}}) = \begin{pmatrix} \Delta_{\mathrm{s}} & 0 \\ 0 & \Delta_1 \end{pmatrix} \star \underbrace{\begin{pmatrix} F_{11}^1 & F_{12}^1 & F_{13}^1 \\ 0 & F_{22}^1 & F_{23}^1 \\ F_{31}^1 & F_{32}^1 & F_{33}^1 \\ F_{41}^1 & F_{42}^1 & F_{43}^1 \end{pmatrix}}_{F^1}
\tag{5.34}
$$

and

$$
\Sigma_2(\Delta_2,\Delta_{\mathrm{s}}) = \begin{pmatrix} \Delta_{\mathrm{s}} & 0 \\ 0 & \Delta_2 \end{pmatrix} \star \underbrace{\begin{pmatrix} F_{11}^2 & 0 & F_{13}^2 & F_{14}^2 \\ F_{21}^2 & F_{22}^2 & F_{23}^2 & F_{24}^2 \\ F_{31}^2 & F_{32}^2 & F_{33}^2 & F_{34}^2 \end{pmatrix}}_{F^2},
\tag{5.35}
$$

respectively, where F^1 and F^2 are assumed to be proper and stable transfer matrices. Now observe that (5.34) and (5.35) can be combined into one linear fractional representation as (5.24), with $\Delta = \mathrm{diag}(\Delta_{\mathrm{s}},\Delta_{\mathrm{s}},\Delta_1,\Delta_2)$ and

$$
G = \begin{pmatrix} G_{\mathrm{ss}} & G_{\mathrm{s}1} & 0 & G_{\mathrm{s}w} & G_{\mathrm{s}u} \\ 0 & G_{11} & 0 & G_{1w} & 0 \\ G_{2\mathrm{s}} & G_{21} & G_{22} & G_{2w} & G_{2u} \\ G_{z\mathrm{s}} & G_{z1} & G_{z2} & G_{zw} & G_{zu} \\ G_{y\mathrm{s}} & G_{y1} & 0 & G_{yw} & 0 \end{pmatrix} = \begin{pmatrix} F_{11}^1 & 0 & F_{12}^1 & 0 & F_{13}^1 & 0 \\ F_{13}^2 F_{31}^1 & F_{11}^2 & F_{13}^2 F_{32}^1 & 0 & F_{13}^2 F_{33}^1 & F_{14}^1 \\ 0 & 0 & F_{22}^1 & 0 & F_{23}^1 & 0 \\ F_{23}^2 F_{31}^1 & F_{21}^2 & F_{23}^2 F_{32}^1 & F_{22}^2 & F_{23}^2 F_{33}^1 & F_{24}^1 \\ F_{33}^2 F_{31}^1 & F_{31}^2 & F_{33}^2 F_{32}^1 & F_{32}^2 & F_{33}^2 F_{33}^1 & F_{34}^1 \\ F_{41}^1 & 0 & F_{42}^1 & 0 & F_{43}^1 & 0 \end{pmatrix}.
$$

Hence, also the class of open-loop gain-scheduling design problems can be viewed as a special case of the general synthesis problem considered in Section 5.4.

5.5.5 Synthesis with uncertain performance weights

Similarly, we can address the design of robust gain-scheduled controllers against uncertain performance weights based on Figure 5.7. These problems emerge e.g. if de-

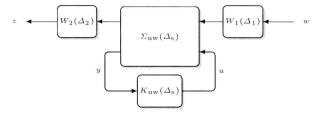

Fig. 5.7 Interconnection for the class of design problems with uncertain performance weights

signing controllers in order to suppress periodic disturbances with time-varying frequencies or by incorporating coloring filters that are subject to parameter variations in order to capture disturbance spectrum uncertainty in stochastic control (See [38, 12] and references therein). Here the uncertain performance weights $W_{\mathrm{u}}(\Delta_{\mathrm{u}})$, $\mathrm{u} \in \{1,2\}$, and the plant $\Sigma_{\mathrm{uw}}(\Delta_{\mathrm{s}})$ are assumed to admit the linear fractional representations

$$W_{\mathrm{u}}(\Delta_{\mathrm{u}}) = \Delta_{\mathrm{u}} \star \underbrace{\begin{pmatrix} W_{11}^{\mathrm{u}} & W_{12}^{\mathrm{u}} \\ W_{21}^{\mathrm{u}} & W_{22}^{\mathrm{u}} \end{pmatrix}}_{W^{\mathrm{u}} \in \mathcal{R}\mathcal{H}_\infty}, \quad \mathrm{u} \in \{1,2\} \quad \text{and} \quad \Sigma_{\mathrm{uw}}(\Delta_{\mathrm{s}}) = \Delta_{\mathrm{s}} \star \underbrace{\begin{pmatrix} F_{11} & F_{12} & F_{13} \\ F_{21} & F_{22} & F_{23} \\ F_{31} & F_{32} & F_{33} \end{pmatrix}}_{F}.$$

Then the overall open-loop plant of the interconnection in Figure 5.7 can be represented by (5.24), for $\Delta = \mathrm{diag}(\Delta_{\mathrm{s}}, \Delta_1, \Delta_2)$ and with

$$G = \begin{pmatrix} G_{ss} & G_{s1} & 0 & G_{sw} & G_{su} \\ 0 & G_{11} & 0 & G_{1w} & 0 \\ G_{2s} & G_{21} & G_{22} & G_{2w} & G_{2u} \\ G_{zs} & G_{z1} & G_{z2} & G_{zw} & G_{zu} \\ G_{ys} & G_{y1} & 0 & G_{yw} & 0 \end{pmatrix} = \begin{pmatrix} F_{11} & F_{12}W_{21}^1 & 0 & F_{12}W_{22}^1 & F_{13} \\ 0 & W_{11}^1 & 0 & W_{12}^1 & 0 \\ W_{12}^2 F_{21} & W_{12}^2 F_{22}W_{21}^1 & W_{11}^2 & W_{12}^2 F_{22}W_{22}^1 & W_{12}^2 F_{23} \\ W_{22}^2 F_{21} & W_{22}^2 F_{22}W_{21}^1 & W_{21}^2 & W_{22}^2 F_{22}W_{22}^1 & W_{22}^2 F_{23} \\ F_{31} & F_{32}W_{21}^1 & 0 & F_{32}W_{22}^1 & F_{33} \end{pmatrix}.$$

Hence, controller synthesis can again be performed in the setting of Section 5.4.

5.5.6 IQC-synthesis with unstable weights

So far, we have not been illustrating the potential benefit of unstable weights. The main purpose of this section is to show how unstable weight can be useful for robust controller synthesis. As has already been extensively discussed, the general robust controller synthesis problem cannot be easily turned into a semi-definite program. This motivates to consider heuristic methods such as, e.g., μ- or IQC-synthesis [143, 40, 11, 28, 146, 144, 63, 208, 198]. These methods proceed by an iteration between the synthesis of an optimal $\mathscr{H}_\infty$-controller, while fixing the involved IQC-multipliers, and finding IQC-multipliers, while fixing the controller. Although the individual steps admit convex solutions, there is no guarantee that the resulting controller is globally optimal. Moreover, all existing methods rely on an intermediate factorization of the involved IQC-multipliers in order to construct stable weights to be used in a new re-weighted plant for designing a robustified controller in a next iteration step. This might cause numerical ill-conditioning in computations. As one of the benefits of the general synthesis framework in this chapter, we now show how to avoid any IQC-multiplier factorization. Since IQC-synthesis is the main theme of Chapter 6, we emphasize that the presentation in the sequel will be concise; in particular, main goal is to illustrate the use of unstable weights.

Let us start by considering the uncertain plant

$$\Sigma(\Delta_1) = \Delta_1 \star \underbrace{\begin{pmatrix} G_{qp} & G_{qw} & G_{qu} \\ \hline G_{zp} & G_{zw} & G_{zu} \\ G_{yp} & G_{yw} & G_{yu} \end{pmatrix}}_{G} \quad \text{with} \quad \Delta_1 \in \boldsymbol{\Delta}_1$$

which is assumed to be well-posed. Here $\boldsymbol{\Delta}_1$ is supposed to have the properties as in Section 5.4 and, similarly, $\boldsymbol{\Pi}_1 = \{\Psi^* P_1 \Psi : P_1 \in \boldsymbol{P}_1\}$ denotes a valid class of IQCs for all $\Delta_1 \in \boldsymbol{\Delta}_1$ with a parameterization as in (5.30). Due to $0 \in \boldsymbol{\Delta}_1$, the left-upper block $\Psi_1^* P_1 \Psi_1$ of any $\Pi_1 \in \boldsymbol{\Pi}_1$ is positive semi-definite on $\mathbb{C}^0$. For many classes it is possible to verify (or to enforce by a perturbation) that

$$\Psi_1^* P_1 \Psi_1 \succ 0 \quad \text{for all} \quad \Psi^* P_1 \Psi \in \boldsymbol{\Pi}_1 \tag{5.36}$$

which is assumed from now on.

We infer again that $\Sigma(\Delta_1) \star K = \Delta_1 \star \mathcal{G}$ with

$$\mathcal{G} = \begin{pmatrix} \mathcal{G}_{qp} & \mathcal{G}_{qw} \\ \mathcal{G}_{zp} & \mathcal{G}_{zw} \end{pmatrix} = \begin{pmatrix} G_{qp} & G_{qw} & G_{qu} \\ \hline G_{zp} & G_{zw} & G_{zu} \\ G_{yp} & G_{yw} & G_{yu} \end{pmatrix} \star K,$$

which leads us to the IQC-analysis result of [116] as presented in Chapter 2: Suppose that K internally stabilizes G and that $\Delta_1 \star \mathcal{G}$ is well-posed for all $\Delta_1 \in \boldsymbol{\Delta}_1$. Then K stabilizes $\Sigma(\Delta_1)$ and achieves $\|\Delta_1 \star \mathcal{G}\|_{i2} < \gamma$ for all $\Delta_1 \in \boldsymbol{\Delta}_1$ if there exists some $P_1 \in \boldsymbol{P}_1$ for which the following FDI holds:

$$
\begin{pmatrix}
\mathcal{G}_{qp} & \mathcal{G}_{qw} \\
I & 0 \\
\hline
\mathcal{G}_{zp} & \mathcal{G}_{zw} \\
0 & I
\end{pmatrix}^{*}
\begin{pmatrix}
\Psi^{*}P_1\Psi & 0 \\
0 & P_\gamma
\end{pmatrix}
\begin{pmatrix}
\mathcal{G}_{qp} & \mathcal{G}_{qw} \\
I & 0 \\
\hline
\mathcal{G}_{zp} & \mathcal{G}_{zw} \\
0 & I
\end{pmatrix} \prec 0. \tag{5.37}
$$

As shown in Chapter 2, the FDI (5.37) translates into a genuine LMI in P_1 and γ by applying the Schur complement and the KYP lemma.

Let us proceed with synthesis. Using (5.36), we have shown in Section 2.10 that (5.37) can equivalently be written as

$$
\mathrm{He}\left[
\begin{pmatrix}
\Psi^{*}P_1\Psi & 0 \\
0 & P_\gamma
\end{pmatrix}
\begin{pmatrix}
-I & 2\mathcal{G}_{qp} & 0 & 2\mathcal{G}_{qw} \\
0 & I & 0 & 0 \\
\hline
0 & 2\mathcal{G}_{zp} & -I & 2\mathcal{G}_{zw} \\
0 & 0 & 0 & I
\end{pmatrix}
\right] \prec 0.
$$

In turn, this can be written as

$$
\begin{pmatrix}
T_z^{*}\mathcal{H}_{22} \\
I
\end{pmatrix}^{*}
\begin{pmatrix}
0 & I \\
I & P_r
\end{pmatrix}
\begin{pmatrix}
T_z^{*}\mathcal{H}_{22} \\
I
\end{pmatrix} \prec 0 \tag{5.38}
$$

for $P_r := \mathrm{diag}(0,0,-2\gamma I,-2\gamma I)$,

$$
T_z^{*} :=
\begin{pmatrix}
\Psi^{*}P_1\Psi & 0 \\
\hline
0 & I \\
0 & 0
\end{pmatrix}
\quad \text{and} \quad
\mathcal{H}_{22} :=
\begin{pmatrix}
-I & 2\mathcal{G}_{qp} & 0 & 2\mathcal{G}_{qw} \\
0 & I & 0 & 0 \\
\hline
0 & 2\mathcal{G}_{zp} & 0 & 2\mathcal{G}_{zw}
\end{pmatrix}.
$$

We observe that this is nothing but (5.4) with $T_w := I$, $T_r := I$ and with void matrices P_e, $\mathcal{H}_{11}$, $\mathcal{H}_{12}$ and $\mathcal{H}_{21}$. For a fixed and possibly unstable weight $T_z \in \mathscr{RL}_\infty$, design problems with this particular structure have been shown to be convex in [157] (see also [118, 69]), and are also covered by the main result of this chapter. This leads to the following robust synthesis algorithm.

Algorithm 5.1 (IQC-Synthesis).

Initialization (iteration step $j = 1$)

1. With Theorem 3.4, design a standard nominal $\mathscr{H}_\infty$-controller K_1 with $\|G \star K_1\|_{\mathrm{i2}} < \gamma_{\mathrm{nom}}$.
2. Construct the closed-loop plant $\mathcal{G}_1 = G \star K_1$ and perform an IQC-analysis: find $P_1 \in \boldsymbol{P}_1$, $\gamma > 0$ for which (5.37) is feasible. Minimization of γ leads to the (almost) optimal bound γ_1 with $\gamma_{\mathrm{nom}} \le \gamma_1$ and $\|\Delta_1 \star \mathcal{G}_1\|_{\mathrm{i2}} < \gamma_1$ for all $\Delta_1 \in \boldsymbol{\Delta}_1$.

Iteration (step $j - 1 \to j$ for $j > 1$)

1. With the multiplier $\Psi^* P_1 \Psi$ from the IQC-analysis for K_{j-1}, design a new robustified controller K_j by applying Theorem 5.1 for the FDI (5.38). Minimization of γ leads to the (almost) optimal bound γ_{rob} with $\gamma_1 \geq \gamma_{j-1} \geq \gamma_{\mathrm{rob}} \geq \gamma_{\mathrm{nom}} > 0$ and $\|\Sigma(\Delta_1) \star K_j\|_{\mathrm{i2}} < \gamma_{\mathrm{rob}}$ for all $\Delta_1 \in \boldsymbol{\Delta}_1$.

2. For the closed-loop $\mathcal{G}_j = G \star K_j$ perform an IQC-analysis: find $P_1 \in \boldsymbol{P}_1$, $\gamma > 0$ for which (5.37) is feasible. Minimization of γ leads to the (almost) optimal bound γ_j with $\gamma_1 \geq \gamma_{j-1} \geq \gamma_{\mathrm{rob}} \geq \gamma_j \geq \gamma_{\mathrm{nom}} > 0$ and $\|\Delta_1 \star \mathcal{G}_j\|_{\mathrm{i2}} < \gamma_j$ for all $\Delta_1 \in \boldsymbol{\Delta}_1$.

3. Terminate if $|\gamma_{j-1} - \gamma_j|$ is small.

The iteration results in a sequence of controllers $K_1, K_2, \ldots, K_{j-1}, K_j$, which guarantee stability and render $\|\Sigma(\Delta_1) \star K_j\|_{\mathrm{i2}}$ less than $\gamma_1 \geq \gamma_2 \geq \cdots \geq \gamma_{j-1} \geq \gamma_j \geq \gamma_{\mathrm{nom}}$ for all uncertainties $\Delta_1 \in \boldsymbol{\Delta}_1$.

In summary, we have formulated an algorithm for the systematic synthesis of robust controllers, which does not require any intermediate factorizations. This illustrates the potential benefit of using unstable weights in a controller/scaling algorithm. The results are an alternative to the ones presented in Chapter 6. We refer the reader to Section 6.7 for a comparison of the two approaches.

5.6 Illustrations

In order to demonstrate the variety of design scenarios that can be handled in a common fashion, we now present two numerical examples from our papers [170] and [197] respectively. Once more we stress that some portions of the text overlap.

5.6.1 Open-loop controller synthesis

Consider again the system interconnection in Figure 5.6 and suppose that the plants (5.34) and (5.35) are given by

$$
\Sigma_1(\delta_1) = \left[\begin{array}{cc|cc} 0 & -1 + 0.95\delta_1 & -2 & 0 \\ 1 & -0.5 & 1 & 0 \\ \hline 1 & 0 & 0 & 0 \\ -1 & 1 & 0 & 0.001 \end{array}\right] \quad \text{and} \quad \Sigma_2(\delta_2) = \left[\begin{array}{cc|cc} -10 & 4 & 1 & 0 \\ -1 + 0.1\delta_2 & 0 & -1 & 1 \\ \hline -1 & 1 & 0 & 0 \end{array}\right]
$$

respectively, for $(\Delta_i z_i)(t) = \delta_i z_i(t)$, $\delta_i \in \alpha[-1,1]$, $i = 1, 2$ and $\alpha \geq 0$.

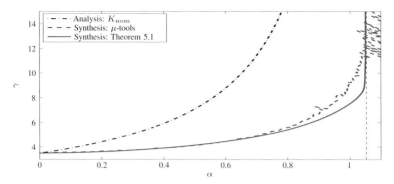

Fig. 5.8 Synthesis and analysis results for increasing α.

The goal is to design a filter K that renders the induced $\mathscr{L}_2$-gain from w to z as small as possible for all $\delta_i \in \alpha[-1,1]$, $i = 1, 2$ and for some given $\alpha \geq 0$. For reasons of comparison we have designed a series of filters for increasing values of α, by (i) considering the μ-tools of Matlab and (ii) solving the semi-definite program in Theorem 5.1 (with a comparable degree-of-freedom in the IQC-multipliers). Finally, we have also performed an IQC-analysis for increasing values of α for a nominal $\mathscr{H}_\infty$-controller K_{nom}. The results are shown in Figure 5.8.

Clearly, the nominal $\mathscr{H}_\infty$-controller K_{nom} cannot keep the worst-case induced $\mathscr{L}_2$-gain γ small, even for small values of α. On the other hand, it is possible to systematically improve the worst-case performance by applying the μ-tools and Theorem 5.1. Note that, even though the μ-tools do not exploit convexity, the global optimum is attained for $\alpha < 0.7$. This fact is verifiable with our results. On the other hand, for $\alpha > 0.7$ the μ-tools either run into numerical problems or do not reach the global optimum. Indeed, observe that the system becomes unstable for $\alpha \geq 1/0.95$ (see vertical dashed line in Figure 5.8). Hence, since we are considering an open-loop design problem, a stabilizing filter does not exist for $\alpha \geq 1/0.95$. This is clearly visible for the synthesis results that were obtain through Theorem 5.1.

5.6.2 Robust observer design

Let us also design various augmented-gain observers as in Section 5.5.3 and based on the unstable uncertain plant

$$
\begin{pmatrix} \dot{x} \\ - \\ e_1 \\ e_2 \end{pmatrix} = \left(\begin{array}{ccccc|cc} 0 & 0 & 10 & 0 & 0 & 0 & 0 \\ 0 & -\frac{\lambda_{\mathrm{s}}}{10} & -\lambda_{\mathrm{s}}^2 & 0 & 10 & 0 & 0 \\ 0 & 1 & 0 & 0 & 0 & 0 & 0 \\ 0 & 0 & 0 & -\frac{\lambda_1}{10} & -\lambda_1^2 & 1 & 0 \\ 0 & 0 & 0 & 1 & 0 & 0 & 0 \\ \hline 0 & 0 & 10 & 0 & 0 & 0 & 0 \\ 1 & 0 & 0 & 0 & 0 & 0 & \frac{1}{20} \end{array} \right) \begin{pmatrix} x \\ - \\ w_1 \\ w_2 \end{pmatrix}.
$$

With fixed constants $\alpha_{\mathrm{s}} = 2$ and $\alpha_1 = 5$, we choose $\lambda_{\mathrm{v}} = \alpha_{\mathrm{v}} \left(1 + \frac{19}{20} \delta_{\mathrm{v}} \right)$, $\mathrm{v} \in \{\mathrm{s}, 1\}$, where Δ_{s} and Δ_1 are defined by $(\Delta_{\mathrm{s}} z_{\mathrm{s}})(t) = \delta_{\mathrm{s}}(t) z_{\mathrm{s}}(t)$, $\delta_{\mathrm{s}}(t) = \sin 2t$, and $(\Delta_1 z_1)(t) = \delta_1 z_1(t)$, $\delta_1 \in [-1, 1]$, respectively; $\delta_{\mathrm{s}}(t)$ is assumed to be online measurable scheduling parameter and δ_1 a time-invariant parametric uncertainty.

We design AGOs that dynamically process the measurement e_2 and the scheduling variable δ_{s} in order to provide an estimate e_3 of the signal $-e_1$ such that the $\mathscr{L}_2$-gain of $w \to e_1 + e_3 = z$ in Figure 5.5 is smaller than $\gamma > 0$. For the following three scenarios we determined AGOs by directly applying Theorem 5.1:

1. A nominal LTI AGO (O_{nom}) with a computed optimal performance-bound of $\gamma = 0.36$ if δ_{s} and δ_1 are assumed to be zero. Note that this scenario can also be handled with standard $\mathscr{H}_\infty$-synthesis techniques such as in [41, 48, 72].

2. A nominal gain-scheduled AGO (O_{nomgs}) with a computed optimal gain-bound of $\gamma = 0.59$ for all $\delta_{\mathrm{s}}(t) \in [-1, 1]$ and if δ_1 vanishes. Let us emphasize that this scenario can alternatively be considered with the synthesis results of [152] as presented in Section 3.7.

3. A robust gain-scheduled AGO (O_{robgs}) with a computed optimal performance level of $\gamma = 1.1$ for all $\delta_{\mathrm{s}}(t) \in [-1, 1]$ and $\delta_1 \in [-1, 1]$. We used the IQC-multiplier 2.5 from Section 2.6.3.1 for the uncertainty δ_1, with constants $\rho = -1$ and $\nu = 1$. Non-dynamic IQC-multipliers ($\nu = 0$) lead to infeasibility of the synthesis LMIs even for large values γ, which illustrates the benefit of simple dynamic IQCs over static ones.

In accordance with Figure 5.5, let us review the designs on the basis of the closed-loop plants $z = \mathcal{G}_{\mathrm{x}}(\Delta_{\mathrm{s}}, \Delta_1) w$ for the corresponding augmented-gain observers O_{x}, $\mathrm{x} \in \{\mathrm{nom}, \mathrm{nomgs}, \mathrm{robgs}\}$. First of all, Figure 5.9 shows the estimation error z for $\delta_1 = 0$ (left) and $\delta_1 = -1$ (right) for a disturbance input $w = \mathrm{col}(w_1, w_2)$ where w_1 is a sinusoid with 'slowly' varying frequency in $[0.1, 10]$rad/sec, and w_2 a random uniformly distributed noise input with an amplitude of 1. We observe that the nominal AGO O_{nom} is not able to estimate the signal $-e_1$, even for $\delta_1 = 0$. On the other hand, with the nominal gain-scheduled AGO O_{nomgs}, the estimation error is small if $\delta_1 = 0$. However, both O_{nom} and O_{nomgs} show a drastic performance degradation for $\delta_1 = -1$. In contrast, it is very nice to see that the robust gain-scheduled AGO O_{robgs} keeps

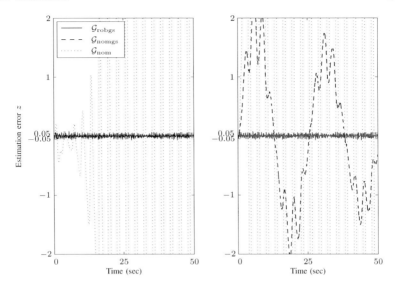

Fig. 5.9 Estimation error for $\delta_1 = 0$ (left) and for $\delta_1 = -1$ (right).

the estimation error small, despite the uncertainty in the system. This behavior is confirmed by simulations with other parameters $\delta_1 \in [-1, 1]$.

The results are consistent with the singular value plots of the transfer matrices from the disturbance input w to the estimated error z in Figure 5.10, if fixing the parameters δ_s and δ_1 to various values in $[-1, 1]$. Indeed, the nominal AGO O_{nom} does not lead to good disturbance attenuation, even if $\delta_1 = 0$. In contrast, the nominal gain-scheduled AGO O_{nomgs} provides a significantly better estimate against disturbances at low frequencies for various points in $\delta_1 \in [-1, 1]$. However, the performance drastically degrades if $\delta_1 \neq 0$. This should be contrasted with the robust gain-scheduled AGO O_{robgs}, which keeps the estimation error small at low frequencies for both scenarios. We would like to stress that all these results have been obtained without the inclusion of performance weights or any sophisticated tuning steps, just in order to illustrate the conceptual possibilities of the design framework proposed in this chapter.

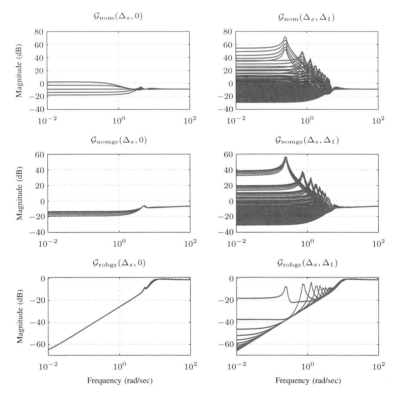

Fig. 5.10 Left column: Singular value plots of $\mathcal{G}_x(\Delta_s, \Delta_1)$ evaluated at $\delta_1 = 0$ and for different values of $\delta_s \in [-1, 1]$. Right column: Singular value plots of $\mathcal{G}_x(\Delta_s, \Delta_1)$ evaluated at different values of $\delta_s \in [-1, 1]$ and $\delta_1 \in [-1, 1]$.

5.7 Summary

As a contribution, and as the main result of this chapter, we presented a generic gain-scheduling synthesis problem and showed how it can be turned into a finite dimensional convex feasibility test in an insightful fashion. This enabled us to effectively address a rich class of design questions related to robust gain-scheduling control in a common framework. Specifically, we provided a solution of the robust gain-scheduled controller synthesis problem for systems without control channel uncertainty and exemplified the ease how to handle a variety of other concrete design scenarios. The framework's flexibility has been illustrated by two numerical examples.

Chapter 6
IQC-synthesis with warm-start options

6.1 Introduction

As we have already briefly discussed in Section 5.5.6, one can resort to a number of useful but non-optimal methods, such as for example μ-synthesis [143, 40, 11, 28, 146, 144, 63, 208], in order to systematically handle the fully general robust controller synthesis problem. Essentially these methods rely on an iteration between the synthesis of an optimal $\mathcal{H}_\infty$-controller, while fixing the involved scalings, and finding scalings by computing an upper-bound of the structured singular value [122] (see also Section 2.8.1), while fixing the controller. Although the individual steps admit convex solutions, there is no guarantee that the resulting controller is globally optimal. In addition, most of the methods require curve-fitting as an intermediate step, though this problem has been addressed and resolved in [146, 144]. Despite these drawbacks, the methods are considered to be reliable and have been successfully applied in many practical applications.

Analogously to the existing μ-tools, we developed an alternative algorithm in Section 5.5.6 for the systematic synthesis of robust controllers based on general dynamic (frequency dependent) IQCs. The suggested algorithm enables us to perform robust controller synthesis for a significantly larger class of uncertainties if compared to the existing methods, just because it relies on general dynamic IQC-multipliers. Indeed, while the classical approaches are restricted to the use of real/complex time-invariant or arbitrarily fast time-varying parametric uncertainties, it has been extensively illustrated in Chapter 2 that the IQC-framework also offers, for example, the possibility to efficiently handle sector-bounded and slope restricted nonlinearities or time-varying parametric uncertainties and uncertain time-varying time-delays, both with bounds on the rate-of-variation.

Apart from curve-fitting all methods have in common that they rely on the immediate factorization of the involved scalings in order to construct stable weights to be

used in a new augmented (re-weighted) plant for designing a robustified controller in
a next iteration step. These immediate factorization steps often cause numerical ill-
conditioning in computations. The key technical twist in Section 5.5.6 was to exploit
Theorem 5.1 and completely avoid any immediate factorization of the scalings.

The key idea of this chapter is to develop a second algorithm for the systematic
synthesis of robust controllers, which is also based on general dynamic IQCs. In contrast to Section 5.5.6, we do proceed in the standard fashion by constructing suitable
factorizations of the IQC-multipliers. However, for general dynamic IQC-multipliers
this is non-trivial. As one of the contributions, we provide an alternative, streamlined and more insightful proof of the factorization in [165] based on the solution of
a general indefinite algebraic Riccati equation (ARE) with a-priori and easily verifiable solvability conditions. As another contribution, we develop a new factorization
for passivity-like IQC-multipliers in order to circumvent a perturbation argument in
[165] which could lead to substantial numerical ill-conditioning in computations. This
involves a novel state-space construction of a one-sided Wiener-Hopf factorization for
tall matrices that is more elementary than those in the literature (see e.g. [57]). As
a third contribution, we reveal the benefit of these factorizations for developing the
second IQC-synthesis algorithm, which also enables us to avoid curve-fitting. More
importantly, and in contrast to all existing approaches, our factorizations are directly constructed by state-space computations, which is the essential ingredient to
use warm-start techniques in the IQC-synthesis algorithm. Depending on the size of
the problem, this can significantly speed up the synthesis process. We emphasize that
the presented results already have been published in [192, 193, 198] and that some
portions of the text overlap.

This chapter is organized as follows. After formally stating the problem in Section 6.2, we briefly recall the robust analysis and nominal controller synthesis LMIs
from Chapter 2 and 3 respectively in Section 6.3. Subsequently, we give an illustrative
example in Section 6.4, which sketches the main technical difficulties of the problem
under consideration. These are resolved in Section 6.5 and 6.6, where we present
our contributions on the factorization of rational matrices and the reformulation of
LMIs. This forms the basis for the main results in Section 6.7: a second algorithm
for the systematic synthesis of robust controllers based on dynamic IQCs. Further,
we will briefly compare the two algorithms and illustrate the main results by means
of two numerical examples in Section 6.8. We conclude the chapter with a summary
in Section 6.9. Some of the proofs are deferred to the appendix.

6.2 Problem formulation

Let us recall from Section 3.2 the linear fractional representation

$$\begin{pmatrix} q \\ z \\ y \end{pmatrix} = \underbrace{\begin{pmatrix} G_{qp} & G_{qw} & G_{qu} \\ G_{zp} & G_{zw} & G_{zu} \\ G_{yp} & G_{yw} & G_{yu} \end{pmatrix}}_{G} \begin{pmatrix} p \\ w \\ u \end{pmatrix}, \quad p = \Delta(q), \tag{6.1}$$

where $p \to q$ represents the uncertainty channel, $w \to z$ the performance channel and $u \to y$ the control channel. As before, the open-loop uncontrolled plant $G \in \mathscr{R}^{(n_q+n_z+n_y) \times (n_p+n_w+n_u)}$ is a real rational and proper transfer matrix that admits a realization of the form

$$G = \begin{pmatrix} G_{qp} & G_{qw} & G_{qu} \\ G_{zp} & G_{zw} & G_{zu} \\ G_{yp} & G_{yw} & G_{yu} \end{pmatrix} = \left[\begin{array}{c|ccc} A & B_p & B_w & B_u \\ \hline C_q & D_{qp} & D_{qw} & D_{qu} \\ C_z & D_{zp} & D_{zw} & D_{zu} \\ C_y & D_{yp} & D_{yw} & 0 \end{array} \right] \quad \text{with} \quad A \in \mathbb{R}^{n_G \times n_G}. \tag{6.2}$$

We assume again that (A, B_u) is stabilizable, that (A, C_y) is detectable and, without loss of generality, that $G_{yu}(\infty) = D_{yu} = 0$. Further, the trouble making component $\Delta : \mathscr{L}_{2e}^{n_q} \to \mathscr{L}_{2e}^{n_p}$ is a bounded and causal operator, which we allow to vary in a given class Δ that satisfies Assumption 2.1 (see Section 2.4.1). We assume that the linear fractional representation (6.1) is well-posed for all $\Delta \in \Delta$.

The ultimate goal of this chapter is the synthesis of a proper LTI system

$$K = \left[\begin{array}{c|c} A_K & B_K \\ \hline C_K & D_K \end{array} \right] \quad \text{with} \quad A_K \in \mathbb{R}^{n_K \times n_K},$$

which dynamically processes the measurement y in order to provide a control input u such that, for all $\Delta \in \Delta$, the system (6.1) is rendered (internally) stable, while a pre-specified quadratic performance criteria is satisfied on the channel $w \to z$, which we assume to be expressed as

$$IQC(\hat{P}, z_1, w_1) + IQC(P_\gamma, z_2, w_2) \le -\epsilon(\|w_1\|_2^2 + \|w_2\|_2^2) \tag{6.3}$$

for all $w_1 \in \mathscr{L}_2^{n_{w_1}}$, for all $w_2 \in \mathscr{L}_2^{n_{w_2}}$ and for some $\epsilon > 0$. Here we partitioned z and w as $z = \mathrm{col}(z_1, z_2)$ and $w = \mathrm{col}(w_1, w_2)$ respectively, while $\hat{P}$ and P_γ are given by

$$\hat{P} = \begin{pmatrix} \hat{P}_{11} & \hat{P}_{12} \\ \hat{P}_{12}^\top & \hat{P}_{22} \end{pmatrix}, \quad \hat{P} \text{ fixed and nonsingular, } \hat{P}_{11} \succcurlyeq 0, \hat{P}_{22} \preccurlyeq 0 \tag{6.4}$$

and

$$P_\gamma = \begin{pmatrix} \gamma^{-1} I_{n_{z_2}} & 0 \\ 0 & -\gamma I_{n_{w_2}} \end{pmatrix}.$$

Quadratic performance is achieved if the latter inequality holds for all trajectories of the closed-loop system $z = (\Delta \star (G \star K))w$. The particular choice of (6.3) will become clearer in the sequel.

6.3 Nominal controller synthesis and robust analysis

As discussed in the introduction, the main goal of this chapter is to develop an algorithm for the systematic synthesis of robust controllers based on the nominal controller synthesis and robust analysis results in Chapter 3 and 2 respectively. In order to improve the readability of the chapter, we will briefly recall the required ingredients in the following two subsections.

6.3.1 Nominal controller synthesis

For the results of this chapter we need a slight generalization of the $\mathscr{H}_\infty$-synthesis results of [48, 72, 162] (which we presented in Section 3.6) in order to be able to handle the quadratic performance criteria (6.3) from above. For this purpose, let us extract from (6.2) the nominal plant

$$
G_{\text{nom}} = \begin{pmatrix} G_{zw} & G_{zu} \\ G_{yw} & G_{yu} \end{pmatrix} = \left[\begin{array}{c|cc} A & B_w & B_u \\ \hline C_z & D_{zw} & D_{zu} \\ C_y & D_{yw} & 0 \end{array} \right] = \left[\begin{array}{c|ccc} A & B_{w_1} & B_{w_2} & B_u \\ \hline C_{z_1} & D_{z_1 w_1} & D_{z_1 w_2} & D_{z_1 u} \\ C_{z_2} & D_{z_2 w_1} & D_{z_2 w_2} & D_{z_2 u} \\ \hline C_y & D_{y w_1} & D_{y w_2} & 0 \end{array} \right]
$$

with a signal-partitioning as in (6.3) and state the following result.

Theorem 6.1. *There exists a controller $K = (A_K, B_K, C_K, D_K)$, which internally stabilizes G_{nom} and renders the performance specification (6.3) on the channel $w \to z$ satisfied if and only if there exist symmetric matrices X, Y for which the following LMIs hold:*

$$
T_y^\top \left[\mathrm{He} \begin{pmatrix} XA & XB_{w_1} & XB_{w_2} & 0 \\ 0 & 0 & 0 & 0 \\ 0 & 0 & -\frac{\gamma}{2} I_{n_{w_2}} & 0 \\ C_{z_2} & D_{z_2 w_1} & D_{z_2 w_2} & -\frac{\gamma}{2} I_{n_{z_2}} \end{pmatrix} + \begin{pmatrix} \star \\ \star \\ \star \end{pmatrix}^\top \hat{P} \begin{pmatrix} C_{z_1} & D_{z_1 w_1} & D_{z_1 w_2} & 0 \\ 0 & I & 0 & 0 \end{pmatrix} \right] T_y \prec 0,
$$

$$
\tag{6.5}
$$

$$T_u^\top \left[\mathrm{He} \begin{pmatrix} YA^\top & YC_{z_1}^\top & YC_{z_2}^\top & 0 \\ 0 & 0 & 0 & 0 \\ 0 & 0 & -\frac{\gamma}{2}I_{n_{z_2}} & 0 \\ B_{w_2}^\top & D_{z_1w_2}^\top & D_{z_2w_2}^\top & -\frac{\gamma}{2}I_{n_{w_2}} \end{pmatrix} - \begin{pmatrix} \star \\ \star \end{pmatrix}^\top \hat{P}^{-1} \begin{pmatrix} 0 & -I & 0 & 0 \\ B_{w_1}^\top & D_{z_1w_1}^\top & D_{z_2w_1}^\top & 0 \end{pmatrix} \right] T_u \prec 0.$$

$$(6.6)$$

$$\begin{pmatrix} X & I \\ I & Y \end{pmatrix} \succ 0, \qquad (6.7)$$

Here, T_y and T_u are basis-matrices of

$$\ker \begin{pmatrix} C_y & D_{yw} & 0_{n_y \times n_z} \end{pmatrix} \quad and \quad \ker \begin{pmatrix} B_u^\top & D_{zu}^\top & 0_{n_u \times n_w} \end{pmatrix}$$

respectively.

Although the particular way of formulating the LMIs (6.5) and (6.6) might seem unusual, we remark that they simplify to the standard form of Theorem 3.4 in Section 3.6 in case $z = z_2$, $w = w_2$ and with void matrices $\hat{P}$, $\hat{P}^{-1}$, B_{w_1}, C_{z_1}, $D_{z_2w_1}$ and $D_{z_1w_2}$.

6.3.2 Robust stability and performance analysis

Let us continue our preparations by recalling the robust analysis results from Section 3.2. As discussed, the controller K stabilizes the closed-loop system (3.2) for all $\Delta \in \mathbf{\Delta}$ and guarantees the induced $\mathscr{L}_2$-gain from w to z to be less than $\gamma > 0$, if the first three assumption in Theorem 3.1 are satisfied and if there exist some $P \in \mathbf{P}$ for which the following FDI is satisfied:

$$\begin{pmatrix} \Psi_1 \mathcal{G}_{qp} + \Psi_2 & \Psi_1 \mathcal{G}_{qw} \\ \hline \mathcal{G}_{zp} & \mathcal{G}_{zw} \\ 0 & I \end{pmatrix}^* \begin{pmatrix} P & 0 \\ \hline 0 & P_\gamma \end{pmatrix} \begin{pmatrix} \Psi_1 \mathcal{G}_{qp} + \Psi_2 & \Psi_1 \mathcal{G}_{qw} \\ \hline \mathcal{G}_{zp} & \mathcal{G}_{zw} \\ 0 & I \end{pmatrix} \prec 0 \text{ on } \mathbb{C}^0. \quad (6.8)$$

By elementary manipulations (Schur) this inequality can be rendered affine in γ as follows:

$$\mathrm{He} \begin{pmatrix} 0 & 0 & 0 \\ 0 & -\frac{\gamma}{2}I_{n_w} & 0 \\ \mathcal{G}_{zp} & \mathcal{G}_{zw} & -\frac{\gamma}{2}I_{n_z} \end{pmatrix} + (\star)^* P \begin{pmatrix} \Psi_1 \mathcal{G}_{qp} + \Psi_2 & \Psi_1 \mathcal{G}_{qw} & 0 \end{pmatrix} \prec 0 \text{ on } \mathbb{C}^0. \quad (6.9)$$

As discussed in Section 3.5, this test can be easily turned into a computations by applying the KYP-lemma. Indeed, if we recall the realizations (3.21) and (3.23), then we can infer that the FDI (6.8) is equivalent to the existence of a symmetric matrix $\mathcal{X} \in \mathbb{S}^{n_\Psi + n_G + n_K}$ for which the following LMI is satisfied:

$$
\begin{pmatrix} \star \\ \star \\ \star \\ \star \\ \star \\ \star \end{pmatrix}^\top
\begin{pmatrix} 0 & \mathcal{X} & 0 & 0 \\ \mathcal{X} & 0 & 0 & 0 \\ 0 & 0 & P & 0 \\ 0 & 0 & 0 & P_\gamma \end{pmatrix}
\begin{pmatrix}
I & 0 & 0 \\
A_\Psi & B_{\Psi 1}\mathcal{C}_q & B_{\Psi 1}\mathcal{D}_{qp} + B_{\Psi 2} & B_{\Psi 1}\mathcal{D}_{qw} \\
0 & \mathcal{A} & \mathcal{B}_p & \mathcal{B}_w \\
\mathcal{C}_\Psi & D_{\Psi 1}\mathcal{C}_q & D_{\Psi 1}\mathcal{D}_{qp} + D_{\Psi 2} & D_{\Psi 1}\mathcal{D}_{qw} \\
0 & \mathcal{C}_z & \mathcal{D}_{zp} & \mathcal{D}_{zw} \\
0 & 0 & 0 & I
\end{pmatrix} \prec 0.
$$

$$(6.10)$$

Here $\mathcal{X}$ is again assumed to be partitioned as in (3.25). If the LMI (6.10) holds we say that it certifies the FDI (6.9), or that $\mathcal{X}$ is a certificate for the FDI (6.9). Further note that (6.10) is a special case of (3.24), if voiding all the block matrices that are related to gain-scheduling. Finally, analogously to (6.8), we can also render the LMI (6.10) affine in γ and somewhat more compactly and elegantly rewrite (6.10) as

$$
\mathrm{He}\left[\begin{pmatrix} \mathcal{X} & 0 \\ 0 & I \end{pmatrix}
\begin{pmatrix}
A_\Psi & B_{\Psi 1}\mathcal{C}_q & B_{\Psi 1}\mathcal{D}_{qp} + B_{\Psi 2} & B_{\Psi 1}\mathcal{D}_{qw} & 0 \\
0 & \mathcal{A} & \mathcal{B}_p & \mathcal{B}_w & 0 \\
0 & 0 & 0 & 0 & 0 \\
0 & 0 & 0 & -\frac{\gamma}{2}I_{n_w} & 0 \\
0 & \mathcal{C}_z & \mathcal{D}_{zp} & \mathcal{D}_{zw} & -\frac{\gamma}{2}I_{n_z}
\end{pmatrix} \right]
$$
$$
+ (\star)^\top P \left(\mathcal{C}_\Psi \quad D_{\Psi 1}\mathcal{C}_q \quad D_{\Psi 1}\mathcal{D}_{qp} + D_{\Psi 2} \quad D_{\Psi 1}\mathcal{D}_{qw} \quad 0 \right) \prec 0. \quad (6.11)
$$

We can now restate the following IQC-analysis result.

Theorem 6.2. *Let the controller K be given and assume that*

1. *K internally stabilizes G (i.e. $\mathcal{A}$ is Hurwitz);*
2. *the closed-loop system (3.2) is well-posed for all $\Delta \in \mathbf{\Delta}$;*
3. *the IQC (3.3) is satisfied for all $\Delta \in \mathbf{\Delta}$ and for all $P \in \mathbf{P}$;*
4. *there exists some $\mathcal{X} \in \mathbb{S}^{n_\Psi + n_G + n_K}$, $P \in \mathbf{P}$ for which the LMI (6.10) holds.*

Then, for all $\Delta \in \mathbf{\Delta}$, the controller K (internally) stabilizes the closed-loop system (3.2) and guarantees the induced $\mathscr{L}_2$-gain from w to z to be less than $\gamma > 0$.

6.4 From analysis to synthesis

For the to be developed algorithm we require, in each iteration step, an appropriate factorization of the IQC-multipliers in order to find suitable stable weights for a new augmented (re-weighted) plant that can be used for the synthesis of a robustified controller in a next iteration step. In contrast to the situation for static (frequency independent) IQC-multipliers, the construction of these weights is rather delicate for dynamic (frequency dependent) IQC-multipliers, as illustrated by means of the following example.

Example 6.1. Consider the IQC-multiplier 2.5 from Section 2.6.3.1 for the class of LTI real diagonally repeated parametric uncertainties. Replacing $\Psi^* P \Psi$ in (6.9) with

$$\Psi^* P \Psi := \begin{pmatrix} \Psi_{11} & 0 \\ 0 & \Psi_{22} \end{pmatrix}^* \begin{pmatrix} P_{11} & P_{12} \\ P_{12}^\top & -P_{11} \end{pmatrix} \begin{pmatrix} \Psi_{11} & 0 \\ 0 & \Psi_{22} \end{pmatrix} \tag{6.12}$$

for $\Psi_{11} := \alpha \psi_\nu \otimes I_{n_q}$ and $\Psi_{22} := \psi_\nu \otimes I_{n_q}$ then yields the following FDI:

$$\mathrm{He} \begin{pmatrix} 0 & 0 & 0 \\ 0 & -\frac{\gamma}{2}I & 0 \\ \mathcal{G}_{zp} & \mathcal{G}_{zw} & -\frac{\gamma}{2}I \end{pmatrix} + \begin{pmatrix} \star \\ \star \\ \star \end{pmatrix}^* \begin{pmatrix} P_{11} & P_{12} \\ P_{12}^\top & -P_{11} \end{pmatrix} \begin{pmatrix} \Psi_{11}\mathcal{G}_{qp} & \Psi_{11}\mathcal{G}_{qw} & 0 \\ \Psi_{22} & 0 & 0 \end{pmatrix} \prec 0 \text{ on } \mathbb{C}^0.$$

$$\tag{6.13}$$

Here Ψ_{22} is tall and thus not invertible; even if being square and invertible on $\mathbb{C}^0$, there is no guarantee that Ψ_{22}^{-1} is stable; moreover, P_{11} is in general indefinite and does not satisfy (6.4). All this prevents us from rewriting (6.13) as a standard quadratic performance constraint. As a remedy, we can factorize (6.12) as

$$\hat\psi^* \hat P \hat\psi := \begin{pmatrix} \hat\Psi_1 & \hat\Psi_3 \\ 0 & \hat\Psi_2 \end{pmatrix}^* \begin{pmatrix} \hat P_{11} & \hat P_{12} \\ \hat P_{12}^\top & \hat P_{22} \end{pmatrix} \begin{pmatrix} \hat\Psi_1 & \hat\Psi_3 \\ 0 & \hat\Psi_2 \end{pmatrix},$$

where $\hat P$ satisfies (6.4), where $\hat\Psi_1$ and $\hat\Psi_3$ are stable and where $\hat\Psi_2$ is minimal phase. Then it is possible to equivalently reformulate (6.13) as

$$\mathrm{He} \begin{pmatrix} 0 & 0 & 0 \\ 0 & -\frac{\gamma}{2}I & 0 \\ \mathcal{G}_{zp} & \mathcal{G}_{zw} & -\frac{\gamma}{2}I \end{pmatrix} + \begin{pmatrix} \star \\ \star \\ \star \end{pmatrix}^* \hat P \begin{pmatrix} \hat\Psi_1\mathcal{G}_{qp} + \hat\Psi_3 & \hat\Psi_1\mathcal{G}_{qw} & 0 \\ \hat\Psi_2 & 0 & 0 \end{pmatrix} \prec 0 \text{ on } \mathbb{C}^0$$

and thus, by congruence, as

$$\mathrm{He} \begin{pmatrix} 0 & 0 & 0 \\ 0 & -\frac{\gamma}{2}I & 0 \\ \mathcal{G}_{zp}\hat\Psi_2^{-1} & \mathcal{G}_{zw} & -\frac{\gamma}{2}I \end{pmatrix} + \begin{pmatrix} \star \\ \star \\ \star \end{pmatrix}^* \hat P \begin{pmatrix} (\hat\Psi_1\mathcal{G}_{qp} + \hat\Psi_3)\hat\Psi_2^{-1} & \hat\Psi_1\mathcal{G}_{qw} & 0 \\ I & 0 & 0 \end{pmatrix} \prec 0 \text{ on } \mathbb{C}^0.$$

This allows us to extract a re-weighted open-loop plant $\hat{\mathcal{G}}_{\mathrm{nom}}$ satisfying

$$\begin{pmatrix} (\hat{\Psi}_1 \mathcal{G}_{qp} + \hat{\Psi}_3)\hat{\Psi}_2^{-1} & \hat{\Psi}_1 \mathcal{G}_{qw} \\ \mathcal{G}_{zp}\hat{\Psi}_2^{-1} & \mathcal{G}_{zw} \end{pmatrix} = \underbrace{\left(\begin{array}{cc|c} (\hat{\Psi}_1 G_{qp} + \hat{\Psi}_3)\hat{\Psi}_2^{-1} & \hat{\Psi}_1 G_{qw} & \hat{\Psi}_1 G_{qu} \\ \hline G_{zp}\hat{\Psi}_2^{-1} & G_{zw} & G_{zu} \\ \hline G_{yp}\hat{\Psi}_2^{-1} & G_{yw} & G_{yu} \end{array} \right)}_{\hat{G}_{\text{nom}}} \star K,$$

which can be used for the synthesis of a robustified controller with Theorem 6.1.

Although Example 6.1 illustrates that the initial IQC-multiplier factorization can be simply replaced with a new one, it is far from trivial to see how this can be done by using state-space arguments for the corresponding LMI (6.11). This is of particular interest since a new certificate for the modified FDI can serve as a suitable initial condition in Theorem 6.1 in order to speed up the controller synthesis, possibly significantly. Indeed, in the sequel, we will present two multiplier-factorizations (which are of independent interest) that will prove to be very useful in the IQC-synthesis algorithm for the synthesis of robust controllers.

6.5 On the factorization of IQC-multipliers

As illustrated, it is relevant to consider the problem of factorizing the IQC-multiplier $\Pi = \Psi^* P \Psi$ as $\Pi = \hat{\Psi}^* \hat{P} \hat{\Psi}$, where $\hat{P}$ satisfies (6.4), where $\hat{\Psi}$ is structured as

$$\hat{\Psi} := \begin{pmatrix} \hat{\Psi}_1 & \hat{\Psi}_3 \\ 0 & \hat{\Psi}_2 \end{pmatrix} \quad \text{with} \quad \hat{\Psi}_1, \hat{\Psi}_2, \hat{\Psi}_2^{-1}, \hat{\Psi}_3 \in \mathcal{RH}_\infty^{\bullet \times \bullet} \tag{6.14}$$

and given that all assumptions in Theorem 6.2 are true. Then the IQC (3.3) and the FDI (6.9) are satisfied with $\Pi = \Psi^* P \Psi$ for all $\Delta \in \boldsymbol{\Delta}$. This implies that Π is, in general, non-negative on $\text{im}(\text{col}(I_{n_q}, 0))$ and that it has n_p negative eigenvalues on $\mathbb{C}^0$. Since the number of positive eigenvalues is not fixed over frequency, it is difficult to make a general statement about the existence of a structured factorization of Π as $\hat{\Psi}^* \hat{P} \hat{\Psi}$. Fortunately, many existing IQC-multipliers, and all the ones discussed in Chapter 2 (with perturbations if necessary), can be either parameterized as

$$\Pi = \Psi^* P \Psi = \begin{pmatrix} \Psi_1 & \Psi_2 \end{pmatrix}^* P \begin{pmatrix} \Psi_1 & \Psi_2 \end{pmatrix} = \begin{pmatrix} \Psi_1^* P \Psi_1 & \Psi_1^* P \Psi_2 \\ \Psi_2^* P \Psi_1 & \Psi_2^* P \Psi_2 \end{pmatrix} \tag{6.15}$$

with $\Pi_{11} = \Psi_1^* P \Psi_1 \succ 0$ on $\mathbb{C}^0$, or as

$$\Pi = \Psi^* P \Psi = \begin{pmatrix} \Psi_1 & \Psi_2 \end{pmatrix}^* P \begin{pmatrix} \Psi_1 & \Psi_2 \end{pmatrix} = \begin{pmatrix} 0 & \Psi_{11}^* P_{12} \Psi_{22} \\ \Psi_{22}^* P_{12}^\top \Psi_{11} & \Psi_{22}^* P_{22} \Psi_{22} \end{pmatrix}, \tag{6.16}$$

with $\Psi = (\Psi_1\ \Psi_2) := \mathrm{diag}(\Psi_{11}, \Psi_{22})$, $\Pi_{11} = 0$ and tall $\Pi_{12}^* := \Psi_{22}^* P_{12}^\top \Psi_{11}$. In view of (6.14) and Theorem 6.1, this motivates the search for either one of the following factorizations:

$$
\hat{\Psi}^* \hat{P} \hat{\Psi} = \begin{pmatrix} \star \\ \star \end{pmatrix}^* \begin{pmatrix} I_{n_q} & 0 \\ 0 & -I_{n_p} \end{pmatrix} \begin{pmatrix} \hat{\Psi}_1 & \hat{\Psi}_3 \\ 0 & \hat{\Psi}_2 \end{pmatrix} = \begin{pmatrix} \hat{\Psi}_1^* \hat{\Psi}_1 & \hat{\Psi}_1^* \hat{\Psi}_3 \\ \hat{\Psi}_3^* \hat{\Psi}_1 & \hat{\Psi}_3^* \hat{\Psi}_3 - \hat{\Psi}_2^* \hat{\Psi}_2 \end{pmatrix}, \quad (6.17a)
$$

$$
\hat{\Psi}^* \hat{P} \hat{\Psi} = \begin{pmatrix} \star \\ \star \end{pmatrix}^* \begin{pmatrix} 0 & I_{n_p} \\ I_{n_p} & 0 \end{pmatrix} \begin{pmatrix} \hat{\Psi}_1 & \hat{\Psi}_3 \\ 0 & \hat{\Psi}_2 \end{pmatrix} = \begin{pmatrix} 0 & \hat{\Psi}_1^* \hat{\Psi}_2 \\ \hat{\Psi}_2^* \hat{\Psi}_1 & \hat{\Psi}_3^* \hat{\Psi}_2 + \hat{\Psi}_2^* \hat{\Psi}_3 \end{pmatrix}. \quad (6.17b)
$$

Even though the factorization of rational matrices has been studied intensively (think e.g. of Wiener-Hopf, canonical, spectral and J-spectral factorizations and see e.g. [16] and references therein), we emphasize that those of this particular form are rarely addressed [55, 56, 165]. On the one hand, it has been shown in [56, 165] that (6.15) can be factorized as (6.17a). However, the results are not applicable for IQC-multipliers of the form (6.16) without perturbations, which might lead to numerical ill-conditioning. On the other hand, it has been shown in [55] how to construct a canonical factorization for generalized positive real transfer functions. This leads to the desired factorization for a very special case of (6.16) in which Π_{12} is square with $\mathrm{He}(\Pi_{12}) \succ 0$ and $\Pi_{22} = 0$. Both limitations are overcome next.

Indeed, depending on whether the left-upper block of the IQC-multiplier is strictly positive or zero on $\mathbb{C}^0$, we now present two constructive procedures for factorizing (6.15) and (6.16) as (6.17a) and (6.17b) respectively. To this end, we need the following two assumptions.

Assumption 6.1. For the parameterization (6.15) there exist at least one $P \in \boldsymbol{P}$ with

$$
\begin{pmatrix} \mathcal{G}_{qp} \\ I \end{pmatrix}^* \Psi^* P \Psi \begin{pmatrix} \mathcal{G}_{qp} \\ I \end{pmatrix} \prec 0 \text{ on } \mathbb{C}^0 \qquad (6.18)
$$

and

$$
\begin{pmatrix} I \\ 0 \end{pmatrix}^* \Psi^* P \Psi \begin{pmatrix} I \\ 0 \end{pmatrix} \succ 0 \text{ on } \mathbb{C}^0.
$$

Assumption 6.2. For the parameterization (6.16) there exist at least one $P \in \boldsymbol{P}$ and an LTI $\Delta \in \boldsymbol{\Delta}$ with

$$
\begin{pmatrix} \mathcal{G}_{qp} \\ I \end{pmatrix}^* \Psi^* P \Psi \begin{pmatrix} \mathcal{G}_{qp} \\ I \end{pmatrix} \prec 0 \text{ on } \mathbb{C}^0, \qquad \begin{pmatrix} I \\ \Delta \end{pmatrix}^* \Psi^* P \Psi \begin{pmatrix} I \\ \Delta \end{pmatrix} \succ 0 \text{ on } \mathbb{C}^0
$$

and

$$
\begin{pmatrix} I \\ \tau\Delta \end{pmatrix}^* \Psi^* P \Psi \begin{pmatrix} I \\ \tau\Delta \end{pmatrix} \not\succcurlyeq 0 \text{ on } \mathbb{C}^0 \ \ \forall \tau \in [0,1).
$$

We emphasize that either one of these assumptions hold for the IQC-multipliers in Chapter 2 (slightly perturb if necessary). Consequently, $\Pi = \Psi^* P \Psi$ is nonsingular and has n_p negative and n_q positive eigenvalues. Moreover, as we have seen in Section 4.4.2, Assumption 6.1 implies that $\Pi = \Psi^* P \Psi$ satisfies the inertia constraints

$$\Psi_1^* P \Psi_1 \succ 0 \text{ on } \mathbb{C}^0, \tag{6.19a}$$

$$\Psi_2^* P \Psi_2 - \Psi_2^* P \Psi_1 \left(\Psi_1^* P \Psi_1\right)^{-1} \Psi_1^* P \Psi_2 \prec 0 \text{ on } \mathbb{C}^0. \tag{6.19b}$$

On the other hand, if Assumption 6.2 holds, we infer that $\Pi_{12}^* = \Psi_{22}^* P_{12}^\top \Psi_{11}$ must have full-column rank on $\mathbb{C}^0$. These facts allow us to state the following generalization of Lemma 4.3.

Lemma 6.1. Let Ψ admit the stable and controllable realization (3.23) and suppose that either Assumption 6.1 or 6.2 holds true. Then there exist matrices

$$\hat{P} = \begin{pmatrix} I_{n_q} & 0 \\ 0 & -I_{n_p} \end{pmatrix} \quad or \quad \hat{P} = \begin{pmatrix} 0 & I_{n_p} \\ I_{n_p} & 0 \end{pmatrix}$$

in case of Assumption 6.1 or 6.2 respectively, and transfer matrices $\hat{\Psi}_1, \hat{\Psi}_2, \hat{\Psi}_3 \in \mathscr{RH}_\infty^{\bullet \times \bullet}$ with $\hat{\Psi}_2^{-1} \in \mathscr{RH}_\infty^{n_p \times n_p}$ as well as a symmetric matrix Z such that $\Pi = \Psi^* P \Psi$ can be factorized as

$$\Psi^* P \Psi = \hat{\Psi}^* \hat{P} \hat{\Psi} \tag{6.20}$$

with

$$\hat{\Psi} = \begin{pmatrix} \hat{\Psi}_1 & \hat{\Psi}_3 \\ 0 & \hat{\Psi}_2 \end{pmatrix} =$$

$$= \left[\begin{array}{c|cc} \hat{A}_\Psi & \hat{B}_{\Psi 1} & \hat{B}_{\Psi 2} \\ \hline \hat{C}_{\Psi 1} & \hat{D}_{\Psi 1} & \hat{D}_{\Psi 3} \\ \hat{C}_{\Psi 2} & 0 & \hat{D}_{\Psi 2} \end{array}\right] = \left[\begin{array}{cc|cc} A_1 & 0 & B_1 & 0 \\ 0 & \hat{A}_2 & 0 & \hat{B}_2 \\ \hline \hat{C}_1 & \hat{C}_3 & \hat{D}_1 & \hat{D}_3 \\ 0 & \hat{C}_2 & 0 & \hat{D}_2 \end{array}\right] = \left[\begin{array}{ccc|cc} A_1 & 0 & 0 & B_1 & 0 \\ 0 & A_{e1} & A_{e2} & 0 & B_e \\ 0 & 0 & A_2 & 0 & B_2 \\ \hline \hat{C}_1 & \hat{C}_{3,1} & \hat{C}_{3,2} & \hat{D}_1 & \hat{D}_3 \\ 0 & \hat{C}_{2,1} & \hat{C}_{2,2} & 0 & \hat{D}_2 \end{array}\right] \tag{6.21}$$

and

$$\begin{pmatrix} \star \\ \star \\ \star \\ \star \\ \star \\ \star \end{pmatrix}^\top \left(\begin{array}{c|c|c|c} 0 & Z & 0 & 0 \\ \hline Z & 0 & 0 & 0 \\ \hline 0 & 0 & \hat{P} & 0 \\ \hline 0 & 0 & 0 & -P \end{array}\right) \left(\begin{array}{c|cc} I & 0 & 0 \\ \hline \hat{A}_\Psi & \hat{B}_{\Psi 1} & \hat{B}_{\Psi 2} \\ \hline \hat{C}_{\Psi 1} & \hat{D}_{\Psi 1} & \hat{D}_{\Psi 3} \\ \hline \hat{C}_{\Psi 2} & 0 & \hat{D}_{\Psi 2} \\ \hline C_\Psi T_1 & D_{\Psi 1} & D_{\Psi 2} \end{array}\right) = 0. \tag{6.22}$$

Here $\hat{A}_\Psi \in \mathbb{R}^{n_{\hat{\Psi}} \times n_{\hat{\Psi}}}$, $\mathrm{eig}(\hat{A}_\Psi) \subset \mathbb{C}^-$ and T_1 is structured as

$$T_1 = \begin{pmatrix} I & 0 & 0 \\ 0 & 0 & I \end{pmatrix} \tag{6.23}$$

in accordance with the block structure of (6.21) and such that $C_\Psi T_1 = \begin{pmatrix} C_1 & 0 & C_2 \end{pmatrix}$.

In order to stay focussed on the main results, and due to its technical and lengthly nature, the proof of Lemma 6.1 has been deferred to Appendix G. Here both factorizations along with the constructive proofs are presented as auxiliary results.

6.6 Warm-start techniques for robust controller synthesis

It is important to note that the previous factorizations are directly constructed by state-space computations (See Appendix G). It has been shown in Chapter 4 (in another context) that this enables us to replace the initial IQC-multiplier $\Psi^* P \Psi$ appearing in the LMI (6.11) by the new factorization $\hat{\Psi}^* \hat{P} \hat{\Psi}$. As a first benefit, and analogously to Example 6.1, this allows us to extract a re-weighted open-loop plant $\hat{G}_{\mathrm{nom}}$, which can be used for the synthesis of a robustified controller with Theorem 6.1. Moreover, as a second advantage, it also becomes possible to use warm-start techniques in the IQC-synthesis algorithm. Depending on the size of the problem, this can significantly speed up the synthesis process. In this section we show how to generate suitable initial conditions for each optimization step in a synthesis/analysis algorithm.

6.6.1 Constructing initial conditions for the synthesis step

Let us start by showing that it is possible to construct initial conditions for the controller synthesis step after the robustness analysis. For this purpose, suppose that all assumptions in Theorem 6.2 are true. Then, given the 'old' IQC-multiplier factorization $\Psi^* P \Psi$ appearing in the FDI (6.9), we can compute the 'new' factorization $\hat{\Psi}^* \hat{P} \hat{\Psi}$ in accordance with Lemma 6.1, and, as illustrated in Example 6.1, simply replace the old factorization by the new one. As a key contribution of this chapter, let us now argue how this can be performed by state-space arguments for the corresponding LMI (6.11). To this end, we need the following intermediate result.

Lemma 6.2. *Suppose that all assumptions in Theorem 6.2 are true with Assumption 6.1 or 6.2. Then there exists a Z with (6.22), some X_e with $\mathrm{He}(X_e T_2 \hat{A}_\Psi T_2^\top) = \mathrm{He}(X_e A_{e1}) \prec 0$ as well as an $\epsilon > 0$ (which can be chosen arbitrarily small) such that*

$$
\text{He}\left[\begin{pmatrix} \hat{\mathcal{X}} & 0 \\ 0 & I \end{pmatrix}\begin{pmatrix} \hat{A}_{\Psi} & \hat{B}_{\Psi 1}\mathcal{C}_q & \hat{B}_{\Psi 1}\mathcal{D}_{qp}+\hat{B}_{\Psi 2} & \hat{B}_{\Psi 1}\mathcal{D}_{qw} & 0 \\ 0 & \mathcal{A} & \mathcal{B}_p & \mathcal{B}_w & 0 \\ 0 & 0 & 0 & 0 & 0 \\ 0 & 0 & 0 & -\frac{\gamma}{2}I_{n_w} & 0 \\ 0 & \mathcal{C}_z & \mathcal{D}_{zp} & \mathcal{D}_{zw} & -\frac{\gamma}{2}I_{n_z} \end{pmatrix}\right] +
$$

$$
+\begin{pmatrix} \star \\ \star \end{pmatrix}^{*}\hat{P}\begin{pmatrix} \hat{C}_{\Psi 1} & \hat{D}_{\Psi 1}\mathcal{C}_q & \hat{D}_{\Psi 1}\mathcal{D}_{qp}+\hat{D}_{\Psi 3} & \hat{D}_{\Psi 1}\mathcal{D}_{qw} & 0 \\ \hat{C}_{\Psi 2} & 0 & \hat{D}_{\Psi 2} & 0 & 0 \end{pmatrix} \prec 0. \quad (6.24)
$$

Here $T_2^{\top} = \begin{pmatrix} 0 & I & 0 \end{pmatrix}^{\top}$ is a basis for the kernel of T_1 and $\hat{\mathcal{X}}$ is, analogously to $\mathcal{X}$, structured as

$$
\hat{\mathcal{X}} = \begin{pmatrix} T_1^{\top}X_{11}T_1 + \epsilon T_2^{\top}X_e T_2 + Z & T_1^{\top}\mathcal{X}_{12} \\ \mathcal{X}_{12}^{\top}T_1 & \mathcal{X}_{22} \end{pmatrix} =
$$

$$
= \begin{pmatrix} X_{11,11}+Z_{11} & Z_{12} & X_{11,12}+Z_{13} & \mathcal{X}_{12,1} \\ Z_{12}^{\top} & \epsilon X_e + Z_{22} & Z_{23} & 0 \\ X_{11,12}^{\top}+Z_{13}^{\top} & Z_{23}^{\top} & X_{11,22}+Z_{33} & \mathcal{X}_{12,2} \\ \mathcal{X}_{12,1}^{\top} & 0 & \mathcal{X}_{12,1}^{\top} & \mathcal{X}_{22} \end{pmatrix} =
$$

$$
= \left(\begin{array}{cc|c} T_1^{\top}X_{11}T_1 + \epsilon T_2^{\top}X_e T_2 + Z & T_1^{\top}\mathcal{X}_{12} & T_1^{\top}U_1 \\ \mathcal{X}_{12}^{\top}T_1 & X_{22} & U_2 \\ \hline U_1^{\top}T_1 & U_2^{\top} & \bar{X} \end{array}\right) = \left(\begin{array}{c|c} \hat{X} & \hat{U} \\ \hline \hat{U}^{\top} & \bar{X} \end{array}\right) \quad (6.25)
$$

with X_{11} and Z being partitioned according to the block structure of Ψ as in (3.23).

Proof. For the proof we will rely on two auxiliary results which are found in Appendix E.1 and F.1 respectively. Since all assumptions in Theorem 6.2 are true with Assumption 6.1 or 6.2, we infer by Lemma 6.1 that there exist a factorization (6.20) as well as a matrix Z with (6.22).

The key of the proof is to merge (6.11) and (6.22) in order to obtain (6.24). The corresponding operation in frequency-domain would be the replacement of the initial factorization $\Psi^{*}P\Psi$ in the FDI (6.9) with the new factorization $\hat{\Psi}^{*}\hat{P}\hat{\Psi}$.

In order to do so, we introduce unobservable dynamics in the LMI (6.11), such that (6.11) and (6.22) are both expressed in terms of the realization matrices $\hat{A}_{\Psi}$, $\hat{B}_{\Psi 1}$ and $\hat{B}_{\Psi 2}$. This can be done by defining the unobservable realization

$$\begin{pmatrix} \Psi_1 & \Psi_2 \end{pmatrix} = \left[\begin{array}{c|cc} \hat{A}_\Psi & \hat{B}_{\Psi 1} & \hat{B}_{\Psi 2} \\ \hline C_\Psi T_1 & D_{\Psi 1} & D_{\Psi 2} \end{array} \right] =$$

$$= \left[\begin{array}{cc|cc} A_1 & 0 & B_1 & 0 \\ 0 & \hat{A}_2 & 0 & \hat{B}_2 \\ \hline C_1 & \begin{pmatrix} 0 & C_2 \end{pmatrix} & D_1 & D_2 \end{array} \right] = \left[\begin{array}{ccc|cc} A_1 & 0 & 0 & B_1 & 0 \\ 0 & A_{e1} & A_{e2} & 0 & B_e \\ 0 & 0 & A_2 & 0 & B_2 \\ \hline C_1 & 0 & C_2 & D_1 & D_2 \end{array} \right] \quad (6.26)$$

and fixing the symmetric matrix X_e such that $\mathrm{He}(X_e T_2 \hat{A}_\Psi T_2^\top) = \mathrm{He}(X_e A_{e1}) \prec 0$. Then it is possible to apply Lemma E.1 in Appendix E.1 and fix an $\epsilon > 0$ (which can be chosen arbitrarily small) in order to infer that (4.11) is equivalent to

$$\left[\mathrm{He} \left[\begin{pmatrix} \tilde{\mathcal{X}} & 0 \\ 0 & I \end{pmatrix} \begin{pmatrix} \hat{A}_\Psi & \hat{B}_{\Psi 1}\mathcal{C}_q & \hat{B}_{\Psi 1}\mathcal{D}_{qp} + \hat{B}_{\Psi 2} & \hat{B}_{\Psi 1}\mathcal{D}_{qw} & 0 \\ 0 & \mathcal{A} & \mathcal{B}_p & \mathcal{B}_w & 0 \\ 0 & 0 & 0 & 0 & 0 \\ 0 & 0 & 0 & -\frac{\gamma}{2}I_{n_w} & 0 \\ 0 & \mathcal{C}_z & \mathcal{D}_{zp} & \mathcal{D}_{zw} & -\frac{\gamma}{2}I_{n_z} \end{pmatrix} \right] + \right.$$

$$\left. + (\star)^\top P \begin{pmatrix} C_\Psi T_1 & D_{\Psi 1}\mathcal{C}_q & D_{\Psi 1}\mathcal{D}_{qp} + D_{\Psi 2} & D_{\Psi 1}\mathcal{D}_{qw} & 0 \end{pmatrix} \right] \prec 0. \quad (6.27)$$

Here $\tilde{\mathcal{X}}$ is structured as

$$\tilde{\mathcal{X}} = \begin{pmatrix} T_1^\top X_{11} T_1 + \epsilon T_2^\top X_e T_2 & T_1^\top \mathcal{X}_{12} \\ \mathcal{X}_{12}^\top T_1 & \mathcal{X}_{22} \end{pmatrix} = \begin{pmatrix} X_{11,11} & 0 & X_{11,12} & \mathcal{X}_{12,1} \\ 0 & \epsilon X_e & 0 & 0 \\ X_{11,12}^\top & 0 & X_{11,22} & \mathcal{X}_{12,2} \\ \mathcal{X}_{12,1}^\top & 0 & \mathcal{X}_{12,1}^\top & \mathcal{X}_{22} \end{pmatrix}.$$

It is now possible to systematically merge (6.27) and (6.22) by applying the so-called gluing lemma [159, Theorem 3] (See also Lemma F.1 in Appendix F.1). This directly yields the inequality (6.24) and proves the claim. ∎

In conclusion, we have replaced the initial factorization $\Psi^* P \Psi$ as appearing in the LMI (6.11) by the new one $\hat{\Psi}^* \hat{P} \hat{\Psi}$ affecting the inequality (6.24) and directly constructed a corresponding certificate (6.25). This allow us to exploit the properties of the factorizations in (6.20), and in particular the fact that

$$\hat{\Psi}_2 = \left[\begin{array}{c|c} \hat{A}_2 & \hat{B}_2 \\ \hline \hat{C}_2 & \hat{D}_2 \end{array} \right] \quad \text{and} \quad \hat{\Psi}_2^{-1} = \left[\begin{array}{c|c} \hat{A}_2 - \hat{B}_2 \hat{D}_2^{-1} \hat{C}_2 & \hat{B}_2 \hat{D}_2^{-1} \\ \hline -\hat{D}_2^{-1} \hat{C}_2 & \hat{D}_2^{-1} \end{array} \right]$$

are both Hurwitz, in order to reformulate the robust analysis LMI into its final desired form. This yields the following key result.

Lemma 6.3. *The certificate* (6.25) *is positive definite and the matrices*

$$\gamma_{\text{init}} := \gamma, \quad X_{\text{init}} := \hat{X}, \quad Y_{\text{init}} := \left(\hat{X} - \hat{U} \bar{X}^{-1} \hat{U}^{\top} \right)^{-1} \quad (6.28)$$

satisfy the synthesis inequalities (6.7), (6.5) *and* (6.6) *of Theorem 6.1 for the re-weighted plant*

$$
\hat{G}_{nom} =
\left[
\begin{array}{c|cc|c}
\hat{A} & \hat{B}_{w1} & \hat{B}_{w2} & \hat{B}_u \\
\hline
\hat{C}_{z1} & \hat{D}_{z_1 w_1} & \hat{D}_{z_1 w_2} & \hat{D}_{z_1 u} \\
\hat{C}_{z2} & \hat{D}_{z_2 w_1} & \hat{D}_{z_2 w_2} & \hat{D}_{z_2 u} \\
\hline
\hat{C}_y & \hat{D}_{y w_1} & \hat{D}_{y w_2} & 0
\end{array}
\right] =
$$

$$
=
\left[
\begin{array}{cc|cc|c}
\hat{A}_\Psi - \hat{L}_1 \hat{D}_{\Psi 2}^{-1} \hat{C}_{\Psi 2} & \hat{B}_{\Psi 1} C_q & \hat{L}_1 \hat{D}_{\Psi 2}^{-1} & \hat{B}_{\Psi 1} D_{qw} & \hat{B}_{\Psi 1} D_{qu} \\
-B_p \hat{D}_{\Psi 2}^{-1} \hat{C}_{\Psi 2} & A & B_p \hat{D}_{\Psi 2}^{-1} & B_w & B_u \\
\hline
\hat{C}_{\Psi 1} - \hat{L}_2 \hat{D}_{\Psi 2}^{-1} \hat{C}_{\Psi 2} & \hat{D}_{\Psi 1} C_q & \hat{L}_2 \hat{D}_{\Psi 2}^{-1} & \hat{D}_{\Psi 1} D_{qw} & \hat{D}_{\Psi 1} D_{qu} \\
-D_{zp} \hat{D}_{\Psi 2}^{-1} \hat{C}_{\Psi 2} & C_q & D_{zp} \hat{D}_{\Psi 2}^{-1} & D_{zw} & D_{zu} \\
\hline
-D_{yp} \hat{D}_{\Psi 2}^{-1} \hat{C}_{\Psi 2} & C_y & D_{yp} \hat{D}_{\Psi 2}^{-1} & D_{yw} & 0
\end{array}
\right], \quad (6.29)
$$

which is stabilizable through u and detectable from y. Here $\hat{L}_1$ *and* $\hat{L}_2$ *are defined through* $\hat{L}_1 = \hat{B}_{\Psi 1} D_{qp} + \hat{B}_{\Psi 2}$ *and* $\hat{L}_2 = \hat{D}_{\Psi 1} D_{qp} + \hat{D}_{\Psi 3}$ *respectively.*

Proof. Let us start by employing Lemma 6.2 and performing the following congruence transformation on (6.24):

$$
\left(
\begin{array}{cc|ccc}
I & 0 & 0 & 0 & 0 \\
0 & I & 0 & 0 & 0 \\
\hline
-\hat{D}_{\Psi 2}^{-1} \hat{C}_{\Psi 2} & 0 & \hat{D}_{\Psi 2}^{-1} & 0 & 0 \\
0 & 0 & 0 & I & 0 \\
0 & 0 & 0 & 0 & I
\end{array}
\right)^{\top}
(6.24)
\left(
\begin{array}{cc|ccc}
I & 0 & 0 & 0 & 0 \\
0 & I & 0 & 0 & 0 \\
\hline
-\hat{D}_{\Psi 2}^{-1} \hat{C}_{\Psi 2} & 0 & \hat{D}_{\Psi 2}^{-1} & 0 & 0 \\
0 & 0 & 0 & I & 0 \\
0 & 0 & 0 & 0 & I
\end{array}
\right). \quad (6.30)
$$

Then we arrive at the final reformulation of (6.11):

$$
\text{He}
\left(
\begin{array}{cccc}
\hat{\mathcal{X}} \hat{\mathcal{A}} & \hat{\mathcal{X}} \hat{\mathcal{B}}_p & \hat{\mathcal{X}} \hat{\mathcal{B}}_w & 0 \\
0 & 0 & 0 & 0 \\
0 & 0 & -\frac{\gamma}{2} I_{n_w} & 0 \\
\hat{\mathcal{C}}_z & \hat{\mathcal{D}}_{zp} & \hat{\mathcal{D}}_{zw} & -\frac{\gamma}{2} I_{n_z}
\end{array}
\right)
+
\left(
\begin{array}{c}
\star \\
\star \\
\star
\end{array}
\right)^{\top}
\hat{P}
\left(
\begin{array}{cccc}
\hat{\mathcal{C}}_q & \hat{\mathcal{D}}_{qp} & \hat{\mathcal{D}}_{qw} & 0 \\
0 & I & 0 & 0
\end{array}
\right)
\prec 0. \quad (6.31)
$$

Here, the calligraphic matrices define the following re-weighted closed-loop system, where the uncertainty channels are re-scaled and adjoined to the performance channel:

$$
\left[
\begin{array}{c|cc}
\hat{\mathcal{A}} & \hat{\mathcal{B}}_p & \hat{\mathcal{B}}_w \\
\hline
\hat{\mathcal{C}}_q & \hat{\mathcal{D}}_{qp} & \hat{\mathcal{D}}_{qw} \\
\hat{\mathcal{C}}_z & \hat{\mathcal{D}}_{zp} & \hat{\mathcal{D}}_{zw}
\end{array}
\right]
:=
\left[
\begin{array}{cc|cc}
\hat{A}_\Psi - \hat{\mathcal{L}}_1 \hat{D}_{\Psi 2}^{-1} \hat{C}_{\Psi 2} & \hat{B}_{\Psi 1}\mathcal{C}_q & \hat{\mathcal{L}}_1 \hat{D}_{\Psi 2}^{-1} & \hat{B}_{\Psi 1}\mathcal{D}_{qw} \\
-\mathcal{B}_p \hat{D}_{\Psi 2}^{-1} \hat{C}_{\Psi 2} & \mathcal{A} & \mathcal{B}_p \hat{D}_{\Psi 2}^{-1} & \mathcal{B}_w \\
\hline
\hat{C}_{\Psi 1} - \hat{\mathcal{L}}_2 \hat{D}_{\Psi 2}^{-1} \hat{C}_{\Psi 2} & \hat{D}_{\Psi 1}\mathcal{C}_q & \hat{\mathcal{L}}_2 \hat{D}_{\Psi 2}^{-1} & \hat{D}_{\Psi 1}\mathcal{D}_{qw} \\
-\mathcal{D}_{zp} \hat{D}_{\Psi 2}^{-1} \hat{C}_{\Psi 2} & \mathcal{C}_z & \mathcal{D}_{zp} \hat{D}_{\Psi 2}^{-1} & \mathcal{D}_{zw}
\end{array}
\right]. \quad (6.32)
$$

Here $\hat{\mathcal{L}}_1 := \hat{B}_{\Psi 1}\mathcal{D}_{qp} + \hat{B}_{\Psi 2}$ and $\hat{\mathcal{L}}_2 := \hat{D}_{\Psi 1}\mathcal{D}_{qp} + \hat{D}_{\Psi 3}$.

Now observe (6.31) implies $\mathrm{He}(\hat{\mathcal{X}}\hat{\mathcal{A}}) \prec 0$. Zooming into the structure of $\hat{\mathcal{A}}$ shows that

$$
\hat{\mathcal{A}} =
\begin{pmatrix}
\hat{A}_\Psi - \hat{\mathcal{L}}_1 \hat{D}_{\Psi 2}^{-1} \hat{C}_{\Psi 2} & \hat{B}_{\Psi 1}\mathcal{C}_q \\
-\mathcal{B}_p \hat{D}_{\Psi 2}^{-1} \hat{C}_{\Psi 2} & \mathcal{A}
\end{pmatrix}
=
\left(
\begin{array}{cc|c}
A_1 & -B_1 \mathcal{D}_{qp} \hat{D}_2^{-1} \hat{C}_2 & B_1 \mathcal{C}_q \\
0 & \hat{A}_2 - \hat{B}_2 \hat{D}_2^{-1} \hat{C}_2 & 0 \\
\hline
0 & -\mathcal{B}_p \hat{D}_2^{-1} \hat{C}_2 & \mathcal{A}
\end{array}
\right). \quad (6.33)
$$

Hence, since A_1, $\hat{A}_2 - \hat{B}_2 \hat{D}_2^{-1} \hat{C}_2$ and $\mathcal{A}$ are all Hurwitz, we conclude that $\hat{\mathcal{A}}$ is Hurwitz and that indeed $\hat{\mathcal{X}}$ is positive definite.

In a next step, note that the calligraphic matrices in (6.32) are defined through

$$
\left(
\begin{array}{c|cc}
\hat{\mathcal{A}} & \hat{\mathcal{B}}_p & \hat{\mathcal{B}}_w \\
\hline
\hat{\mathcal{C}}_q & \hat{\mathcal{D}}_{qp} & \hat{\mathcal{D}}_{qw} \\
\hat{\mathcal{C}}_z & \hat{\mathcal{D}}_{zp} & \hat{\mathcal{D}}_{zw}
\end{array}
\right) =
$$

$$
= \left(
\begin{array}{cccc|cc}
\hat{A}_\Psi - \hat{L}_1 \hat{D}_{\Psi 2}^{-1} \hat{C}_{\Psi 2} & \hat{B}_{\Psi 1}C_q & 0 & \hat{L}_1 \hat{D}_{\Psi 2}^{-1} & \hat{B}_{\Psi 1}D_{qw} \\
-B_p \hat{D}_{\Psi 2}^{-1} \hat{C}_{\Psi 2} & A & 0 & B_p \hat{D}_{\Psi 2}^{-1} & B_w \\
0 & 0 & 0 & 0 & 0 \\
\hline
\hat{C}_{\Psi 1} - \hat{L}_2 \hat{D}_{\Psi 2}^{-1} \hat{C}_{\Psi 2} & \hat{D}_{\Psi 1}C_q & 0 & \hat{L}_2 \hat{D}_{\Psi 2}^{-1} & \hat{D}_{\Psi 1}D_{qw} \\
-D_{zp} \hat{D}_{\Psi 2}^{-1} \hat{C}_{\Psi 2} & C_q & 0 & D_{zp} \hat{D}_{\Psi 2}^{-1} & D_{zw}
\end{array}
\right)
+
\left(
\begin{array}{cc}
0 & \hat{B}_{\Psi 1}D_{qu} \\
0 & B_u \\
I & 0 \\
\hline
0 & \hat{D}_{\Psi 1}D_{qu} \\
0 & D_{zu}
\end{array}
\right) \times
$$

$$
\times \begin{pmatrix} A_K & B_K \\ C_K & D_K \end{pmatrix}
\begin{pmatrix}
0 & 0 & I & 0 & 0 \\
-D_{yp} \hat{D}_{\Psi 2}^{-1} \hat{C}_{\Psi 2} & C_y & 0 & D_{yp} \hat{D}_{\Psi 2}^{-1} & D_{yw}
\end{pmatrix},
$$

This allows us to extract the re-weighted open-loop plant (6.29) from (6.32) that is used for the synthesis of a next controller based on Theorem 6.1.

Again, since A_1 and $\hat{A}_2 - \hat{B}_2 \hat{D}_2^{-1} \hat{C}_2$ are Hurwitz, and in accordance with the requirements, we conclude that (6.29) is indeed stabilizable through u and detectable from y. Recalling (6.25), the elimination of the controller variables from (6.31) with Lemma C.2 finally reveals that (6.28) indeed satisfy (6.7), (6.5) and (6.6) for the plant (6.29). ∎

Lemma 6.3 nicely illustrates that the re-weighted open-loop plant (6.29) can be used for the synthesis of a new 'robustified' controller based on Theorem 6.1. Moreover, the matrices (6.28) can be used as an initial condition for the LMI-solvers in order to initiate a warm-start.

6.6.2 Constructing initial conditions for the analysis step

In complete analogy to the previous arguments, but in the reversed order, it is also possible to generate initial conditions for the analysis step after controller synthesis. Indeed, given the re-weighted plant (6.29), suppose that we obtained some $\hat{X}$, $\hat{Y}$ and a $\tilde{\gamma} \leq \gamma$ for which the LMIs in Theorem 6.1 are feasible. Then we can construct the positive definite matrix

$$
\hat{\mathcal{X}} := \left(\begin{array}{c|c} \hat{X} & \hat{X} - \hat{Y} \\ \hline \hat{X} - \hat{Y} & \hat{X} - \hat{Y} \end{array} \right) = \left(\begin{array}{c|c} \hat{X} & \hat{U} \\ \hline \hat{U}^{\top} & \bar{X} \end{array} \right) = \left(\begin{array}{cc|c} \hat{X}_{11} & \hat{X}_{12} & \hat{U}_1 \\ \hline \hat{X}_{12}^{\top} & \hat{X}_{22} & \hat{U}_2 \\ \hline \hat{U}_1^{\top} & \hat{U}_2^{\top} & \bar{X} \end{array} \right) =
$$

$$
= \left(\begin{array}{ccc|c} \hat{X}_{11,11} & \hat{X}_{11,12} & \hat{X}_{11,13} & \hat{\mathcal{X}}_{12,1} \\ \hat{X}_{11,12}^{\top} & \hat{X}_{11,22} & \hat{X}_{11,23} & \hat{\mathcal{X}}_{12,2} \\ \hat{X}_{11,13}^{\top} & \hat{X}_{11,23}^{\top} & \hat{X}_{11,33} & \hat{\mathcal{X}}_{12,3} \\ \hline \hat{\mathcal{X}}_{12,1}^{\top} & \hat{\mathcal{X}}_{12,2}^{\top} & \hat{\mathcal{X}}_{12,3}^{\top} & \hat{\mathcal{X}}_{22} \end{array} \right) = \left(\begin{array}{c|c} \hat{X}_{11} & \hat{\mathcal{X}}_{12} \\ \hline \hat{\mathcal{X}}_{12}^{\top} & \hat{\mathcal{X}}_{22} \end{array} \right) \succ 0, \quad (6.34)
$$

which is partitioned in accordance with the block-structure of (6.25), and a controller $\tilde{K} = (\tilde{A}_K, \tilde{B}_K, \tilde{C}_K, \tilde{D}_K)$ for which the (nominal) analysis LMI (6.31) holds with (6.25), $K = (A_K, B_K, C_K, D_K)$ and γ replaced by (6.34), $\tilde{K} = (\tilde{A}_K, \tilde{B}_K, \tilde{C}_K, \tilde{D}_K)$ and $\tilde{\gamma}$ respectively. This leads to the following key result.

Lemma 6.4. *The analysis LMI* (6.11) *holds with*

$$
\gamma_{\text{init}} := \tilde{\gamma}, \quad \mathcal{X}_{\text{init}} := \tilde{\mathcal{X}}, \quad P_{\text{init}} := P \tag{6.35}
$$

(i.e. P from the previous analysis) and (3.2) *replaced by*

$$
\tilde{\mathcal{G}} = \left(\begin{array}{cc} \tilde{\mathcal{G}}_{qp} & \tilde{\mathcal{G}}_{qw} \\ \tilde{\mathcal{G}}_{zp} & \tilde{\mathcal{G}}_{zw} \end{array} \right) = \left[\begin{array}{c|cc} \tilde{\mathcal{A}} & \tilde{\mathcal{B}}_p & \tilde{\mathcal{B}}_w \\ \hline \tilde{\mathcal{C}}_q & \tilde{\mathcal{D}}_{qp} & \tilde{\mathcal{D}}_{qw} \\ \tilde{\mathcal{C}}_z & \tilde{\mathcal{D}}_{zp} & \tilde{\mathcal{D}}_{zw} \end{array} \right] =
$$

$$
= \left[\begin{array}{cc|cc} A + B_u \tilde{D}_K C_y & B_u \tilde{C}_K & B_p + B_u \tilde{D}_K D_{yp} & B_w + B_u \tilde{D}_K D_{yw} \\ \tilde{B}_K C_y & \tilde{A}_K & \tilde{B}_K D_{yp} & \tilde{B}_K D_{yw} \\ \hline C_q + D_{qu} \tilde{D}_K C_y & D_{qu} \tilde{C}_K & D_{qp} + D_{qu} \tilde{D}_K D_{yp} & D_{qw} + D_{qu} \tilde{D}_K D_{yw} \\ C_z + D_{zu} \tilde{D}_K C_y & D_{zu} \tilde{C}_K & D_{zp} + D_{zu} \tilde{D}_K D_{yp} & D_{zw} + D_{zu} \tilde{D}_K D_{yw} \end{array} \right]. \tag{6.36}
$$

Here

$$\tilde{\mathcal{X}} = \begin{pmatrix} T_1(\hat{X}_{11} - Z)T_1^\top & T_1\hat{\mathcal{X}}_{12} \\ \hat{\mathcal{X}}_{12}^\top T_1^\top & \hat{\mathcal{X}}_{22} \end{pmatrix} - \begin{pmatrix} (\hat{X}_{11} - Z)T_2^\top \\ \hat{\mathcal{X}}_{12}^\top T_2^\top \end{pmatrix} \left(T_2(\hat{X}_{11} - Z)T_2^\top \right)^{-1} (\star)^\top =$$

$$\begin{pmatrix} \hat{X}_{11,11} - Z_{11} & \hat{X}_{11,13} - Z_{13} & \hat{\mathcal{X}}_{12,1} \\ \hat{X}_{11,13}^\top - Z_{13}^\top & \hat{X}_{11,33} - Z_{33} & \hat{\mathcal{X}}_{12,3} \\ \hat{\mathcal{X}}_{12,1}^\top & \hat{\mathcal{X}}_{12,3}^\top & \hat{\mathcal{X}}_{22} \end{pmatrix} - \begin{pmatrix} \hat{X}_{11,12} - Z_{12} \\ \hat{X}_{11,23}^\top - Z_{23}^\top \\ \hat{\mathcal{X}}_{12,2}^\top \end{pmatrix} \left(\hat{X}_{11,22} - Z_{22} \right)^{-1} (\star)^\top .$$

$$(6.37)$$

Proof. As discussed, the inequality (6.31) is satisfied with $\tilde{K} = (\tilde{A}_K, \tilde{B}_K, \tilde{C}_K, \tilde{D}_K)$, $\hat{\mathcal{X}}$ and $\tilde{\gamma}$. Hence, we can undo the congruence transformation (6.30) in order obtain the inequality (6.24), which now holds with (6.34), (6.36) and $\tilde{\gamma}$. Subsequently, it is possible to replace the 'new' factorization $\hat{\Psi}^* \hat{P} \hat{\Psi}$ with the 'old' factorization $\Psi^* P \Psi$ by reversing the arguments in the proof of Lemma 6.2. This yields the inequality

$$\mathrm{He}\left[\begin{pmatrix} \hat{X}_{11} - Z & \hat{\mathcal{X}}_{12} & 0 \\ \hat{\mathcal{X}}_{12}^\top & \hat{\mathcal{X}}_{22} & 0 \\ 0 & 0 & I \end{pmatrix} \begin{pmatrix} \hat{A}_\Psi & \hat{B}_{\Psi 1}\mathcal{C}_q & \hat{B}_{\Psi 1}\mathcal{D}_{qp} + \hat{B}_{\Psi 2} & \hat{B}_{\Psi 1}\mathcal{D}_{qw} & 0 \\ 0 & \tilde{\mathcal{A}} & \tilde{\mathcal{B}}_p & \tilde{\mathcal{B}}_w & 0 \\ 0 & 0 & 0 & 0 & 0 \\ 0 & 0 & 0 & -\frac{\gamma}{2}I_{n_w} & 0 \\ 0 & \tilde{\mathcal{C}}_z & \tilde{\mathcal{D}}_{zp} & \tilde{\mathcal{D}}_{zw} & -\frac{\gamma}{2}I_{n_z} \end{pmatrix} \right]$$

$$+ (\star)^\top P \left(C_\Psi T_1 \quad D_{\Psi 1}\tilde{\mathcal{C}}_q \quad D_{\Psi 1}\tilde{\mathcal{D}}_{qp} + D_{\Psi 2} \quad D_{\Psi 1}\tilde{\mathcal{D}}_{qw} \quad 0 \right) \prec 0, \quad (6.38)$$

where

$$\begin{pmatrix} \hat{X}_{11} - Z & \hat{\mathcal{X}}_{12} \\ \hat{\mathcal{X}}_{12}^\top & \hat{\mathcal{X}}_{22} \end{pmatrix} = \begin{pmatrix} \hat{X}_{11,11} - Z_{11} & \hat{X}_{11,12} - Z_{12} & \hat{X}_{11,13} - Z_{13} & \hat{\mathcal{X}}_{12,1} \\ \hat{X}_{11,12}^\top - Z_{12}^\top & \hat{X}_{11,22} - Z_{22} & \hat{X}_{11,23} - Z_{23} & \hat{\mathcal{X}}_{12,2} \\ \hat{X}_{11,13}^\top - Z_{13}^\top & \hat{X}_{11,23}^\top - Z_{23}^\top & \hat{X}_{11,33} - Z_{33} & \hat{\mathcal{X}}_{12,3} \\ \hat{\mathcal{X}}_{12,1}^\top & \hat{\mathcal{X}}_{12,2}^\top & \hat{\mathcal{X}}_{12,3}^\top & \hat{\mathcal{X}}_{22} \end{pmatrix} .$$

Now recall the realization (6.26) and observe that (6.38) contains exactly the same unobservable dynamics with which we extended the system (6.11) in order to obtain (6.27). Hence, by Lemma E.1 (see Appendix E.1), we can remove the unobservable dynamics from (6.38) and infer that (6.37) satisfies the analysis LMI (6.11) for $\tilde{K} = (\tilde{A}_K, \tilde{B}_K, \tilde{C}_K, \tilde{D}_K)$, which only involves the IQC-multiplier $\Psi^* P \Psi$ from the previous analysis step. ∎

In conclusion, we can use the matrices (6.35) as an initial condition for the analysis of $\tilde{K} = (\tilde{A}_K, \tilde{B}_K, \tilde{C}_K, \tilde{D}_K)$ based on Theorem 6.2.

6.7 IQC-synthesis with warm-start options

We now have introduced all the ingredients to formulate the following algorithm for the systematic synthesis of robust controllers based on general dynamic IQC-multipliers with warm-start options.

Algorithm 6.1 (IQC-Synthesis).

Initialization (iteration step $j = 1$)

1. Perform a standard $\mathscr{H}_\infty$-synthesis by applying Theorem 3.4. Feasibility leads to a nominal controller K_1 with $\|G \star K_1\|_{i2} < \gamma_{\mathrm{nom}}$.
2. Construct the closed-loop plant $\mathcal{G}_1 := G \star K_1$ and perform an IQC-analysis by applying Theorem 6.2: find $\mathcal{X} \in \mathbb{S}^\bullet$, $P \in \boldsymbol{P}$, $\gamma > 0$ for which (6.11) is feasible. Minimization of γ leads to the (almost) optimal bound γ_1 with $\gamma_1 \geq \gamma_{\mathrm{nom}}$ and $\|\Delta \star \mathcal{G}_1\|_{i2} < \gamma_1$.

Iteration (step $j - 1 \to j$ for $j > 1$)

3. Construct the re-weighted plant (6.29) as well as the initial conditions of (6.28) and design a new robustified controller K_j by applying Theorem 6.1.
4. Construct the closed-loop plant $\mathcal{G}_j = G \star K_j$ together with the initial conditions of (6.35) and perform an IQC-analysis by applying Theorem 6.2: find $\mathcal{X} \in \mathbb{S}^\bullet$, $P \in \boldsymbol{P}$, $\gamma > 0$ for which (6.11) is feasible. Minimization of γ leads to the (almost) optimal bound γ_j with $\gamma_1 \geq \gamma_{j-1} \geq \gamma_j \geq \gamma_{\mathrm{nom}}$ and $\|\Delta \star \mathcal{G}_j\|_{i2} < \gamma_j$.
5. Terminate if $|\gamma_{j-1} - \gamma_j|$ is small.

The iteration results in a sequence of controllers $K_1, K_2, \ldots, K_{j-1}, K_j$, which guarantee stability and render $\|\Delta \star \mathcal{G}_j\|_{i2}$ less than $\gamma_1 \geq \gamma_2 \geq \cdots \geq \gamma_{j-1} \geq \gamma_j \geq \gamma_{\mathrm{nom}}$ for all uncertainties $\Delta_1 \in \boldsymbol{\Delta}_1$.

In summary we have formulated an algorithm for the systematic synthesis of robust controllers based on general dynamic IQC-multipliers. We have resolved the difficulties of factorizing IQC-multipliers into the special form (6.17), which enabled us to suitably reformulate the analysis and synthesis LMIs. The reformulated LMIs yielded the re-weighted open-loop plant (6.29), which can be used for the synthesis of a new robustified controller in Theorem 6.1, as well as the initial conditions (6.28) and (6.35), which can be used for warm-start purposes.

The McMillan degree of the controller K is determined by the McMillan degree of the IQC-multiplier $\Pi = \hat{\Psi}^* \hat{P} \hat{\Psi}$ and of the open-loop plant G as $n_K := n_G + n_{\hat{\Psi}}$. If comparing the results with the alternative procedure in Section 5.5.6, which avoids any IQC-multiplier factorization, we conclude that this is at most n_Ψ smaller than the McMillan of the controller obtained by Algorithm 5.1. Hence, the price to be paid for avoiding the factorization of $\Psi^* P \Psi$ as $\hat{\Psi}^* \hat{P} \hat{\Psi}$ is a higher McMillan degree of the controller.

We emphasize that the implementation of the presented tools is numerically rather delicate. Though it is expected that the alternative approach in Section 5.5.6 is numerically more reliable, a thorough comparison of the numerical behavior between the two approaches still needs to be performed. Further, it remains to be investigated whether the previous warm-start techniques extend to Algorithm 5.1 as well. Let us finally note that the presented techniques can be easily generalized to gain-scheduling by adopting the results of [121, 174, 152], which we presented in Section 3.7.

Remark 6.1. In case of infeasibility in Step 2 of Algorithm 6.1 one can incorporate a parameter $\tau \in [0, 1]$ and find the largest $\hat{\tau}$ for which the analysis LMI of Theorem 6.2 with respect to the uncertainty set $\hat{\tau}\boldsymbol{\Delta}$ is feasible. Moreover, if $\hat{\tau} < 1$, one can try to increase τ during the iteration in Step 4 of the algorithm.

Remark 6.2. Note that it is straightforward to combine the two factorizations that we presented in Section 6.5 in order to handle block diagonally repeated uncertainties in Algorithm 6.1.

6.8 Illustrations

In order to demonstrate the capabilities of Algorithm 6.1, we now present the two numerical examples from our recent paper [198]. We emphasize that some portions of the text overlap.

6.8.1 A mixed sensitivity design problem with parametric and delay uncertainties

We first applied Algorithm 6.1 for a mixed sensitivity design problem as shown in Figure 6.1. The uncertain plant and the disturbance model are given by

$$G(s, \delta) = \frac{4}{s^2 + 0.1(1 + 0.5\delta)s + (1 + 0.5\delta)^2} \quad \text{and} \quad G_{\mathrm{d}}(s) = \frac{10}{s + 0.1}$$

respectively, with the time-invariant uncertain parameter $\delta \in [-1, 1]$. In addition, the measurement delay-time α is assumed to be not larger than 0.025.

With the performance weights chosen as

$$W_{\mathrm{e}}(s) = \frac{2(s + 3.674)^2}{3(s + 0.03)^2}, \quad W_{\mathrm{u}}(s) = \frac{s + 10}{s + 10^4} \quad \text{and} \quad W_{\mathrm{d}} = 1,$$

the synthesis objectives are, respectively, to track and reject the commanded reference signal r and the disturbance signal d at low frequencies, while penalizing control at

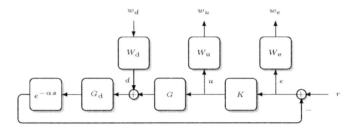

Fig. 6.1 Block diagram of the mixed sensitivity design problem

high frequencies. For this purpose, we employ Algorithm 6.1 with the IQC-multipliers for LTI parametric uncertainties and an uncertain time-delay as presented in Section 2.6.3.1 and 2.6.5.1 respectively. Here we used the first basis-function in (2.11) for different lengths $\nu \in \mathbb{N}_0$ and pole-location $\rho = -1$.

We have designed a nominal $\mathscr{H}_\infty$-controller K_{nom} as well as two robust controllers K_{sta} and K_{dyn} by applying Theorem 3.4 and Algorithm 6.1 respectively; K_{sta} is obtained with static IQC-multipliers (i.e. $\nu = 0$), while K_{dyn} has been designed for dynamic IQC-multipliers with $\nu = 1$.

In Figure 6.2 time-domain simulation results are shown for all three controllers in the nominal case with $\delta = 0$ and $\alpha = 0$ (nom. response), and for 25 random samples of $\delta \in [-1, 1]$ and $\alpha = 0.025$ (pert. response). As can be seen in Figure 6.2(a), the nominal controller K_{nom} tracks the reference and rejects the disturbance well, while its performance degrades drastically in the simulation with uncertainties. On the other hand, it is nice to see in Figure 6.2(b) and 6.2(c) that robust performance can be significantly improved through robust designs based on Algorithm 6.1. The figures also clearly illustrate the benefit of dynamics IQC-multipliers, leading to a significantly better performance in terms of settling time.

This is also quantitatively confirmed by means of the root mean square (rms) values of the nominal and worst simulated tracking errors given in Table 6.1. Furthermore, in contrast to the controllers K_{nom} and K_{sta}, the controller K_{dyn} is better able to suppress the shifting resonance peaks of G around $\omega = 1$ (rad/sec) caused by the parametric uncertainty δ, while it still retains a high gain at low frequencies. This can be seen in Figure 6.3(a).

Let us also address some computational details. Figure 6.3(b) shows the γ-values that were obtained after each optimization step. As can be seen, the γ-levels stay approximately constant after a few iterations only and the limiting values are consistent with the time-domain simulation results. Figure 6.3(b) also clearly illustrates the benefit of dynamics in the IQC-multipliers, since the robust $\mathscr{L}_2$-gain level $\gamma_{\mathrm{dyn}} = 3.19$ achieved with K_{dyn} is much closer to the nominal performance level $\gamma_{\mathrm{nom}} = 1.60$ if compared with K_{sta}, which only guarantees $\gamma_{\mathrm{sta}} = 8.67$.

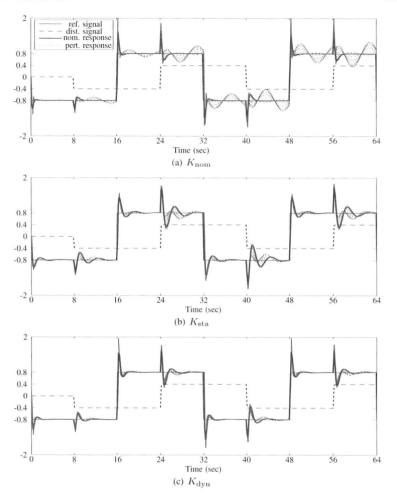

Fig. 6.2 Time-domain simulation results for random samples δ_i of $\delta \in [-1, 1]$

Last but not least, Table 6.2 shows the computation times for different controllers that were obtained through different synthesis methods. It can be extracted that it is indeed possible to significantly speed up the synthesis process by using warm-start methods; the bigger the size of the optimization problem, the more time profit (percentage-wise) can be achieved. Let us remark at this point that the problem considered in this example can also be handled by the μ-tools of Matlab [11]. As a

	K_{nom}	K_{sta}	K_{dyn}
nominal case	0.151	0.197	0.169
worst simulated case	0.258	0.217	0.197

Table 6.1 rms-values of the nominal and worst simulated tracking errors

	γ	States	Cold-start (sec)	Warm-start (sec)	Time profit (%)
K_{sta}	8,67	6	7.36	6.02	18
K_{dyn}	3.19	14	131.30	89.38	32
$K_{\mu 50}$ (50 grid-points)	2.8	18	22.13	-	-
$K_{\mu 500}$ (500 grid-points)	2.8	18	107.51	-	-

Table 6.2 Synthesis time for different controllers

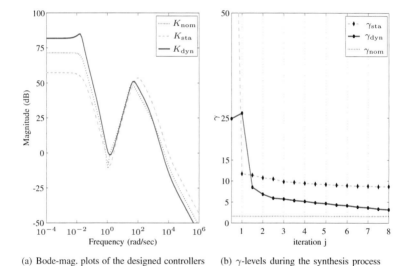

(a) Bode-mag. plots of the designed controllers   (b) γ-levels during the synthesis process

Fig. 6.3 Additional simulation results

matter of fact, it is no surprise that a comparable degree-of-freedom in the multipliers leads to (almost) identical controllers. As opposed to the IQC-tools, the μ-tools rely on gridding and approximation techniques. Typical for this strategy is that one can accelerate the synthesis by decreasing the number of grid-points. However, this also increases the chance to miss critical frequencies. Table 6.2 shows the computation times for the μ-synthesis based on 50 as well as 500 grid points and a comparable degree-of-freedom in the IQC-multipliers, if compared to the controller K_{dyn}. (Note that, although there are no theoretical restrictions, the μ-tools of Matlab (R2011b) do not allow to use static multipliers, which prevented a comparison with K_{sta}.) Clearly, increasing the number of grid-points significantly increases the computation time, yet without implying guarantees for all frequencies. This issue causes no trouble in the proposed IQC methodology.

6.8.2 A distillation process with a control channel nonlinearity

In this section we illustrate the flexibility of the IQC-framework by a design example that cannot be handled with the μ-tools. For this purpose, consider the distillation process from [178], extended with a control channel nonlinearity, as shown in Figure 6.4. Here the constant diagonal matrix β and the plant G are given by

$$\beta = \begin{pmatrix} \beta_1 & 0 \\ 0 & \beta_2 \end{pmatrix}, \quad \beta_1 \geq 0, \quad \beta_2 \geq 0 \quad \text{and} \quad G(s) = \frac{1}{75s+1} \begin{pmatrix} 87.8 & -86.4 \\ 108.2 & -109.6 \end{pmatrix}$$

respectively. The dead-zone function $\Delta_{\mathrm{dz}}(\cdot)$ is defined as

$$\underbrace{\begin{pmatrix} p_1 \\ p_2 \end{pmatrix}}_{p} = \underbrace{\begin{pmatrix} \Delta_{\mathrm{dz}1}(q_1) \\ \Delta_{\mathrm{dz}2}(q_2) \end{pmatrix}}_{\Delta_{\mathrm{dz}}(q)},$$

where

$$p_i = \Delta_{\mathrm{dz}i}(q_i) = q_i - \min\{|q_i|, \alpha_i\}\mathrm{sign}(q_i), \quad \alpha_i > 0, \quad i = 1, 2.$$

Note that for $\beta = I$ the interconnection within the dashed box in Figure 6.4 simplifies into a standard repeated saturation nonlinearity.

With the performance weights

$$W_{\mathrm{e}}(s) = \frac{s + 0.1}{2s + 10^{-5}} I_2 \quad \text{and} \quad W_{\mathrm{u}}(s) = \frac{s + 10}{s + 100} I_2,$$

the synthesis objective is to track the commanded reference signal r at low frequencies, while penalizing the control action at high frequencies. We employ Algorithm 6.1 with the IQC-multipliers for sector-bounded and slope-restricted nonlinearities as discussed in Section 2.6.7 and for the following two scenarios:

1. $\beta_1 = 0.6$, $\beta_2 = 0.3$, $\alpha_1 = \alpha_2 = 0.05$,
2. $\beta_1 = \beta_2 = 0.95$, $\alpha_1 = \alpha_2 = 0.15$.

Apart from a nominal $\mathscr{H}_\infty$-controller, K_{nom}, we have designed three robust controllers (i.e. K_{sc}, K_{fb} and K_{zf}) based on the diagonal multipliers of Section 2.6.8.1, the full-block multiplier circle criterion of Section 2.6.8.2 and the diagonal multipliers

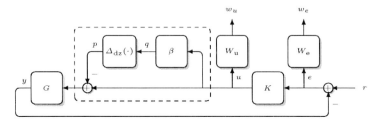

Fig. 6.4 Block diagram of the distillation process

combined with the Zames-Falb multiplier parameterization of Section 2.6.8.3. For the latter, we employed the first basis-function in (2.11) for basis-length $\nu = 1$ and pole-location $\rho = -1$.

In Figure 6.5 and 6.6 the responses are shown for Scenario 1 and 2 respectively, for a simultaneous unit-step in both reference channels and in case of $\beta = 0$ (left-column) and $\beta \neq 0$ (right-column). Again the nominal controller, K_{nom}, shows good tracking for $\beta = 0$, but the performance degrades drastically if $\beta \neq 0$. On the other hand, it is nice to see that by giving up on nominal performance a little, robust performance can be significantly improved with the designs based on Algorithm 6.1. It remains hard to assess which of the robust controllers performs best, since the step responses are rather similar. One might judge in favor of the Zames-Falb multiplier-based controller K_{zf} since it performs somewhat better in one channel at the expense of more overshoot in the other. Finally, similar conclusions can be drawn from Figure 6.7, which shows the γ-values obtained after each optimization step. Clearly, for both scenarios, all controller designs approximately guarantee the same worst-case performance levels.

6.9 Summary

In this chapter we generalized the classical μ-synthesis tools to the IQC-framework. We developed an alternative algorithm for the systematic synthesis of robust controllers based on an iteration standard nominal controller synthesis and IQC-analysis with general dynamic IQC-multipliers. This offers the possibility to efficiently handle a more diverse class of uncertainties and nonlinearities. As key technical tools, we presented new insights into the factorization of dynamic IQC-multipliers and the related manipulations on linear matrix inequalities. In contrast to most of the existing methods, the proposed algorithm not only avoids frequency-gridding and curve-fitting, but it also enabled us to realize warm-starts for each of the optimization steps in the iteration. As illustrated by the numerical examples, this can significantly speed up the synthesis process.

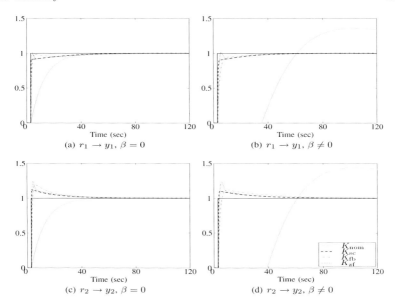

Fig. 6.5 Time-domain simulation results for Scenario 1.

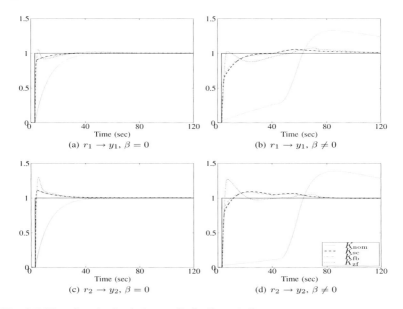

Fig. 6.6 Time-domain simulation results for Scenario 2.

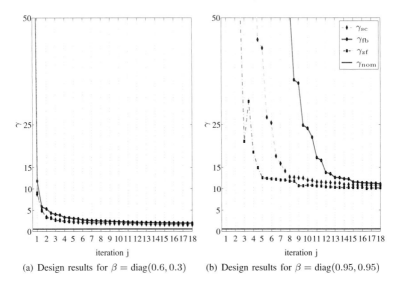

(a) Design results for $\beta = \mathrm{diag}(0.6, 0.3)$ (b) Design results for $\beta = \mathrm{diag}(0.95, 0.95)$

Fig. 6.7 γ-levels during the synthesis process.

Chapter 7
Conclusions and recommendations

7.1 Conclusions

In this thesis we have made several steps towards a general framework for the systematic analysis of uncertain systems and the design of robust controllers based on integral quadratic constraints (IQCs). To achieve the main goal of this thesis, we have first provided an easy-to-access tutorial introduction on the most important IQC-analysis tools. These are supported by an extensive survey on the formulation and parameterization of IQC-multipliers as well as some illustrative numerical examples. Further we contributed to the IQC-analysis tools with some new and useful technical insights, as precisely indicated in Chapters 1 and 2.

The analysis tools formed a solid basis for the further development of the IQC-tools on the synthesis side. To achieve this goal we have presented various results and generalizations on the design of robust gain-scheduling controllers by convex optimization techniques. This includes

- an LMI solution for the robust gain-scheduled estimation problem. For plants with a linear fractional dependency on both the uncertainties as well as the scheduling variables, we have shown how to convexify the problem if the properties uncertainties and scheduling variables are described by general dynamic and static full-block IQC-multipliers respectively. For further details we refer the reader to Chapter 4.

- an LMI solution for a more general generic feasibility problem, which encompasses a rich class of concrete design scenarios. For example, we have illustrated that robust gain-scheduled controller synthesis is convex under the rough hypothesis that the control channel is not affected by uncertainties. Further, we have shown how also other problems can be handled with ease, like e.g. robust gain-scheduled observer design and robust controller synthesis with unstable weights. More details and a comprehensive overview of other design scenarios that are covered are found in Chapter 5.

- a 'μ-like' synthesis algorithm for the general robust controller synthesis problem. The suggested algorithm enables us to perform robust controller synthesis for a significantly larger class of uncertainties if compared to the existing methods. In addition, we have shown how new insights on the factorization of IQC-multipliers allow us to use warm-start techniques. This can speed up the synthesis process. These results are found in Chapter 6.

Summing up, we have made various significant contributions on both the analysis as well as the synthesis side. Altogether, we have unveiled that the IQC-tools are not only very useful for the analysis of a large class of uncertain systems, but also for the systematic synthesis of robust controllers for a diverse class of practically useful design scenarios.

7.2 Recommendations for future research

In order to continue the endeavor to establish a well-known and widely applied framework the systematic analysis of uncertain systems and the synthesis of robust controllers based on IQCs, the following problems and challenges are suggested for future research.

- As already briefly discussed at the end of Section 2.2, the modeling of uncertain dynamical systems and the reformulation of these systems in the standard LFT-form is a nontrivial task. Despite various useful results (see e.g [210, 30, 21, 62, 61, 185] and references therein), there still is a need for systematic tools that allow us to efficiently obtain linear fractional representations, which are of high fidelity on the one hand, and which have a 'low-dimension' on the other hand. Here 'low-dimension' means: i) the nominal model has a small McMillan degree, and ii) the uncertainty blocks have only a few number of repetitions.
- As discussed in Remark 2.6 in Section 2.5.1, it is unknown how to properly select basis functions for the IQC-multiplier parameterizations. It has been illustrated in Section 2.9 that this might cause numerical issues as well as a higher computational loads in the computations. For this purpose, further research is needed in order to be able to systematically choose 'good' basis functions and/or more efficient alternative IQC-multiplier parameterizations.
- As conjectured in Section 2.10, the alternative dissipativity based proof of the IQC-theorem opens the way to merge frequency-domain techniques with time-domain conditions known from Lyapunov theory. This might lead to new and interesting applications. Unfortunately, in order to find the Lyapunov function that proves robust stability of the closed-loop uncertain system, we had to solve an indefinite

algebraic Riccati equation (ARE). Therefore the constructed Lyapunov function does not solely consists of original data, which prevents us from obtaining useful results. To date, it is unknown how to characterize the solution of the ARE through an LMI. Further research should point out whether or not this problem can be resolved.

- As indicated in the discussions after Theorems 4.2 and 5.1 in Chapters 4 and 5 respectively, the results on gain-scheduling in this thesis are limited to static (frequency independent) full-block multipliers. For general dynamic (frequency dependent) IQC-multipliers the nominal gain-scheduling synthesis problem is still open, while full results for dynamic D and D/G-scalings have been recently presented in [167, 157, 158, 161]. Although the nominal gain-scheduling controller synthesis problem with general dynamic IQC-multipliers is rather tough, we emphasize that a full solution could lead to numerous new and interesting applications. Further, it would be very interesting to see whether or not it is possible to merge the results in Chapter 4 and 5 with the ones presented in [167, 157, 158, 161].

- Recent years have witnessed various results on structured and distributed controller synthesis (see e.g. [13, 34, 42, 106, 107, 78, 45, 201, 94, 25, 156, 160]). Some of the applied techniques are very similar to the ones presented in this thesis and might be straightforwardly merged with our synthesis results in order to design structured or distributed robust gain-scheduling controllers. For example, it is a natural conjecture that the $\mathscr{H}_\infty$-synthesis results in [160] for nested system interconnections can be unified with the synthesis results in Chapters 4 and 5.

- Although we have presented a long list of concrete design scenarios that are covered by Problem 5.1 in Chapter 5, the author believes that the full potential of this result has still to be discovered. Especially the use of unstable weights in practice has to be further explored.

- As discussed in Chapter 6 the developed IQC-synthesis algorithm is numerically rather delicate. Although it is expected that the alternative results in Section 5.5.5 are numerically more reliable, a thorough comparison of two approaches still has to be carried out. Moreover, it remains a conjecture that the warm-start techniques in Chapter 6 extend to the results in Section 5.5.5.

- In spite of various solvers for semi-definite programs (see e.g. [50, 10, 179, 184, 128, 60, 187, 109, 8, 58, 59] and references therein), it remains a rather delicate issue to solve LMI-problems that consist of a large number of decision variables. For this purpose, there is a need for better LMI-solvers that are specialized for control, and which exploit the extra structure in the optimization problem.

- Despite the long list of available IQC-multipliers, there is still a need for new IQC-multipliers that capture properties of uncertainties and/or nonlinearities that are not yet covered. For example, in the case of saturation nonlinearities, it is unknown how to effectively take into account saturation bounds with IQCs. As solution would lead to a very competitive alternative for the existing Lyapunov approaches.

- It is well-known that many IQC-tools can be straightforwardly applied for discrete-time system too. However, it remains a conjecture that the LMI-solutions in Chapter 4 and 5 have a complete discrete-time analogue as well. We refer the reader to [93] for more insights in this matter.

- Although some IQC-toolboxes are available for the systematic analysis of uncertain systems (see e.g. [115]), there is no IQC-toolbox available for both analysis as well as synthesis purposes. Only recently, we have started to developed such a toolbox. However, more work needs to be done before a user-friendly version can be released.

Appendices

Appendix A
Explanation of symbols

In the following sections dimensions are indicated by the nonnegative integers n and m. If the dimensions are unspecified, we frequently use the symbol •, which is also often dropped.

A.1 Vector spaces and matrices

In the following enumeration a is a real scaler, and A, B are real or complex matrices of appropriate dimensions.

$\mathbb{N}$	Set of positive integers
$\mathbb{N}_0$	Set of positive integers including zero
$\mathbb{R}$	Set of real numbers
$\mathbb{R}^{n \times m}$	Set of real-valued matrices of dimension $n \times m$
$\mathbb{C}$	Set of complex numbers
$\mathbb{C}^-$	Open left-half complex plane
$\mathbb{C}^0$	Extended imaginary axis ($i\mathbb{R} \cup \{\infty\}$)
$\mathbb{C}^+$	Open right-half complex plane
$\mathbb{C}^{n \times m}$	Set of complex-valued matrices of dimension $n \times m$
$\mathbb{S}^n$	Set of symmetric matrices of dimension $n \times n$
I_n	Identity matrix of dimension $n \times n$
$0_{n \times m}$	Zero matrix of dimension $n \times m$
$A^\top$, A^*	Transpose and conjugate transpose of A
A^{-1}	Inverse of the square and nonsingular matrix A
$\mathrm{He}(A)$	$A + A^*$
$A \succ (\prec) B$	A, B Hermitian, $A - B$ is positive (negative) definite

$A \succcurlyeq (\preccurlyeq)B$ A, B Hermitian, $A - B$ is positive (negative) semi-definite
$A \otimes B$ Kronecker product of A and B
$\det(A)$ Determinant of A
$\ker(A)$, $\operatorname{im}(A)$ Kernel and image of A
$\operatorname{diag}(A_1, \ldots, A_n)$ Block diagonal matrix with $A_1, \ldots, A_n$ on its diagonal
$\operatorname{col}(A_1, \ldots, A_n)$ $(A_1^\top, \ldots, A_n^\top)^\top$
$\operatorname{eig}(A)$ Eigenvalues of A
$\operatorname{trace}(A)$ Trace of A
$\operatorname{sgn}(a)$ Sign of the real scalar a (i.e. $\operatorname{sgn}(a) = a/|a|$)

A.2 Function spaces

$\mathscr{L}_1^n$ Space of vector-valued absolute integrable functions on $(-\infty, \infty)$. The subspaces that are defined on $(-\infty, 0]$ and $[0, \infty)$ are denoted by $\mathscr{L}_1^n(-\infty, 0]$ and $\mathscr{L}_1^n[0, \infty)$ respectively.

$\mathscr{L}_{2e}^n$ Space of vector-valued locally square integrable functions on $[0, \infty)$.

$\mathscr{L}_2^n \subset \mathscr{L}_{2e}^n$ Subspace of vector-valued locally square integrable functions on $[0, \infty)$ with finite energy. The corresponding space for functions defined on $(-\infty, \infty)$ is denoted by $\mathscr{L}_2^n(-\infty, \infty)$.

$\mathscr{L}_\infty^n$ Space of vector-valued measurable and essentially bounded functions on $(-\infty, \infty)$.

As usual $\mathscr{L}_2$ spaces defined on $(-\infty, \infty)$ (and, hence, also those on $[0, \infty)$) are equipped with the standard inner product and norm, as respectively given by

$$\langle v, w \rangle = \int_{-\infty}^{\infty} v(t)^\top w(t) dt \ \text{ and } \ \|w\|_2 = \sqrt{\int_{-\infty}^{\infty} \|w(t)\|^2 dt}.$$

Here

$$\|w(t)\| := \sqrt{\sum_{i=1}^{n} |w_i(t)|^2} \ \text{ with } \ w = \operatorname{col}(w_1, \ldots, w_n)$$

denotes the standard Euclidian-norm. For any $w \in \mathscr{L}_2^n$ one can determine the Fourier transform $\hat{w}$ which is a $\mathbb{C}^n$-valued function defined on the extended imaginary axis $\mathbb{C}^0$ and which has the property

$$\int_{-\infty}^{\infty} \hat{w}(i\omega)^* \hat{w}(i\omega) d\omega \ \text{ is finite.}$$

In fact, for signals in $\mathscr{L}_2^n$ it is well-known that Parseval's theorem states that

$$\int_0^\infty v(t)^\top w(t)dt = \frac{1}{2\pi}\int_{-\infty}^\infty \hat{v}(i\omega)^* \hat{w}(i\omega)d\omega.$$

Sometimes we need the so-called truncation operator, which is defined for each $T > 0$ and assigns to any signal $w \in \mathscr{L}_2^n$ the signal w_T as follows:

$$w_T(t) := \begin{cases} w(t) & \text{for } t \in [0,T], \\ 0 & \text{for } t \in (T,\infty). \end{cases}$$

For example, a map (operator) $G : \mathscr{L}_{2e}^n \to \mathscr{L}_{2e}^m$ is said to be causal if it satisfies

$$G(w)_T = G(w_T) \quad \forall T > 0, \quad w \in \mathscr{L}_{2e}^n.$$

In addition, if there exist a constant $\gamma \in \mathbb{R}$ with

$$\|G(w)_T\|_2 \le \gamma \|w_T\|_2 \quad \forall T > 0, \quad w \in \mathscr{L}_{2e}^n,$$

then the map $G : \mathscr{L}_{2e}^n \to \mathscr{L}_{2e}^m$ is said to be bounded. The infimal $\gamma \ge 0$ with the latter property is called the induced $\mathscr{L}_2$-gain of the map $G : \mathscr{L}_{2e}^n \to \mathscr{L}_{2e}^m$ and can as well be expressed as

$$\|G\|_{i2} := \sup_{T>0,\ w\in\mathscr{L}_{2e}^n,\ \|w_T\|_2 \ne 0} \frac{\|G(w)_T\|_2}{\|w_T\|_2}.$$

A.3 Other function spaces

In the following enumeration, $s := \lambda + i\omega \in \mathbb{C}$ is a complex number with $\lambda, \omega \in \mathbb{R}$.

$\mathscr{R}^{\bullet\times\bullet}$ The space of all real rational and proper transfer matrices $G(s)$.

$\mathscr{H}_\infty^{\bullet\times\bullet}$ The space of all complex valued matrix functions $G(s)$, which are analytic for $s \in \mathbb{C}^+$ and satisfy

$$\|G\|_\infty := \sup_{\lambda>0} \|G(\lambda + i\omega)\| < \infty. \tag{A.1}$$

Here $\|G\|_\infty$ is called the $\mathscr{H}_\infty$-norm of G.

$\mathscr{R}\mathscr{H}_\infty^{\bullet\times\bullet} \subset \mathscr{H}_\infty^{\bullet\times\bullet}$ The subspace of all real rational and proper transfer matrices $G(s)$ without poles in $\mathbb{C}^0 \cup \mathbb{C}^+$. For transfer matrices $G \in \mathscr{R}\mathscr{H}_\infty^{\bullet\times\bullet}$ the norm (A.1) is equivalent to

$$\|G\|_\infty := \sup_{\omega\in\mathbb{R}} \|G(i\omega)\|.$$

$\mathscr{H}_2^{\bullet\times\bullet}$ The space of all complex valued matrix functions $G(s)$, which are analytic for $s \in \mathbb{C}^+$ and satisfy

$$\|G\|_{H_2} := \sqrt{\frac{1}{2\pi}\sup_{\lambda>0}\ \operatorname{trace}\int_{-\infty}^{\infty} G(\lambda+i\omega)^* G(\lambda+i\omega)d\omega} < \infty.$$

(A.2)

Here $\|G\|_{H_2}$ is called the $\mathscr{H}_2$-norm of G.

$\mathscr{R}\mathscr{H}_2^{\bullet\times\bullet} \subset \mathscr{H}_2^{\bullet\times\bullet}$ The subspace of all real rational and proper transfer matrices $G(s)$ without poles in $\mathbb{C}^0 \cup \mathbb{C}^+$ and which satisfy $G(\infty)=0$. For transfer matrices $G \in \mathscr{R}\mathscr{H}_2^{\bullet\times\bullet}$ the norm (A.2) is equivalent to

$$\|G\|_{H_2} := \sqrt{\frac{1}{2\pi}\operatorname{trace}\int_{-\infty}^{\infty} G(i\omega)^* G(i\omega)d\omega}.$$

$\mathscr{R}\mathscr{L}_\infty^{\bullet\times\bullet}$ The space of all real rational and proper transfer matrices $G(s)$ without poles on $\mathbb{C}^0$.

A.4 Miscellaneous

For the given operators F, G and H the upper and lower linear fractional transformations (LFTs)

$$F \star \underbrace{\begin{pmatrix} G_{11} & G_{12} \\ G_{21} & G_{22} \end{pmatrix}}_{G} \quad \text{and} \quad \underbrace{\begin{pmatrix} G_{11} & G_{12} \\ G_{21} & G_{22} \end{pmatrix}}_{G} \star H$$

are defined as

$$G_{22} + G_{21}F\left(I - G_{11}F\right)^{-1}G_{12} \quad \text{and} \quad G_{11} + G_{12}H\left(I - G_{22}H\right)^{-1}G_{21}$$

respectively, under the assumption that the inverses exist; in addition, if the inverses are also causal, the LFTs are said to be well-posed. Realizations of real rational and proper transfer matrices $G \in \mathscr{R}^{\bullet\times\bullet}$ are denoted by

$$G = \left[\begin{array}{c|c} A & B \\ \hline C & D \end{array}\right] \quad \text{or} \quad G = (A,B,C,D)$$

and $G(i\omega)^*$ is defined as $G(-i\omega)^\top$. If $G = G^* \in \mathscr{R}\mathscr{L}_\infty^{\bullet\times\bullet}$, then

$$G \prec 0 \text{ on } \mathbb{C}^0$$

means that G satisfies the frequency domain inequality (FDI)

$$G(i\omega) \prec 0 \quad \forall \omega \in \mathbb{R} \cup \{\infty\}.$$

Finally, objects that can be inferred by symmetry are indicated by $\star$. For example, we frequently abbreviate

$$G^* PG \quad \text{and} \quad \begin{pmatrix} G_{11} & G_{12} \\ G_{12}^* & G_{22} \end{pmatrix} \quad \text{as} \quad (\star)^* PG \quad \text{and} \quad \begin{pmatrix} G_{11} & G_{12} \\ \star & G_{22} \end{pmatrix}$$

respectively.

Appendix B
Abbreviations

Throughout this work we make use of the abbreviations

ARE	Algebraic Riccati Equation
ARI	Algebraic Riccati Inequality
FDI	Frequency Domain Inequality
IQC	Integral Quadratic Constraint
KYP	Kalman Yakubovich Popov
LFT	Linear Fractional Transformation
LMI	Linear Matrix Inequality
LPV	Linear Parameter Varying
LTI	Linear Time Invariant
LTV	Linear Time Varying
MIMO	Multi Input Multi Output
SISO	Single Input Single Output
SOS	Sum-of-squares
SSV	Structured Singular Value

Appendix C
Dualization and elimination

Lemma C.1 ([65]). *Let* $P = P^* \in \mathbb{C}^{(n_1+n_2)\times(n_1+n_2)}$, $A \in \mathbb{C}^{(n_1+n_2)\times n_1}$ *and* $B \in \mathbb{C}^{(n_1+n_2)\times n_2}$ *suppose that* $(A \; B)$ *and* P *are nonsingular. Then*

$$A^* P A \prec 0 \quad and \quad B^* P B \succcurlyeq 0$$

are equivalent to

$$A_\perp^* P^{-1} A_\perp \succ 0 \quad and \quad B_\perp^* P^{-1} B_\perp \preccurlyeq 0,$$

where the columns of $A_\perp$ *and* $B_\perp$ *form a basis of the orthogonal complement of the image of* A *and* B *respectively.*

Lemma C.2 ([65]). *For arbitrary matrices* $A \in \mathbb{R}^{x_1 \times n}$, $B \in \mathbb{R}^{x_2 \times m}$, $C \in \mathbb{R}^{n \times m}$ *and any nonsingular matrix* $P = P^\top$ *with* m *negative and* n *positive eigenvalues, consider the matrix inequality*

$$\begin{pmatrix} I \\ A^\top X B + C \end{pmatrix}^\top P \begin{pmatrix} I \\ A^\top X B + C \end{pmatrix} \prec 0 \qquad (C.1)$$

or the equivalent version

$$\begin{pmatrix} -B^\top X^\top A - C^\top \\ I \end{pmatrix}^\top P^{-1} \begin{pmatrix} -B^\top X^\top A - C^\top \\ I \end{pmatrix} \succ 0 \qquad (C.2)$$

in the unstructured matrix unknown $X \in \mathbb{R}^{x_1 \times x_2}$. *Then (C.1) has a solution if and only if*

$$\begin{pmatrix} B^\perp \\ CB^\perp \end{pmatrix}^\top P \begin{pmatrix} B^\perp \\ CB^\perp \end{pmatrix} \prec 0, \quad and \quad \begin{pmatrix} -C^\top A^\perp \\ A^\perp \end{pmatrix}^\top P^{-1} \begin{pmatrix} -C^\top A^\perp \\ A^\perp \end{pmatrix} \succ 0.$$

Here $A^\perp$ *and* $B^\perp$ *are arbitrary basis matrices of the kernels of* A *and* B *respectively.*

Appendix D
A particular version of the KYP-Lemma

Lemma D.1. *Let $P \in \mathbb{S}^{\bullet}$ and let $G \in \mathscr{R}\mathscr{H}_{\infty}^{\bullet \times \bullet}$ admit the realization (A, B, C, D) with $A \in \mathbb{R}^{n_x \times n_x}$, $\mathrm{eig}(A) \subset \mathbb{C}^{-}$ and (A, B) controllable. Then the following two statements are equivalent:*

1. $G^* P G = 0$ on $\mathbb{C}^0$.
2. There exist a matrix $X \in \mathbb{S}^{n_x}$ such that

$$
\begin{pmatrix} I & 0 \\ A & B \\ C & D \end{pmatrix}^{\top} \begin{pmatrix} 0 & X & 0 \\ X & 0 & 0 \\ 0 & 0 & P \end{pmatrix} \begin{pmatrix} I & 0 \\ A & B \\ C & D \end{pmatrix} = 0.
$$

Proof. To show $1 \Rightarrow 2$, let us exploit the realization $G = (A, B, C, D)$ in order to infer that

$$
G^* P G = \left[\begin{array}{cc|c} -A^{\top} & C^{\top} P C & C^{\top} P D \\ 0 & A & B \\ \hline -B^{\top} & D^{\top} P C & D^{\top} P D \end{array} \right].
$$

Since $G(i\omega)^* P G(i\omega) = 0$ holds for $w = \infty$ the latter implies that $D^{\top} P D = 0$. Moreover, with A being Hurwitz, the Lyapunov equation $A^{\top} X + X A + C^{\top} P C = 0$ has a unique solution X. Hence, by similarity we have

$G^*PG =$

$$= \left[\begin{array}{cc|c} -A^\top & C^\top PC & C^\top PD \\ \hline 0 & A & B \\ \hline -B^\top & D^\top PC & D^\top PD \end{array} \right] = \left[\begin{array}{cc|c} -A^\top & A^\top X + XA + C^\top PC & C^\top PD + XB \\ \hline 0 & A & B \\ \hline -B^\top & D^\top PC + B^\top X & D^\top PD \end{array} \right] =$$

$$= \left[\begin{array}{c|c} -A^\top & C^\top PD + XB \\ \hline -B^\top & \frac{1}{2} D^\top PD \end{array} \right] + \left[\begin{array}{c|c} A & B \\ \hline D^\top PC + B^\top X & \frac{1}{2} D^\top PD \end{array} \right] = 0.$$

Clearly, this implies that

$$\left[\begin{array}{c|c} A & B \\ \hline D^\top PC + B^\top X & \frac{1}{2} D^\top PD \end{array} \right] = 0$$

and, hence, since the pair (A, B) is controllable, that $D^\top PC + B^\top X = 0$. We conclude that

$$\left(\begin{array}{cc} A^\top X + XA + C^\top PC & C^\top PD + XB \\ D^\top PC + B^\top X & D^\top PD \end{array} \right) = \left(\begin{array}{c} \star \\ \star \\ \star \end{array} \right)^\top \left(\begin{array}{ccc} 0 & X & 0 \\ X & 0 & 0 \\ 0 & 0 & P \end{array} \right) \left(\begin{array}{cc} I & 0 \\ A & B \\ C & D \end{array} \right) = 0.$$

To show $2 \Rightarrow 1$, observe that the FDI for $w = \infty$ directly follows from the right-lower block of the LMI. For $\omega \in \mathbb{R}$ (and recalling that A is Hurwitz such that $i\omega$ is no eigenvalue of A) observe that

$$0 = \left(\begin{array}{c} \star \\ \star \\ \star \end{array} \right)^\top \left(\begin{array}{cc} I & 0 \\ A & B \\ C & D \end{array} \right)^\top \left(\begin{array}{ccc} 0 & X & 0 \\ X & 0 & 0 \\ 0 & 0 & P \end{array} \right) \left(\begin{array}{cc} I & 0 \\ A & B \\ C & D \end{array} \right) \left(\begin{array}{c} (i\omega I - A)^{-1} B \\ I \end{array} \right) =$$

$$= \left(\begin{array}{c} \star \\ \star \\ \star \end{array} \right)^\top \left(\begin{array}{ccc} 0 & X & 0 \\ X & 0 & 0 \\ 0 & 0 & P \end{array} \right) \left(\begin{array}{c} (i\omega I - A)^{-1} B \\ i\omega (i\omega I - A)^{-1} B \\ C(i\omega I - A)^{-1} B + D \end{array} \right) = G(i\omega)^* PG(i\omega).$$

■

Introducing and removing dynamics in LMIs

E.1 Introducing and removing unobservable modes in LMIs

Consider the two KYP-LMIs

$$
\begin{pmatrix} I & 0 \\ A_{\mathrm{o}} & B_{\mathrm{o}} \\ C_{\mathrm{o}} & D_{\mathrm{o}} \end{pmatrix}^{\top}
\begin{pmatrix} 0 & X_{\mathrm{o}} & 0 \\ X_{\mathrm{o}} & 0 & 0 \\ 0 & 0 & P \end{pmatrix}
\begin{pmatrix} I & 0 \\ A_{\mathrm{o}} & B_{\mathrm{o}} \\ C_{\mathrm{o}} & D_{\mathrm{o}} \end{pmatrix} \prec 0,
\tag{E.1}
$$

$$
\begin{pmatrix} I & 0 & 0 \\ 0 & I & 0 \\ A_{\mathrm{u}} & A_{\mathrm{uo}} & B_{\mathrm{u}} \\ 0 & A_{\mathrm{o}} & B_{\mathrm{o}} \\ 0 & C_{\mathrm{o}} & D_{\mathrm{o}} \end{pmatrix}^{\top}
\begin{pmatrix} 0 & 0 & X_{\mathrm{uu}} & X_{\mathrm{uo}} & 0 \\ 0 & 0 & X_{\mathrm{uo}}^{\top} & X_{\mathrm{oo}} & 0 \\ X_{\mathrm{uu}} & X_{\mathrm{uo}} & 0 & 0 & 0 \\ X_{\mathrm{uo}}^{\top} & X_{\mathrm{oo}} & 0 & 0 & 0 \\ 0 & 0 & 0 & 0 & P \end{pmatrix}
\begin{pmatrix} I & 0 & 0 \\ 0 & I & 0 \\ A_{\mathrm{u}} & A_{\mathrm{uo}} & B_{\mathrm{u}} \\ 0 & A_{\mathrm{o}} & B_{\mathrm{o}} \\ 0 & C_{\mathrm{o}} & D_{\mathrm{o}} \end{pmatrix} \prec 0
\tag{E.2}
$$

for transfer matrices with realizations that differ by unobservable modes. The following lemma, which was (literally) taken from our paper [198], describes a useful relation between their solution sets.

Lemma E.1. *Let X_{o} satisfy* (E.1). *If $X_{\mathrm{u}} = X_{\mathrm{u}}^{\top}$ is chosen with $\mathrm{He}(X_{\mathrm{u}} A_{\mathrm{u}}) \prec 0$, there exists some $\epsilon > 0$ (that can be chosen arbitrarily small) such that* (E.2) *holds for*

$$
\begin{pmatrix} X_{\mathrm{uu}} & X_{\mathrm{uo}} \\ X_{\mathrm{uo}}^{\top} & X_{\mathrm{oo}} \end{pmatrix} := \begin{pmatrix} \epsilon X_{\mathrm{u}} & 0 \\ 0 & X_{\mathrm{o}} \end{pmatrix}.
\tag{E.3}
$$

Conversely, (E.2) *implies that X_{uu} is invertible and that* (E.1) *is satisfied for*

$$X_\mathrm{o} := X_\mathrm{oo} - X_\mathrm{uo}^\top X_\mathrm{uu}^{-1} X_\mathrm{uo} \tag{E.4}$$

Proof. Since (E.1) is strictly negative definite, there exists a sufficiently small $\epsilon > 0$ such that

$$\begin{pmatrix} I & 0 \\ A_\mathrm{o} & B_\mathrm{o} \\ C_\mathrm{o} & D_\mathrm{o} \end{pmatrix}^\top \begin{pmatrix} 0 & X_\mathrm{o} & 0 \\ X_\mathrm{o} & 0 & 0 \\ 0 & 0 & P \end{pmatrix} \begin{pmatrix} I & 0 \\ A_\mathrm{o} & B_\mathrm{o} \\ C_\mathrm{o} & D_\mathrm{o} \end{pmatrix}$$
$$- \epsilon(\star)^\top \Big(\mathrm{He}(X_\mathrm{u} A_\mathrm{u}) \Big)^{-1} \begin{pmatrix} X_\mathrm{u} A_\mathrm{uo} & X_\mathrm{u} B_\mathrm{u} \end{pmatrix} \prec 0.$$

The first statement follows by applying the Schur complement. For the second statement observe that (E.2) implies $\mathrm{He}(X_\mathrm{uu} A_\mathrm{u}) \prec 0$ which in turn shows that X_uu is invertible. By a simple congruence transformation, (E.2) can be written as

$$\begin{pmatrix} I & 0 & 0 \\ 0 & I & 0 \\ A_\mathrm{u} & \star & \star \\ 0 & A_\mathrm{o} & B_\mathrm{o} \\ 0 & C_\mathrm{o} & D_\mathrm{o} \end{pmatrix}^\top \begin{pmatrix} 0 & 0 & X_\mathrm{uu} & 0 & 0 \\ 0 & 0 & 0 & X_\mathrm{o} & 0 \\ X_\mathrm{uu} & 0 & 0 & 0 & 0 \\ 0 & X_\mathrm{o} & 0 & 0 & 0 \\ 0 & 0 & 0 & 0 & P \end{pmatrix} \begin{pmatrix} I & 0 & 0 \\ 0 & I & 0 \\ A_\mathrm{u} & \star & \star \\ 0 & A_\mathrm{o} & B_\mathrm{o} \\ 0 & C_\mathrm{o} & D_\mathrm{o} \end{pmatrix} \prec 0$$

with X_o as in (E.4). A cancelation of the $(1,1)$-block proves the second statement. ∎

E.2 Introducing and removing uncontrollable modes in LMIs

Consider the two KYP-LMIs

$$\begin{pmatrix} I & 0 \\ A_\mathrm{o} & B_\mathrm{o} \\ C_\mathrm{o} & D_\mathrm{o} \end{pmatrix}^\top \begin{pmatrix} 0 & X_\mathrm{o} & 0 \\ X_\mathrm{o} & 0 & 0 \\ 0 & 0 & P \end{pmatrix} \begin{pmatrix} I & 0 \\ A_\mathrm{o} & B_\mathrm{o} \\ C_\mathrm{o} & D_\mathrm{o} \end{pmatrix} \prec 0, \tag{E.5}$$

$$\begin{pmatrix} I & 0 & 0 \\ 0 & I & 0 \\ A_\mathrm{u} & 0 & 0 \\ A_\mathrm{ou} & A_\mathrm{o} & B_\mathrm{o} \\ C_\mathrm{u} & C_\mathrm{o} & D_\mathrm{o} \end{pmatrix}^\top \begin{pmatrix} 0 & 0 & X_\mathrm{uu} & X_\mathrm{uo} & 0 \\ 0 & 0 & X_\mathrm{uo}^\top & X_\mathrm{oo} & 0 \\ X_\mathrm{uu} & X_\mathrm{uo} & 0 & 0 & 0 \\ X_\mathrm{uo}^\top & X_\mathrm{oo} & 0 & 0 & 0 \\ 0 & 0 & 0 & 0 & P \end{pmatrix} \begin{pmatrix} I & 0 & 0 \\ 0 & I & 0 \\ A_\mathrm{u} & 0 & 0 \\ A_\mathrm{ou} & A_\mathrm{o} & B_\mathrm{o} \\ C_\mathrm{u} & C_\mathrm{o} & D_\mathrm{o} \end{pmatrix} \prec 0 \tag{E.6}$$

for transfer matrices with realizations that differ by uncontrollable modes. The following lemma describes a useful relation between their solution sets.

Lemma E.2. *Let X_{o} satisfy* (E.5). *If $X_{\mathrm{u}} = X_{\mathrm{u}}^{\top}$ is chosen as* $\mathrm{He}(X_{\mathrm{u}}A_{\mathrm{u}}) \prec 0$, *there exists some $\epsilon > 0$ (that can be chosen arbitrarily small) such that* (E.6) *holds for*

$$\begin{pmatrix} X_{\mathrm{uu}} & X_{\mathrm{uo}} \\ X_{\mathrm{uo}}^{\top} & X_{\mathrm{oo}} \end{pmatrix} := \begin{pmatrix} \frac{1}{\epsilon}X_{\mathrm{u}} & 0 \\ 0 & X_{\mathrm{o}} \end{pmatrix}, \tag{E.7}$$

Conversely, (E.6) *implies that* (E.5) *is satisfied for X_{oo}.*

Proof. Since (E.5) is strictly negative definite, there exists a sufficiently small $\epsilon > 0$ such that both $\mathrm{He}(X_{\mathrm{u}}A_{\mathrm{u}}) + \epsilon C_{\mathrm{u}}^{\top}PC_{\mathrm{u}} \prec 0$ and

$$\begin{pmatrix} I & 0 \\ A_{\mathrm{o}} & B_{\mathrm{o}} \\ C_{\mathrm{o}} & D_{\mathrm{o}} \end{pmatrix}^{\top} \begin{pmatrix} 0 & X_{\mathrm{o}} & 0 \\ X_{\mathrm{o}} & 0 & 0 \\ 0 & 0 & P \end{pmatrix} \begin{pmatrix} I & 0 \\ A_{\mathrm{o}} & B_{\mathrm{o}} \\ C_{\mathrm{o}} & D_{\mathrm{o}} \end{pmatrix}$$
$$- \epsilon (\star)^{\top} \left(\mathrm{He}(X_{\mathrm{u}}A_{\mathrm{u}}) + \epsilon C_{\mathrm{u}}^{\top}PC_{\mathrm{u}} \right)^{-1} \left(A_{\mathrm{ou}}^{\top}X_{\mathrm{o}} + C_{\mathrm{u}}^{\top}PC_{\mathrm{o}} \quad C_{\mathrm{u}}^{\top}PD_{\mathrm{o}} \right) \prec 0,$$

persist to hold. It follows that

$$\begin{pmatrix} I & 0 \\ A_{\mathrm{o}} & B_{\mathrm{o}} \\ C_{\mathrm{o}} & D_{\mathrm{o}} \end{pmatrix}^{\top} \begin{pmatrix} 0 & X_{\mathrm{o}} & 0 \\ X_{\mathrm{o}} & 0 & 0 \\ 0 & 0 & P \end{pmatrix} \begin{pmatrix} I & 0 \\ A_{\mathrm{o}} & B_{\mathrm{o}} \\ C_{\mathrm{o}} & D_{\mathrm{o}} \end{pmatrix}$$
$$- (\star)^{\top} \left(\mathrm{He}\left(\tfrac{1}{\epsilon}X_{\mathrm{u}}A_{\mathrm{u}}\right) + C_{\mathrm{u}}^{\top}PC_{\mathrm{u}} \right)^{-1} \left(A_{\mathrm{ou}}^{\top}X_{\mathrm{o}} + C_{\mathrm{u}}^{\top}PC_{\mathrm{o}} \quad C_{\mathrm{u}}^{\top}PD_{\mathrm{o}} \right) \prec 0.$$

The first statement now directly follows by applying the Schur complement. The second statement is trivial.

∎

Appendix F
Operations on FDIs and their corresponding LMIs

F.1 Replacing multiplier factorizations

The two frequency domain conditions

$$G^* \Psi_\mathrm{o}^* P_\mathrm{o} \Psi_\mathrm{o} G \prec 0 \text{ on } \mathbb{C}^0 \quad \text{and} \quad \Psi_\mathrm{o}^* P_\mathrm{o} \Psi_\mathrm{o} = \Psi_\mathrm{n}^* P_\mathrm{n} \Psi_\mathrm{n} \text{ on } \mathbb{C}^0 \tag{F.1}$$

trivially imply

$$G^* \Psi_\mathrm{n}^* P_\mathrm{n} \Psi_\mathrm{n} G \prec 0 \text{ on } \mathbb{C}^0. \tag{F.2}$$

The following lemma is a special case of [159, Theorem 3] and describes how certificates for (F.1) can be combined to obtain one for (F.2).

Lemma F.1. *Let* G, Ψ_o *and* Ψ_n *admit the realizations* $G = (A, B, C, D)$, $\Psi_\mathrm{o} = (A_\Psi, B_\Psi, C_{\Psi_\mathrm{o}}, D_{\Psi_\mathrm{o}})$ *and* $\Psi_\mathrm{n} = (A_\Psi, B_\Psi, C_{\Psi_\mathrm{n}}, D_{\Psi_\mathrm{n}})$ *respectively, and suppose that* X *and* Z *satisfy*

$$\begin{pmatrix} \star \\ \star \\ \star \\ \star \\ \star \end{pmatrix}^\top \left(\begin{array}{cc|cc|c} 0 & 0 & X_{11} & X_{12} & 0 \\ 0 & 0 & X_{12}^\top & X_{22} & 0 \\ \hline X_{11} & X_{12} & 0 & 0 & 0 \\ X_{12}^\top & X_{22} & 0 & 0 & 0 \\ \hline 0 & 0 & 0 & 0 & P_\mathrm{o} \end{array} \right) \left(\begin{array}{cc|c} I & 0 & 0 \\ 0 & I & 0 \\ \hline A_\Psi & B_\Psi C & B_\Psi D \\ 0 & A & B \\ \hline C_{\Psi_\mathrm{o}} & D_{\Psi_\mathrm{o}} C & D_{\Psi_\mathrm{o}} D \end{array} \right) \prec 0 \tag{F.3}$$

and

$$
\begin{pmatrix} I & 0 \\ A_\Psi & B_\Psi \\ \hline C_{\Psi \mathrm{n}} & D_{\Psi \mathrm{n}} \\ C_{\Psi \mathrm{o}} & D_{\Psi \mathrm{o}} \end{pmatrix}^\top
\begin{pmatrix} 0 & Z & 0 & 0 \\ Z & 0 & 0 & 0 \\ \hline 0 & 0 & P_\mathrm{n} & 0 \\ 0 & 0 & 0 & -P_\mathrm{o} \end{pmatrix}
\begin{pmatrix} I & 0 \\ A_\Psi & B_\Psi \\ \hline C_{\Psi \mathrm{n}} & D_{\Psi \mathrm{n}} \\ C_{\Psi \mathrm{o}} & D_{\Psi \mathrm{o}} \end{pmatrix} = 0. \tag{F.4}
$$

Then these two inequalities imply

$$
\begin{pmatrix} \star \\ \star \\ \hline \star \\ \star \\ \hline \star \end{pmatrix}^\top
\begin{pmatrix} 0 & 0 & X_{11}+Z & X_{12} & 0 \\ 0 & 0 & X_{12}^\top & X_{22} & 0 \\ \hline X_{11}+Z & X_{12} & 0 & 0 & 0 \\ X_{12}^\top & X_{22} & 0 & 0 & 0 \\ \hline 0 & 0 & 0 & 0 & P_\mathrm{n} \end{pmatrix}
\begin{pmatrix} I & 0 & 0 \\ 0 & I & 0 \\ \hline A_\Psi & B_\Psi C & B_\Psi D \\ 0 & A & B \\ \hline C_{\Psi \mathrm{n}} & D_{\Psi \mathrm{n}} C & D_{\Psi \mathrm{n}} D \end{pmatrix} \prec 0. \tag{F.5}
$$

Proof. By augmenting (F.4) and performing a congruence transformation with $\operatorname{diag}(I,(C\ D))$ we obtain

$$
\begin{pmatrix} \star \\ \star \\ \hline \star \\ \star \\ \hline \star \\ \star \end{pmatrix}^\top
\begin{pmatrix} 0 & 0 & Z & 0 & 0 & 0 \\ 0 & 0 & 0 & 0 & 0 & 0 \\ \hline Z & 0 & 0 & 0 & 0 & 0 \\ 0 & 0 & 0 & 0 & 0 & 0 \\ \hline 0 & 0 & 0 & 0 & P_\mathrm{n} & 0 \\ 0 & 0 & 0 & 0 & 0 & -P_\mathrm{o} \end{pmatrix}
\begin{pmatrix} I & 0 & 0 \\ 0 & I & 0 \\ \hline A_\Psi & B_\Psi C & B_\Psi D \\ 0 & A & B \\ \hline C_{\Psi \mathrm{n}} & D_{\Psi \mathrm{n}} C & D_{\Psi \mathrm{n}} D \\ C_{\Psi \mathrm{o}} & D_{\Psi \mathrm{o}} C & D_{\Psi \mathrm{o}} D \end{pmatrix}. \tag{F.6}
$$

Substituting (F.6) in (F.3) then directly yields (F.5).

∎

F.2 Congruence transformations

Consider the two FDIs

$$
G_1^* P G_1 \prec 0 \ \text{on} \ \mathbb{C}^0 \quad \text{and} \quad G_2^* G_2 \succ 0 \ \text{on} \ \mathbb{C}^0, \tag{F.7}
$$

and observe that they trivially imply

$$
G_2^* G_1^* P G_1 G_2 \prec 0 \ \text{on} \ \mathbb{C}^0. \tag{F.8}
$$

Moreover, if G_2 is nonsingular on $\mathbb{C}^0$, then the converse is true as well.

In this section we discuss how to perform the corresponding dynamical congruence transformation in the state-space. For this purpose, let G_1 and G_2 admit the realizations (A_1, B_1, C_1, D_1) and (A_2, B_2, C_2, D_2) respectively, with $\mathrm{eig}(A_i) \cap \mathbb{C}^0 = \emptyset$, $i = 1, 2$. By the KYP-lemma the FDIs in (F.7) and (F.8) are equivalent to the existence of some symmetric matrices X_1, X_2 and

$$
X = \begin{pmatrix} X_{11} & X_{12} \\ X_{12}^\top & X_{22} \end{pmatrix}
$$

for which the following LMIs hold true

$$
\begin{pmatrix} I & 0 \\ A_1 & B_1 \\ C_1 & D_1 \end{pmatrix}^\top \begin{pmatrix} 0 & X_1 & 0 \\ X_1 & 0 & 0 \\ 0 & 0 & P \end{pmatrix} \begin{pmatrix} I & 0 \\ A_1 & B_1 \\ C_1 & D_1 \end{pmatrix} \prec 0, \tag{F.9}
$$

$$
\begin{pmatrix} I & 0 \\ A_2 & B_2 \\ C_2 & D_2 \end{pmatrix}^\top \begin{pmatrix} 0 & -X_2 & 0 \\ -X_2 & 0 & 0 \\ 0 & 0 & I \end{pmatrix} \begin{pmatrix} I & 0 \\ A_2 & B_2 \\ C_2 & D_2 \end{pmatrix} \succ 0, \tag{F.10}
$$

$$
\begin{pmatrix} \star \\ \star \\ \star \\ \star \\ \star \end{pmatrix}^\top \left(\begin{array}{cc:cc:c} 0 & 0 & X_{11} & X_{12} & 0 \\ 0 & 0 & X_{12}^\top & X_{22} & 0 \\ \hdashline X_{11} & X_{12} & 0 & 0 & 0 \\ X_{12}^\top & X_{22} & 0 & 0 & 0 \\ \hdashline 0 & 0 & 0 & 0 & P \end{array} \right) \left(\begin{array}{cc:c} I & 0 & 0 \\ 0 & I & 0 \\ \hdashline A_1 & B_1 C_2 & B_1 D_2 \\ 0 & A_2 & B_2 \\ \hdashline C_1 & D_1 C_2 & D_1 D_2 \end{array} \right) \succ 0. \tag{F.11}
$$

We can now state the following result.

Lemma F.2. *Suppose there exist $X_1 = X_1^\top$ and $X_2 = X_2^\top$ for which respectively (F.9) and (F.10) hold true. Then (F.8) is certified by $X := \mathrm{diag}(X_1, X_2)$. Conversely, if $G_2 = (A_2, B_2, C_2, D_2)$ is invertible on $\mathbb{C}^0$ and if there exist some $X = X^\top$ for which (F.11) holds true, then (F.9) is satisfied with $X_1 := X_{11} - X_{12} X_{22}^{-1} X_{12}^\top$.*

Proof. Observe that (F.9) can be rewritten as

$$
Q_1 := D_1^\top P D_1 - (\star)^\top \left(\mathrm{He}(X_1 A_1) + C_1^\top P C_1 \right)^{-1} \left(X_1 B_1 + C_1^\top P D_1 \right) \prec 0.
$$

Since $Q_1 \prec 0$ there exists some square and invertible T_1 such that $Q_1 = -T_1^\top T_1$. Moreover, since $\mathrm{eig}(A_2) \cap \mathbb{C}^0 = \emptyset$, X_2 is invertible and by Lemma C.1 we can infer that (F.10) is equivalent to

$$\begin{pmatrix} -A_2^\top & -C_2^\top \\ I & 0 \\ -B_2^\top & -D_2^\top \end{pmatrix}^\top \begin{pmatrix} 0 & -X_2 & 0 \\ -X_2 & 0 & 0 \\ 0 & 0 & I \end{pmatrix}^{-1} \begin{pmatrix} -A_2^\top & -C_2^\top \\ I & 0 \\ -B_2^\top & -D_2^\top \end{pmatrix} \succ 0.$$

A simple congruence transformation with $\mathrm{diag}(I, T_1^\top)$ then yields

$$\begin{pmatrix} -A_2^\top & -C_2^\top T_1^\top \\ I & 0 \\ -B_2^\top & -D_2^\top T_1^\top \end{pmatrix}^\top \begin{pmatrix} 0 & -X_2 & 0 \\ -X_2 & 0 & 0 \\ 0 & 0 & I \end{pmatrix}^{-1} \begin{pmatrix} -A_2^\top & -C_2^\top T_1^\top \\ I & 0 \\ -B_2^\top & -D_2^\top T_1^\top \end{pmatrix} \succ 0$$

and thus

$$\begin{pmatrix} I & 0 \\ A_2 & B_2 \\ C_2 & D_2 \end{pmatrix}^\top \begin{pmatrix} 0 & X_2 & 0 \\ X_2 & 0 & 0 \\ 0 & 0 & Q_1 \end{pmatrix} \begin{pmatrix} I & 0 \\ A_2 & B_2 \\ C_2 & D_2 \end{pmatrix} \prec 0.$$

Moreover, after applying a Schur complement we obtain

$$\begin{pmatrix} \mathrm{He}(X_1 A_1) + C_1^\top P C_1 & (X_1 B_1 + C_1^\top P D_1) C_2 & (X_1 B_1 + C_1^\top P D_1) D_2 \\ \star & \mathrm{He}(X_2 A_2) + C_2^\top (D_1^\top P D_1) C_2 & B_2 X_2 + C_2 (D_1^\top P D_1) D_2 \\ \star & \star & D_2^\top (D_1^\top P D_1) D_2 \end{pmatrix} \prec 0,$$

which is nothing but (F.11) with $X := \mathrm{diag}(X_1, X_2)$.

Conversely, since G_2 is invertible on $\mathbb{C}^0$, we have that $G_2(\infty) = D_2$ is nonsingular. Hence, we can apply the congruence transformation

$$\begin{pmatrix} 0 & I & 0 \\ I & 0 & 0 \\ -D_2^{-1}C_2 & 0 & D_2^{-1} \end{pmatrix}^\top (F.11) \begin{pmatrix} 0 & I & 0 \\ I & 0 & 0 \\ -D_2^{-1}C_2 & 0 & D_2^{-1} \end{pmatrix},$$

which yields the inequality

$$\begin{pmatrix} \star \\ \star \\ \star \\ \star \\ \star \end{pmatrix}^\top \left(\begin{array}{cc|cc|c} 0 & 0 & X_{22} & X_{12}^\top & 0 \\ 0 & 0 & X_{12} & X_{11} & 0 \\ \hline X_{22} & X_{12}^\top & 0 & 0 & 0 \\ X_{12} & X_{11} & 0 & 0 & 0 \\ \hline 0 & 0 & 0 & 0 & P \end{array} \right) \left(\begin{array}{cc|c} I & 0 & 0 \\ 0 & I & 0 \\ \hline A_2 - B_2 D_2^{-1}C_2 & 0 & B_2 D_2^{-1} \\ \hline 0 & A_1 & B_1 \\ 0 & C_1 & D_1 \end{array} \right) \succ 0.$$

(F.12)

By Lemma E.1 we finally infer that (F.12) implies that (F.9) holds with $X_1 := X_{11} - X_{12} X_{22}^{-1} X_{12}^\top$.

∎

Appendix G
On the factorization of rational matrix functions

In this appendix we address the factorization of structured symmetric transfer matrices

$$\Pi = \begin{pmatrix} \Pi_{11} & \Pi_{12} \\ \Pi_{12}^* & \Pi_{22} \end{pmatrix} \in \mathscr{RL}_\infty^{(n_1+n_2) \times (n_1+n_2)}$$

into the following form:

$$\Pi = \hat{\Psi}^* \hat{P} \hat{\Psi} = \begin{pmatrix} \hat{\Psi}_1 & \hat{\Psi}_3 \\ 0 & \hat{\Psi}_2 \end{pmatrix}^* \begin{pmatrix} P_{11} & P_{12} \\ P_{12}^\top & P_{22} \end{pmatrix} \begin{pmatrix} \hat{\Psi}_1 & \hat{\Psi}_3 \\ 0 & \hat{\Psi}_2 \end{pmatrix}. \tag{G.1}$$

Here the middle matrix $\hat{P}$ is nonsingular and satisfies the constraints $P_{11} \succcurlyeq 0$ and $P_{22} \preccurlyeq 0$, while $\hat{\Psi}$ is upper triangular and satisfies the minimal requirements that $\hat{\Psi}$ and the partial inverse

$$\begin{pmatrix} \hat{\Psi}_1 & \hat{\Psi}_3 \hat{\Psi}_2^{-1} \\ 0 & \hat{\Psi}_2^{-1} \end{pmatrix}$$

are both stable, and, whenever possible, that $\hat{\Psi}$ is square and invertible and has a stable inverse. We emphasize that most parts in this section have already appeared (sometimes literally) in [198].

The factorization of real rational matrix functions that are bounded on the imaginary axis has been studied intensively and is of independent interest (see e.g. [57, 16] and references therein). However, we stress that those of this particular form are rarely addressed [55, 56, 165] and are useful for reasons described in [198, 156, 196] as well as in Chapter 4 and 6. Throughout this appendix we assume that Π admits the initial from

$$\Pi = \Psi^* P \Psi$$

with some real symmetric matrix $P \in \mathbb{S}^{\bullet}$ and a stable, typically tall outer-factor $\Psi \in \mathscr{RH}_{\infty}^{\bullet \times (n_1 + n_2)}$. Note that any such factorization always exists and is non-unique [156, Lemma 2]. Moreover, the number of positive and negative eigenvalues of an arbitrary transfer matrix $\Pi = \Pi^* \in \mathscr{RL}_{\infty}^{(n_1 + n_2) \times (n_1 + n_2)}$ need not be fixed over frequency. This makes it hard to formulate a general statement about the existence of a structured factorization of Π as (G.1). Fortunately, under some additional (practically relevant) assumptions on Π, it is possible to proceed as we will see in the sequel.

G.1 Factorization I

Let us partition the outer-factor Ψ as $\Psi = \begin{pmatrix} \Psi_1 & \Psi_2 \end{pmatrix} \in \mathscr{RH}_{\infty}^{\bullet \times (n_1 + n_2)}$, such that

$$\Pi = \Psi^* P \Psi = \begin{pmatrix} \Psi_1 & \Psi_2 \end{pmatrix}^* P \begin{pmatrix} \Psi_1 & \Psi_2 \end{pmatrix} = \begin{pmatrix} \Psi_1^* P \Psi_1 & \Psi_1^* P \Psi_2 \\ \Psi_2^* P \Psi_1 & \Psi_2^* P \Psi_2 \end{pmatrix}, \tag{G.2}$$

and assume that (G.2) satisfies the inertia constraints

$$\Psi_1^* P \Psi_1 \succ 0 \text{ on } \mathbb{C}^0, \tag{G.3a}$$

$$\Psi_2^* P \Psi_2 - \Psi_2^* P \Psi_1 (\Psi_1^* P \Psi_1)^{-1} \Psi_1^* P \Psi_2 \prec 0 \text{ on } \mathbb{C}^0. \tag{G.3b}$$

Then $\Pi = \Psi^* P \Psi \in \mathscr{RL}_{\infty}^{(n_1 + n_2) \times (n_1 + n_2)}$ has n_1 positive and n_2 negative eigenvalues and can be factorized as

$$\Pi = \hat{\Psi}^* \hat{P} \hat{\Psi} := \begin{pmatrix} \hat{\Psi}_1 & \hat{\Psi}_3 \\ 0 & \hat{\Psi}_2 \end{pmatrix}^* \begin{pmatrix} I_{n_1} & 0 \\ 0 & -I_{n_2} \end{pmatrix} \begin{pmatrix} \hat{\Psi}_1 & \hat{\Psi}_3 \\ 0 & \hat{\Psi}_2 \end{pmatrix}, \tag{G.4}$$

where $\hat{\Psi}$ is square and invertible and satisfies one of the following properties

1. $\hat{\Psi}_1 \in \mathscr{RH}_{\infty}$ and $\hat{\Psi}^{-1} = \begin{pmatrix} \hat{\Psi}_1^{-1} & -\hat{\Psi}_1^{-1} \hat{\Psi}_3 \hat{\Psi}_2^{-1} \\ 0 & \hat{\Psi}_2^{-1} \end{pmatrix} \in \mathscr{RH}_{\infty}$ \hfill (G.5a)

2. $\hat{\Psi} = \begin{pmatrix} \hat{\Psi}_1 & \hat{\Psi}_3 \\ 0 & \hat{\Psi}_2 \end{pmatrix} \in \mathscr{RH}_{\infty}$ and $\begin{pmatrix} \hat{\Psi}_1 & \hat{\Psi}_3 \hat{\Psi}_2^{-1} \\ 0 & \hat{\Psi}_2^{-1} \end{pmatrix} \in \mathscr{RH}_{\infty}$ \hfill (G.5b)

3. $\hat{\Psi}_1 \in \mathscr{RH}_{\infty}$ and $\hat{\Psi}^{-*} = \begin{pmatrix} \hat{\Psi}_1^{-*} & 0 \\ -\hat{\Psi}_2^{-*} \hat{\Psi}_3^* \hat{\Psi}_1^{-*} & \hat{\Psi}_2^{-*} \end{pmatrix} \in \mathscr{RH}_{\infty}$ \hfill (G.5c)

Note that factorizations of this particular form have also been addressed in [56, 165]. Here we stress that [56] restricts itself to frequency domain techniques and that the proof in the sequel is significantly simplified if compared to [165].

In order to perform the factorization (G.4), it is well-known that we need to solve an indefinite algebraic Riccati equation (ARE) (see e.g. [16]). Unfortunately, for general indefinite AREs simple conditions for the existence of stabilizing (or anti-stabilizing) solutions do not exist. For the problem under consideration we exploit the particular structure in order to guarantee the existence such solutions under mild assumptions.

As a preparation, let us assume that (G.3a) is valid and that Ψ_1 and Ψ_2 are realized as

$$
\Psi_i = \left[\begin{array}{c|c} A_i & B_i \\ \hline C_i & D_i \end{array} \right], \quad \mathrm{eig}(A_i) \subset \mathbb{C}^-, \quad (A_i, B_i) \text{ controllable}, \quad i = 1, 2.
$$

By the KYP-lemma, we infer that there exists some symmetric matrix Y_{11} for which the following LMI holds true:

$$
\begin{pmatrix} I & 0 \\ A_1 & B_1 \\ C_1 & D_1 \end{pmatrix}^\top \begin{pmatrix} 0 & Y_{11} & 0 \\ Y_{11} & 0 & 0 \\ 0 & 0 & P \end{pmatrix} \begin{pmatrix} I & 0 \\ A_1 & B_1 \\ C_1 & D_1 \end{pmatrix} \succ 0. \tag{G.6}
$$

This implies that there exist some square and non-singular $\hat{D}_1$ such that $D_1^\top P D_1 = \hat{D}_1^\top \hat{D}_1 \succ 0$. With any such factorization, (G.6) is equivalent to the algebraic Riccati inequality

$$
A_1^\top Y_{11} + Y_{11} A_1 + C_1^\top P C_1 - (\star)^\top \hat{D}_1^{-1} \hat{D}_1^{-\top} \left(B_1^\top Y_{11} + D_1^\top P C_1 \right) \succ 0. \tag{G.7}
$$

Since the pair (A_1, B_1) is controllable and $\hat{D}_1^\top \hat{D}_1 \succ 0$, the corresponding ARE has a unique stabilizing solution Z_{11}, i.e. $A_1 - B_1 \hat{D}_1^{-1} \hat{C}_1$ is Hurwitz for $\hat{C}_1 = \hat{D}_1^{-\top} \left(B_1^\top Z_{11} + D_1^\top P C_1 \right)$. Note that Z_{11} has the additional property $Z_{11} \succ Y_{11}$ for all solutions Y_{11} of (G.7). For the square and invertible

$$
\hat{\Psi}_1 = \left[\begin{array}{c|c} A_1 & B_1 \\ \hline \hat{C}_1 & \hat{D}_1 \end{array} \right]
$$

we infer that

$$
\Psi_1^* P \Psi_1 = \hat{\Psi}_1^* \hat{\Psi}_1
$$

and that $\hat{\Psi}_1^{-1}$ is stable.

Let us continue our preparations by defining the transfer matrix $\hat{\Psi}_3 = \hat{\Psi}_1^{-*} \Psi_1^* P \Psi_2$ which admits the realization

$$\hat{\Psi}_3 = \left[\begin{array}{cc|c} -A_1^\top + \hat{C}_1^\top \hat{D}_1^{-\top} B_1^\top & \hat{P}_1 C_2 & \hat{P}_1 D_2 \\ \hline 0 & A_2 & B_2 \\ \hline -\hat{D}_1^{-\top} B_1^\top & \hat{P}_2 C_2 & \hat{P}_2 D_2 \end{array}\right] \tag{G.8}$$

with $\hat{P}_1 = (C_1^\top - \hat{C}_1^\top \hat{D}_1^{-\top} D_1^\top)P$ and $\hat{P}_2 = \hat{D}_1^{-\top} D_1^\top P$ respectively. In order to be able to solve the subsequent indefinite algebraic Riccati equation, we need to work with a controllable realization of $\hat{\Psi}_3$. If (G.8) has uncontrollable modes, one can always perform a simple state-coordinate change in order to arrive at a controllable realization that is structured as follows:

$$\hat{\Psi}_3 = \left[\begin{array}{c|c} \hat{A}_2 & \hat{B}_2 \\ \hline \hat{C}_3 & \hat{D}_3 \end{array}\right] = \left[\begin{array}{cc|cc} \star & 0 & 0 & 0 \\ \star & \star & \star & \star \\ \hline 0 & 0 & A_2 & B_2 \\ \hline \star & \star & \star & \hat{P}_2 D_2 \end{array}\right] = \left[\begin{array}{cc|c} \star & \star & \star \\ 0 & A_2 & B_2 \\ \hline \star & \star & \hat{P}_2 D_2 \end{array}\right]. \tag{G.9}$$

Indeed, with some nonsingular matrix T we can display the uncontrollable modes as follows:

$$\left(\begin{array}{cc|c} -A_1^\top + \hat{C}_1^\top \hat{D}_1^{-\top} B_1^\top & \hat{P}_1 C_2 & \hat{P}_1 D_2 \\ \hline 0 & A_2 & B_2 \\ \hline -\hat{D}_1^{-\top} B_1^\top & \hat{P}_2 C_2 & \hat{P}_2 D_2 \end{array}\right) \underbrace{\left(\begin{array}{cc|c} T_1 & T_2 & T_3 & 0 \\ 0 & 0 & I & 0 \\ 0 & 0 & 0 & I \end{array}\right)}_{T} =$$

$$= \underbrace{\left(\begin{array}{cc|c} T_1 & T_2 & T_3 & 0 \\ 0 & 0 & I & 0 \\ 0 & 0 & 0 & I \end{array}\right)}_{T} \left(\begin{array}{cc|cc} \star & 0 & 0 & 0 \\ \star & \star & \star & \star \\ \hline 0 & 0 & A_2 & B_2 \\ \hline \star & \star & \star & \hat{P}_2 D_2 \end{array}\right).$$

Let us now come back to the transfer matrix (G.2) and the corresponding constraints (G.3). Thanks to our preparations, we can define the controllable realization

$$\Psi = \left[\begin{array}{c|c} A_\Psi & B_\Psi \\ \hline C_\Psi & D_\Psi \end{array}\right] = \left[\begin{array}{cc|cc} A_1 & 0 & B_1 & 0 \\ 0 & \hat{A}_2 & 0 & \hat{B}_2 \\ \hline C_1 & C_2 J_r & D_1 & D_2 \end{array}\right] = \left[\begin{array}{ccc|cc} A_1 & 0 & 0 & B_1 & 0 \\ 0 & \star & \star & 0 & \star \\ 0 & 0 & A_2 & 0 & B_2 \\ \hline C_1 & 0 & C_2 & D_1 & D_2 \end{array}\right], \tag{G.10}$$

where $J_r = \left(\begin{array}{cc} 0 & I \end{array}\right)$, and consider the following indefinite algebraic Riccati equation:

$$A_\Psi^\top Z_\Psi + Z_\Psi A_\Psi + C_\Psi^\top P C_\Psi - (\star)^\top \left(D_\Psi^\top P D_\Psi\right)^{-1}\left(B_\Psi^\top Z_\Psi + D_\Psi^\top P C_\Psi\right) = 0. \quad \text{(G.11)}$$

Due to the particular way of realizing Ψ we have the following result.

Lemma G.1. *Suppose that Π satisfies the inertia constraints in (G.3) and let Ψ be realized as in (G.10) with (A_Ψ, B_Ψ) being controllable. Then the indefinite ARE (G.11) has a (unique) stabilizing solution $Z_{\Psi-}$, i.e., $A_\Psi - B_\Psi(D_\Psi^\top P D_\Psi)^{-1}(B_\Psi^\top Z_{\Psi-} + D_\Psi^\top P C_\Psi)$ is Hurwitz.*

Proof. By the construction of $\hat{\Psi}_3$, it is possible to write (G.3b) as $\tilde{\Psi}^* \tilde{P}\tilde{\Psi} \prec 0$, with $\tilde{P} = \mathrm{diag}(P, -I)$ and $\tilde{\Psi} = \mathrm{col}(\Psi_2, \hat{\Psi}_3) = (\hat{A}_2, \hat{B}_2, \tilde{C}_2, \tilde{D}_2)$, where $\tilde{C}_2 = \mathrm{col}(C_2 J_\mathrm{r}, \hat{C}_3)$ and $\tilde{D}_2 = \mathrm{col}(D_2, \hat{D}_3)$. In complete analogy to the above-mentioned preparatory steps, we can apply the KYP-lemma in order to certify the FDI $\tilde{\Psi}^* \tilde{P}\tilde{\Psi} \prec 0$ by the symmetric matrix Y_{22}. Subsequently, we can express $\tilde{D}_2^\top \tilde{P}\tilde{D}_2 \prec 0$ as $-\hat{D}_2^\top \hat{D}_2 \prec 0$, where $\hat{D}_2$ is square and invertible. With any such factorization we obtain the ARI

$$Y_{22}\hat{A}_2 + \hat{A}_2^\top Y_{22} + \tilde{C}_2^\top \tilde{P}\tilde{C}_2 + (\star)^\top \hat{D}_2^{-1}\hat{D}_2^{-\top}\left(\hat{B}_2^\top Y_{22} + \hat{D}_2^\top \tilde{P}\tilde{C}_2\right) \prec 0. \quad \text{(G.12)}$$

Since $(\hat{A}_2, \hat{B}_2)$ is controllable and $\hat{D}_2^\top \hat{D}_2 \succ 0$ the corresponding ARE has a unique stabilizing solution Z_{22}, with the property $Z_{22} \prec Y_{22}$ for all solutions Y_{22} of (G.12). With $\hat{C}_2 = -\hat{D}_2^{-\top}(\hat{B}_2^\top Z_{22} + \hat{D}_2^\top \tilde{P}\tilde{C}_2)$ we conclude that $\hat{A}_2 - \hat{B}_2\hat{D}_2^{-1}\hat{C}_2$ is Hurwitz. Moreover, with $\hat{\Psi}_2 = (\hat{A}_2, \hat{B}_2, \hat{C}_2, \hat{D}_2)$, we infer that $\tilde{\Psi}^* \tilde{P}\tilde{\Psi} = -\hat{\Psi}_2^*\hat{\Psi}_2$ where $\hat{\Psi}_2$ is square, invertible and $\hat{\Psi}_2^{-1}$ stable. With $Z_{12} = -\begin{pmatrix} T_2 & T_3 \end{pmatrix}$ it is finally a matter of direct verification that (G.11) is satisfied for

$$Z_{\Psi-} = \begin{pmatrix} Z_{11} & Z_{12} \\ Z_{12}^\top & Z_{22} \end{pmatrix} \quad \text{(G.13)}$$

and that $A_\Psi - B_\Psi(D_\Psi^\top P D_\Psi)^{-1}(B_\Psi^\top Z_{\Psi-} + D_\Psi^\top P C_\Psi)$ is Hurwitz. ∎

Under the hypotheses of Lemma G.1 we now exploit (G.11) and rewrite it as

$$\begin{pmatrix} I & 0 & 0 \\ \hline A_1 & 0 & B_1 & 0 \\ 0 & \hat{A}_2 & 0 & \hat{B}_2 \\ \hline \hat{C}_1 & \hat{C}_3 & \hat{D}_1 & \hat{D}_3 \\ 0 & \hat{C}_2 & 0 & \hat{D}_2 \\ \hline C_1 & C_2 J_\mathrm{r} & D_1 & D_2 \end{pmatrix}^\top \begin{pmatrix} 0 & \mathcal{Z} & 0 & 0 \\ \mathcal{Z} & 0 & 0 & 0 \\ \hline 0 & 0 & P & 0 \\ 0 & 0 & 0 & -P \end{pmatrix} \begin{pmatrix} I & 0 & 0 \\ \hline A_1 & 0 & B_1 & 0 \\ 0 & \hat{A}_2 & 0 & \hat{B}_2 \\ \hline \hat{C}_1 & \hat{C}_3 & \hat{D}_1 & \hat{D}_3 \\ 0 & \hat{C}_2 & 0 & \hat{D}_2 \\ \hline C_1 & C_2 J_\mathrm{r} & D_1 & D_2 \end{pmatrix} \quad \text{(G.14)}$$

for $\mathcal{Z} := -Z_{\Psi-}$. Standard computations reveal that (G.14) implies (G.4) with (G.5a) for the outer-factor

$$\hat{\Psi} = \begin{pmatrix} \hat{\Psi}_1 & \hat{\Psi}_3 \\ 0 & \hat{\Psi}_2 \end{pmatrix} = \left[\begin{array}{cc|cc} A_1 & 0 & B_1 & 0 \\ 0 & \hat{A}_2 & 0 & \hat{B}_2 \\ \hline \hat{C}_1 & \hat{C}_3 & \hat{D}_1 & \hat{D}_3 \\ 0 & \hat{C}_2 & 0 & \hat{D}_2 \end{array} \right]. \qquad \text{(G.15)}$$

Hence, $\hat{\Psi}$ is square and invertible, while $\hat{\Psi}_1$ and $\hat{\Psi}^{-1}$ are stable, just because Z_{Ψ_-} is the stabilizing solution of (G.11). Now observe that one can also construct an anti-stabilizing solution Z_{Ψ_+} as well as mixed versions of the indefinite ARE (G.11), just by choosing either the stabilizing or anti-stabilizing solutions of the AREs that correspond to (G.7) and (G.12). In this fashion we can achieve (G.4) with all the desired properties of $\hat{\Psi}$ in accordance with (G.5).

G.2 Factorization II

For the second factorization we consider the case that the left-upper block of $\Pi \in \mathscr{RL}_{\infty}^{(n_1+n_2)\times(n_1+n_2)}$ is not positive definite, but zero (i.e. $\Pi_{11} = 0$). Then we can partition P and Ψ as

$$P = \begin{pmatrix} 0 & P_{12} \\ P_{12}^{\top} & P_{22} \end{pmatrix} \in \mathbb{R}^{\bullet \times \bullet} \text{ and } \Psi = \begin{pmatrix} \Psi_{11} & 0 \\ 0 & \Psi_{22} \end{pmatrix} \in \mathscr{RH}_{\infty}^{\bullet \times (n_1+n_2)},$$

such that Π now reads as

$$\Pi = \begin{pmatrix} \star \\ \star \end{pmatrix}^* \begin{pmatrix} 0 & P_{12} \\ P_{12}^{\top} & P_{22} \end{pmatrix} \begin{pmatrix} \Psi_{11} & 0 \\ 0 & \Psi_{22} \end{pmatrix} = \begin{pmatrix} 0 & \Psi_{11}^* P_{12}\Psi_{22} \\ \Psi_{22}^* P_{12}^{\top}\Psi_{11} & \Psi_{22}^* P_{22}\Psi_{22} \end{pmatrix}. \quad \text{(G.16)}$$

In addition, we assume that Π satisfies the inertia constraints

$$\begin{pmatrix} G_1 \\ I \end{pmatrix}^* \Pi \begin{pmatrix} G_1 \\ I \end{pmatrix} \prec 0, \quad \begin{pmatrix} I \\ G_2 \end{pmatrix}^* \Pi \begin{pmatrix} I \\ G_2 \end{pmatrix} \succ 0 \qquad \text{(G.17)}$$

and

$$\begin{pmatrix} I \\ \tau G_2 \end{pmatrix}^* \Pi \begin{pmatrix} I \\ \tau G_2 \end{pmatrix} \succcurlyeq 0 \text{ on } \mathbb{C}^0 \quad \forall \tau \in [0,1), \qquad \text{(G.18)}$$

for some given $G_1 \in \mathscr{RH}_{\infty}^{n_1 \times n_2}$ and $G_2 \in \mathscr{RH}_{\infty}^{n_2 \times n_1}$. Consequently, Π has n_1 positive and n_2 negative eigenvalues and can be factorized as

$$\Pi = \hat{\Psi}^* \hat{P} \hat{\Psi} := \begin{pmatrix} \hat{\Psi}_1 & \hat{\Psi}_3 \\ 0 & \hat{\Psi}_2 \end{pmatrix}^* \begin{pmatrix} 0 & I \\ I & 0 \end{pmatrix} \begin{pmatrix} \hat{\Psi}_1 & \hat{\Psi}_3 \\ 0 & \hat{\Psi}_2 \end{pmatrix}, \tag{G.19}$$

where $\hat{\Psi}$ satisfies the following properties

$$\hat{\Psi} = \begin{pmatrix} \hat{\Psi}_1 & \hat{\Psi}_3 \\ 0 & \hat{\Psi}_2 \end{pmatrix} \in \mathcal{RH}_\infty \quad \text{and} \quad \begin{pmatrix} \hat{\Psi}_1 & \hat{\Psi}_3 \hat{\Psi}_2^{-1} \\ 0 & \hat{\Psi}_2^{-1} \end{pmatrix} \in \mathcal{RH}_\infty \tag{G.20}$$

Note that any Π with these particular properties can also be factorized as (G.4) with (G.5) in Section G.1, by slightly perturbing the left-upper block of Π as $\Pi_{11} = \epsilon I \succ 0$. However, this might lead to numerical ill-conditioning. On the other hand, it has been shown in [55] how to construct a canonical factorization for generalized positive real transfer functions. This leads to the desired factorization for a very special case of (G.16) in which Π_{12} is square with $\mathrm{He}(\Pi_{12}) \succ 0$ and $\Pi_{22} = 0$. Both limitations are overcome in the next technical part of this appendix.

Since $\Pi \in \mathcal{RL}_\infty^{(n_1+n_2) \times (n_1+n_2)}$ has n_1 positive and n_2 negative eigenvalues we infer that Π is nonsingular and $\Pi_{12}^* = \Psi_{22}^* P_{12}^\top \Psi_{11}$ has full column-rank on $\mathbb{C}^0$. In addition to that, (G.17)-(G.18) implies that $G_2(I - G_1 G_2)^{-1}$ is LTI and stable. This will be exploited in order guarantee the existence of the one-sided Wiener-Hopf factorization $\Pi_{12}^* = \Psi_{22}^* P_{12}^\top \Psi_{11} = \hat{\Psi}_2^* \hat{\Psi}_1$, where $\hat{\Psi}_1$ is stable and $\hat{\Psi}_2$ is minimal-phase [57].

Before we proceed, let us collect some preparatory facts. First note that, since Π_{12}^* has full column-rank on $\mathbb{C}^0$, there exists some nonsingular $\hat{D}_2$ such that $\Pi_{12}(\infty)\hat{D}_2^{-1} = (I \ 0) =: \hat{D}_1^\top$. If we assume that Ψ_{11} and Ψ_{22} admit the realizations

$$\Psi_{ii} = \left[\begin{array}{c|c} A_i & B_i \\ \hline C_{ii} & D_{ii} \end{array} \right], \quad \mathrm{eig}(A_i) \subset \mathbb{C}^-, \quad (A_i, B_i) \text{ controllable}, \quad i = 1, 2,$$

then it is possible to realize $\hat{D}_2^{-\top} \Pi_{12}^*$ as

$$\hat{D}_2^{-\top} \Pi_{12}^* = \left[\begin{array}{ccc|c} -A_2^\top & C_{22}^\top P_{12}^\top C_{11} & & C_{22}^\top P_{12}^\top D_{11} \\ 0 & & A_1 & B_1 \\ \hline -\hat{D}_2^{-\top} B_2^\top & \hat{D}_2^{-\top} D_{22}^\top P_{12}^\top C_{11} & & \hat{D}_1 \end{array} \right] =: \left[\begin{array}{c|c} A_\Phi & B_\Phi \\ \hline C_{\Phi 1} & I \\ C_{\Phi 2} & 0 \end{array} \right]. \tag{G.21}$$

Now suppose that $\mathcal{T}_+$ denotes the anti-stable generalized eigenspace of A_Φ. Moreover, let $\mathcal{T}_-$ denote the largest controlled invariant output-nulling subspace with internal stability; this means that it satisfies $(A_\Phi - B_\Phi C_{\Phi 1})\mathcal{T}_- \subset \mathcal{T}_-$ and $C_{\Phi 2}\mathcal{T}_- = \{0\}$ such that the eigenvalues of $(A_\Phi - B_\Phi C_{\Phi 1})|_{\mathcal{T}_-}$ are contained in $\mathbb{C}^-$ while the remaining eigenvalues of $A_\Phi - B_\Phi C_{\Phi 1}$ belong to $\mathbb{C}^+$. For constructing the Wiener-Hopf factorization of Π_{12}^* we need the following fact.

Lemma G.2. *Suppose that Π is structured as* (G.16) *and satisfies* (G.17)-(G.18). *Then*

$$\mathcal{T}_+ \cap \mathcal{T}_- = \{0\}.$$

Proof. Observe (G.17) implies that

$$\begin{pmatrix} \hat{D}_2 G_2 (I - G_1 G_2)^{-1} \\ \hat{D}_2^{-\top} \Pi_{12}^* \end{pmatrix}^* \begin{pmatrix} 0 & I \\ I & 0 \end{pmatrix} \begin{pmatrix} \hat{D}_2 G_2 (I - G_1 G_2)^{-1} \\ \hat{D}_2^{-\top} \Pi_{12}^* \end{pmatrix} \succ 0 \text{ on } \mathbb{C}^0 \quad \text{(G.22)}$$

and consider the following minimal realizations

$$\hat{D}_2 G_2 (I - G_1 G_2)^{-1} = \left[\begin{array}{c|c} A_\Lambda & B_\Lambda \\ \hline C_{\Lambda 1} & D_{\Lambda 1} \\ C_{\Lambda 2} & D_{\Lambda 2} \end{array} \right], \quad \hat{D}_2^{-\top} \Pi_{12}^* = \left[\begin{array}{c|c} A_\Phi & B_\Phi \\ \hline C_{\Phi 1} & I \\ C_{\Phi 2} & 0 \end{array} \right].$$

By the KYP-lemma we can infer that (G.22) is equivalent to the existence of some symmetric matrix X for which the following matrix inequality holds:

$$\begin{pmatrix} \star \\ \star \\ \hline \star \\ \star \\ \hline \star \\ \star \\ \hline \star \\ \star \end{pmatrix}^\top \left(\begin{array}{cc|cc|cc|cc} 0 & 0 & X_{11} & X_{12} & 0 & 0 & 0 & 0 \\ 0 & 0 & X_{12}^\top & X_{22} & 0 & 0 & 0 & 0 \\ \hline X_{11} & X_{12} & 0 & 0 & 0 & 0 & 0 & 0 \\ X_{12}^\top & X_{22} & 0 & 0 & 0 & 0 & 0 & 0 \\ \hline 0 & 0 & 0 & 0 & 0 & I & 0 \\ 0 & 0 & 0 & 0 & 0 & 0 & I \\ \hline 0 & 0 & 0 & 0 & I & 0 & 0 & 0 \\ 0 & 0 & 0 & 0 & 0 & I & 0 & 0 \end{array} \right) \left(\begin{array}{cc|c} I & 0 & 0 \\ 0 & I & 0 \\ \hline A_\Phi & 0 & B_\Phi \\ 0 & A_\Lambda & B_\Lambda \\ \hline 0 & C_{\Lambda 1} & D_{\Lambda 1} \\ 0 & C_{\Lambda 2} & D_{\Lambda 2} \\ \hline C_{\Phi 1} & 0 & I \\ C_{\Phi 2} & 0 & 0 \end{array} \right) \succ 0. \quad \text{(G.23)}$$

Also note that applying the congruence transformation

$$\left(\begin{array}{c|c} I & 0 & 0 \\ \hline -X_{22}^{-1} X_{12}^\top & I & 0 \\ \hline -C_{\Phi 1} & 0 & I \end{array} \right)^\top \text{(G.23)} \left(\begin{array}{c|c} I & 0 & 0 \\ \hline -X_{22}^{-1} X_{12}^\top & I & 0 \\ \hline -C_{\Phi 1} & 0 & I \end{array} \right)$$

yields

$$
\begin{pmatrix} \star \\ \star \\ \star \\ \star \\ \star \\ \star \\ \star \\ \star \end{pmatrix}^{\!\top}
\left(\begin{array}{cc|cc|cc|cc}
0 & 0 & \hat{X}_{11} & 0 & 0 & 0 & 0 & 0 \\
0 & 0 & 0 & X_{22} & 0 & 0 & 0 & 0 \\ \hline
\hat{X}_{11} & 0 & 0 & 0 & 0 & 0 & 0 & 0 \\
0 & X_{22} & 0 & 0 & 0 & 0 & 0 & 0 \\ \hline
0 & 0 & 0 & 0 & 0 & 0 & I & 0 \\
0 & 0 & 0 & 0 & 0 & 0 & 0 & I \\ \hline
0 & 0 & 0 & 0 & I & 0 & 0 & 0 \\
0 & 0 & 0 & 0 & 0 & I & 0 & 0
\end{array}\right)
\left(\begin{array}{cc|c}
I & 0 & 0 \\
0 & I & 0 \\ \hline
\hat{A}_{\Phi} & 0 & B_{\Phi} \\
\star & A_{\Lambda} & B_{\Lambda} \\ \hline
\star & C_{\Lambda 1} & D_{\Lambda 1} \\
\star & C_{\Lambda 2} & D_{\Lambda 2} \\ \hline
0 & 0 & I \\
C_{\Phi 2} & 0 & 0
\end{array}\right) \succ 0,
$$

with $\hat{X}_{11} := X_{11} - X_{12} X_{22}^{-1} X_{12}^{\top}$, $\hat{A}_{\Phi} := A_{\Phi} - B_{\Phi} C_{\Phi 1}$ and $\star$ of being no interest. Let T_{+}, T_{-} be basis-matrices of $\mathcal{T}_{+}$, $\mathcal{T}_{-}$. Then there exist A_{+} (anti-Hurwitz) and A_{-} (Hurwitz) with $A_{\Phi} T_{+} = T_{+} A_{+}$, $\hat{A}_{\Phi} T_{-} = T_{-} A_{-}$ and $C_{\Phi 2} T_{-} = 0$. Now note that (G.23) implies $\mathrm{He}(X_{11} A_{\Phi}) \succ 0$ and $\mathrm{He}(X_{22} A_{\Lambda}) \succ 0$. Since A_{Λ} is stable, we infer that $X_{22} \prec 0$. Moreover, since A_{+} is anti-Hurwitz we have $\mathrm{He}(X_{11} A_{\Phi}) \succ 0 \Rightarrow \mathrm{He}(T_{+}^{\top} X_{11} A_{\Phi} T_{+}) \succ 0 \Rightarrow \mathrm{He}(T_{+}^{\top} X_{11} T_{+} A_{+}) \succ 0 \Rightarrow T_{+}^{\top} X_{11} T_{+} \succ 0$. Also observe that (G.2) implies

$$
\mathrm{He}\left[\begin{pmatrix} \hat{X}_{11} & 0 \\ 0 & X_{22} \end{pmatrix}\begin{pmatrix} \hat{A}_{\Phi} & 0 \\ \star & A_{\Lambda} \end{pmatrix} + \begin{pmatrix} \star & \star \\ C_{\Lambda 1}^{\top} & C_{\Lambda 2}^{\top} \end{pmatrix}\begin{pmatrix} 0 & 0 \\ C_{\Phi 2} & 0 \end{pmatrix}\right] \succ 0. \tag{G.24}
$$

Hence, by right- and left-multiplying (G.24) with $\mathrm{col}(T_{-},0)$ and its transpose and recalling $C_{\Phi 2} T_{-} = 0$ we can infer $\mathrm{He}(T_{-}^{\top} \hat{X}_{11} \hat{A}_{\Phi} T_{-}) \succ 0 \Rightarrow \mathrm{He}(T_{-}^{\top} \hat{X}_{11} T_{-} A_{-}) \succ 0 \Rightarrow T_{-}^{\top} \hat{X}_{11} T_{-} \prec 0$. Due to $X_{22} \prec 0$ and $T_{-}^{\top} \hat{X}_{11} T_{-} = T_{-}^{\top}(X_{11} - X_{12} X_{22}^{-1} X_{12}^{\top}) T_{-} \prec 0$ we have $T_{-}^{\top} X_{11} T_{-} \prec 0$. In summary we get $T_{+}^{\top} X_{11} T_{+} \succ 0$ and $T_{-}^{\top} X_{11} T_{-} \prec 0$, which reveals that $\mathrm{im}(T_{+}) \cap \mathrm{im}(T_{-}) = \{0\}$. ∎

Since $-A_{2}^{\top}$ in (G.21) is anti-stable, we can now choose basis-matrices T_{+}, T_{-} of $\mathcal{T}_{+}$, $\mathcal{T}_{-}$ and an extension T_{e} such that

$$
\left(T_{+} \; T_{e} \; T_{-}\right) = \begin{pmatrix} I & -V^{\top} U \\ 0 & U \end{pmatrix} \quad \text{with} \quad \det(U) \neq 0. \tag{G.25}
$$

Subsequently applying a state-coordinate change with (G.25) then yields the following realization:

$$\hat{D}_2^{-\top} \Pi_{12}^* =$$

$$= \left[\begin{array}{ccc|c} -A_2^\top & C_{22}^\top P_{12}^\top C_{11} U^{-1} - A_2^\top V^\top - V^\top U A_1 U^{-1} & C_{22}^\top P_{12}^\top D_{11} - V^\top U B_1 \\ 0 & U A_1 U^{-1} & U B_1 \\ \hline -\hat{D}_2^{-\top} B_2^\top & \hat{D}_2^{-\top}(D_{22}^\top P_{12}^\top C_{11} U^{-1} - B_2^\top V^\top) & \hat{D}_1 \end{array} \right] =$$

$$=: \left[\begin{array}{ccc|c} \tilde{A}_{11} & \tilde{A}_{12} & \tilde{B}_1 \tilde{D}^\top \tilde{C}_3 & \tilde{B}_1 \\ \hline 0 & \tilde{A}_{22} & \tilde{B}_2 \tilde{D}^\top \tilde{C}_3 & \tilde{B}_2 \\ 0 & \tilde{A}_{32} & \tilde{A}_{33} & \tilde{B}_3 \\ \hline \tilde{C}_1 & \tilde{C}_2 & \tilde{C}_3 & \tilde{D} \end{array} \right] = \left[\begin{array}{ccc|c} \tilde{A}_{11} & \tilde{A}_{12} & \tilde{B}_1 \tilde{C}_{13} & \tilde{B}_1 \\ \hline 0 & \tilde{A}_{22} & \tilde{B}_2 \tilde{C}_{13} & \tilde{B}_2 \\ 0 & \tilde{A}_{32} & \tilde{A}_{33} & \tilde{B}_3 \\ \hline \tilde{C}_{11} & \tilde{C}_{12} & \tilde{C}_{13} & I \\ \tilde{C}_{21} & \tilde{C}_{22} & 0 & 0 \end{array} \right] . \quad \text{(G.26)}$$

The particular structure of the latter realization is induced by the properties of $\mathcal{T}_-$ as will become clear in the proof of the following key Wiener-Hopf factorization result.

Lemma G.3. *Consider the realization* (G.26). *There exist matrices* $L = \text{col}(L_1, L_2)$ *and* $F = \text{col}(F_1, F_2)$, *with*

$$\lambda\left(\begin{pmatrix} \tilde{A}_{11} & \tilde{A}_{12} \\ 0 & \tilde{A}_{22} \end{pmatrix} + \begin{pmatrix} -\tilde{B}_1 & L_1 \\ -\tilde{B}_2 & L_2 \end{pmatrix} \begin{pmatrix} \tilde{C}_{11} & \tilde{C}_{12} \\ \tilde{C}_{21} & \tilde{C}_{22} \end{pmatrix} \right) \subset \mathbb{C}^+,$$

$$\lambda\left(\tilde{A}_{22} - \begin{pmatrix} \tilde{B}_2 & -L_2 \end{pmatrix} \begin{pmatrix} \tilde{C}_{12} - F_1 \\ \tilde{C}_{22} - F_2 \end{pmatrix} \right) \subset \mathbb{C}^+ \qquad \text{(G.27)}$$

such that

$$\hat{D}_2^{-\top} \Pi_{12}^* =$$

$$\left[\begin{array}{cc|cc} \tilde{A}_{11} & \tilde{A}_{12} - \tilde{B}_1 \tilde{C}_{12} + L_1 \tilde{C}_{22} + \tilde{B}_1 F_1 - L_1 F_2 & \tilde{B}_1 & -L_1 \\ 0 & \tilde{A}_{22} - \tilde{B}_2 \tilde{C}_{12} + L_2 \tilde{C}_{22} + \tilde{B}_2 F_1 - L_2 F_2 & \tilde{B}_2 & -L_2 \\ \hline \tilde{C}_{11} & F_1 & I & 0 \\ \tilde{C}_{21} & F_2 & 0 & I \end{array} \right] \left[\begin{array}{cc|c} \tilde{A}_{22} & \tilde{B}_2 \tilde{C}_{13} & \tilde{B}_2 \\ \tilde{A}_{32} & \tilde{A}_{33} & \tilde{B}_3 \\ \hline \tilde{C}_{12} - F_1 & \tilde{C}_{13} & I \\ \tilde{C}_{22} - F_2 & 0 & 0 \end{array} \right]$$

$$=: \left[\begin{array}{cc|c} \tilde{A}_{11} & \tilde{A}_{12} - \hat{L}_1(\tilde{C}_2 - F) & \hat{L}_1 \\ 0 & \tilde{A}_{22} - \hat{L}_2(\tilde{C}_2 - F) & \hat{L}_2 \\ \hline \tilde{C}_1 & F & I \end{array} \right] \left[\begin{array}{cc|c} \tilde{A}_{22} & \tilde{B}_2 \tilde{D}^\top \tilde{C}_3 & \tilde{B}_2 \\ \tilde{A}_{32} & \tilde{A}_{33} & \tilde{B}_3 \\ \hline \tilde{C}_2 - F & \tilde{C}_3 & \tilde{D} \end{array} \right] . \quad \text{(G.28)}$$

Here the left-factor is anti-minimal-phase, while the right-factor is stable.

Proof. For (G.26) observe that

$$\lambda(\tilde{A}_{11}) \subset \mathbb{C}^+ \quad \text{and} \quad \lambda\begin{pmatrix} \tilde{A}_{22} & \tilde{B}_2\tilde{C}_{13} \\ \tilde{A}_{32} & \tilde{A}_{33} \end{pmatrix} \subset \mathbb{C}^-. \tag{G.29}$$

At this point we exploit the definition of $\mathcal{T}_-$ in order to infer that

$$\begin{pmatrix} \tilde{A}_{11} & \tilde{A}_{12} & \tilde{B}_1\tilde{C}_{13} & \vdots & \tilde{B}_1 \\ 0 & \tilde{A}_{22} & \tilde{B}_2\tilde{C}_{13} & \vdots & \tilde{B}_2 \\ 0 & \tilde{A}_{32} & \tilde{A}_{33} & \vdots & \tilde{B}_3 \\ \hdashline \tilde{C}_{11} & \tilde{C}_{12} & \tilde{C}_{13} & \vdots & I \\ \tilde{C}_{21} & \tilde{C}_{22} & 0 & \vdots & 0 \end{pmatrix} \begin{pmatrix} I & 0 & 0 & \vdots & 0 \\ 0 & I & 0 & \vdots & 0 \\ 0 & 0 & I & \vdots & 0 \\ \hdashline -\tilde{C}_{11} & -\tilde{C}_{12} & -\tilde{C}_{13} & \vdots & I \end{pmatrix} =$$

$$= \begin{pmatrix} \tilde{A}_{11} - \tilde{B}_1\tilde{C}_{11} & \tilde{A}_{12} - \tilde{B}_1\tilde{C}_{12} & 0 & \vdots & \tilde{B}_1 \\ -\tilde{B}_2\tilde{C}_{11} & \tilde{A}_{22} - \tilde{B}_2\tilde{C}_{12} & 0 & \vdots & \tilde{B}_2 \\ -\tilde{B}_3\tilde{C}_{11} & \tilde{A}_{32} - \tilde{B}_3\tilde{C}_{12} & \tilde{A}_{33} - \tilde{B}_3\tilde{C}_{13} & \vdots & \tilde{B}_3 \\ \hdashline 0 & 0 & 0 & \vdots & I \\ \tilde{C}_{21} & \tilde{C}_{22} & 0 & \vdots & 0 \end{pmatrix}$$

has the properties

$$\lambda\begin{pmatrix} \tilde{A}_{11} - \tilde{B}_1\tilde{C}_{11} & \tilde{A}_{12} - \tilde{B}_1\tilde{C}_{12} \\ -\tilde{B}_2\tilde{C}_{11} & \tilde{A}_{22} - \tilde{B}_2\tilde{C}_{12} \end{pmatrix} \subset \mathbb{C}^+ \quad \text{and} \quad \lambda(\tilde{A}_{33} - \tilde{B}_3\tilde{C}_{13}) \subset \mathbb{C}^-.$$

Now observe that the pairs

$$\left(\begin{pmatrix} \tilde{A}_{11} - \tilde{B}_1\tilde{C}_{11} & \tilde{A}_{12} - \tilde{B}_1\tilde{C}_{12} \\ -\tilde{B}_2\tilde{C}_{11} & \tilde{A}_{22} - \tilde{B}_2\tilde{C}_{12} \end{pmatrix}, \begin{pmatrix} \tilde{C}_{21} & \tilde{C}_{22} \end{pmatrix} \right) \quad \text{and} \quad \left(\tilde{A}_{22}, \tilde{B}_2 \right)$$

are anti-detectable and controllable respectively. The first fact is obvious; the second is proven by contradiction. Suppose there is $\lambda \in \mathbb{C}$ and some vector $x_2 \neq 0$ with $x_2^*(\tilde{A}_{22} - \lambda I \quad \tilde{B}_2) = 0$. We infer

$$\begin{pmatrix} x_2^* & 0 \end{pmatrix} \begin{pmatrix} \tilde{A}_{22} - \lambda I & \tilde{B}_2\tilde{C}_{13} \\ \tilde{A}_{32} & \tilde{A}_{33} - \lambda I \end{pmatrix} = 0$$

and

$$\begin{pmatrix} 0 & x_2^* \end{pmatrix} \begin{pmatrix} \tilde{A}_{11} - \tilde{B}_1\tilde{C}_{11} - \lambda I & \tilde{A}_{12} - \tilde{B}_1\tilde{C}_{12} \\ -\tilde{B}_2\tilde{C}_{11} & \tilde{A}_{22} - \tilde{B}_2\tilde{C}_{12} - \lambda I \end{pmatrix} = 0.$$

This implies $\text{Re}(\lambda) < 0$ (by (G.29)) and $\text{Re}(\lambda) > 0$, a contradiction. Hence, it is possible to choose some $L = \text{col}(L_1, L_2)$ and some $F = \text{col}(F_1, F_2)$ such that (G.27) is satisfied.

It is now a matter of direct verification that an anti-stable left-inverse of (G.26) can be defined as follows:

$$
\left[
\begin{array}{ccc|cc}
\tilde{A}_{11} - \tilde{B}_1\tilde{C}_{11} + L_1\tilde{C}_{21} & \tilde{A}_{12} - \tilde{B}_1\tilde{C}_{12} + L_1\tilde{C}_{22} & 0 & \tilde{B}_1 & -L_1 \\
-\tilde{B}_2\tilde{C}_{11} + L_2\tilde{C}_{21} & \tilde{A}_{22} - \tilde{B}_2\tilde{C}_{12} + L_2\tilde{C}_{22} & 0 & \tilde{B}_2 & -L_2 \\
-\tilde{B}_3\tilde{C}_{11} & \tilde{A}_{32} - \tilde{B}_3\tilde{C}_{12} & \tilde{A}_{33} - \tilde{B}_3\tilde{C}_{13} & \tilde{B}_3 & 0 \\
\hline
-\tilde{C}_{11} & -\tilde{C}_{12} & -\tilde{C}_{13} & I & 0 \\
-\tilde{C}_{21} & -\tilde{C}_{22} & 0 & 0 & I
\end{array}
\right].
$$

This motivates to left-multiply (G.26) with the anti-stable transfer matrix

$$
\left[
\begin{array}{cc|cc}
\tilde{A}_{11} - \tilde{B}_1\tilde{C}_{11} + L_1\tilde{C}_{21} & \tilde{A}_{12} - \tilde{B}_1\tilde{C}_{12} + L_1\tilde{C}_{22} & \tilde{B}_1 & -L_1 \\
-\tilde{B}_2\tilde{C}_{11} + L_2\tilde{C}_{21} & \tilde{A}_{22} - \tilde{B}_2\tilde{C}_{12} + L_2\tilde{C}_{22} & \tilde{B}_2 & -L_2 \\
\hline
-\tilde{C}_{11} & -F_1 & I & 0 \\
-\tilde{C}_{21} & -F_2 & 0 & I
\end{array}
\right]
\tag{G.30}
$$

which yields the stable right-factor in (G.28). On the other hand, note that the inverse of (G.30) is nothing but the left-factor in (G.28), which is itself anti-stable by inspection.

∎

We stress that this construction is inspired from [57] but drastically simplified if compared to existing approaches in the literature. Based on (G.26) and (G.28), it is straightforward to construct the factorization $\Pi_{12}^* = \hat{\Psi}_2^*\hat{\Psi}_1$ as follows:

$$
\Pi_{12}^* =
$$

$$
\left[
\begin{array}{cc|c}
-\tilde{A}_{22}^\top + (\tilde{C}_2 - F)^\top \hat{L}_2^\top & \tilde{A}_{12}^\top - (\tilde{C}_2 - F)^\top \hat{L}_1^\top & F^\top \hat{D}_2 \\
0 & A_2 & B_2 \\
\hline
-\hat{L}_2^\top & \hat{L}_1^\top & \hat{D}_2
\end{array}
\right]^*
\left[
\begin{array}{c|c}
A_1 & B_1 \\
\hline
\left(\tilde{C}_2 - F \quad \tilde{C}_3\right)U & \hat{D}_1
\end{array}
\right]
$$

$$
=:
\left[
\begin{array}{c|c}
\hat{A}_2 & \hat{B}_2 \\
\hline
\hat{C}_2 & \hat{D}_2
\end{array}
\right]^*
\left[
\begin{array}{c|c}
A_1 & B_1 \\
\hline
\hat{C}_1 & \hat{D}_1
\end{array}
\right]
= \hat{\Psi}_2^*\hat{\Psi}_1.
$$

Now observe that, in order to factorize $\Pi = \Psi^* P \Psi$ as (G.19), it remains to show that there exists a stable $\hat{\Psi}_3$ such that $\Psi_{22}^* P_{22} \Psi_{22} = \mathrm{He}(\hat{\Psi}_2^*\hat{\Psi}_3)$. This brings us to the following observation.

Lemma G.4. *For the constructed $\hat{\Psi}_2$ there exists a stable $\hat{\Psi}_3$ with realization*

$$
\hat{\Psi}_3 =
\left[
\begin{array}{c|c}
\hat{A}_2 & \hat{B}_2 \\
\hline
\hat{C}_3 & \hat{D}_3
\end{array}
\right]
:=
\left[
\begin{array}{c|c}
\hat{A}_2 & \hat{B}_2 \\
\hline
\hat{D}_2^{-\top}(D_{22}^\top P_{22}(C_{22}J_{\mathrm{r}} - \frac{1}{2}D_{22}\hat{D}_2^{-1}\hat{C}_{22}) - \hat{B}_2^\top Z) & \frac{1}{2}\hat{D}_2^{-\top}D_{22}^\top P_{22}D_{22}
\end{array}
\right]
$$

such that $\Psi_{22}^* P_{22} \Psi_{22} = \mathrm{He}(\hat{\Psi}_3^* \hat{\Psi}_2)$. Here Z satisfies the Lyapunov equation

$$\check{C}_2^\top P_{22} \check{C}_2 - Z \check{A}_2 - \check{A}_2^\top Z = 0, \qquad (\text{G.31})$$

where the matrices $\check{A}_2$ and $\check{C}_2$ are given by $\check{A}_2 := \hat{A}_2 - \hat{B}_2 \hat{D}_2^{-1} \hat{C}_2$, $\check{C}_2 := C_{22} J_{\mathrm{r}} - D_{22} \hat{D}_2^{-1} \hat{C}_2$.

Proof. Let us observe that $\Psi_{22}^* P_{22} \Psi_{22} = \mathrm{He}(\hat{\Psi}_2^* \hat{\Psi}_3)$ is equivalent to $\hat{\Psi}_2^{-*} \Psi_{22}^* P_{22} \Psi_{22} \hat{\Psi}_2^{-1} = \mathrm{He}(\hat{\Psi}_3 \hat{\Psi}_2^{-1})$. Also note that

$$\Psi_{22} \hat{\Psi}_2^{-1} = \left[\begin{array}{c|c} \hat{A}_2 - \hat{B}_2 \hat{D}_2^{-1} \hat{C}_2 & \hat{B}_2 \hat{D}_2^{-1} \\ \hline C_{22} J_{\mathrm{r}} - D_{22} \hat{D}_2^{-1} \hat{C}_2 & D_2 \hat{D}_2^{-1} \end{array} \right] = \left[\begin{array}{c|c} \check{A}_2 & \check{B}_2 \\ \hline \check{C}_2 & \check{D}_2 \end{array} \right].$$

Hence, $\hat{\Psi}_2^{-*} \Psi_{22}^* P_{22} \Psi_{22} \hat{\Psi}_2^{-1}$ admits the realization

$$\left[\begin{array}{cc|c} -\check{A}_2^\top & \check{C}_2^\top P_{22} \check{C}_2 & \check{C}_2^\top P_{22} \check{D}_2 \\ 0 & \check{A}_2 & \check{B}_2 \\ \hline -\check{B}_2^\top & \check{D}_2^\top P_{22} \check{C}_2 & \check{D}_2^\top P_{22} \check{D}_2 \end{array} \right].$$

After solving the Lyapunov equation (G.31) and performing an elementary state-coordinate change, $\hat{\Psi}_2^{-*} \Psi_{22}^* P_{22} \Psi_{22} \hat{\Psi}_2^{-1}$ is seen to equal

$$\left[\begin{array}{cc|c} -\check{A}_2^\top & 0 & \check{C}_2^\top P_{22} \check{D}_2 - Z \check{B}_2 \\ 0 & \check{A}_2 & \check{B}_2 \\ \hline -\check{B}_2^\top & \check{D}_2^\top P_{22} \check{C}_2 - \check{B}_2^\top Z & \check{D}_2^\top P_{22} \check{D}_2 \end{array} \right] =$$

$$= \mathrm{He} \left[\begin{array}{c|c} \check{A}_2 & \check{B}_2 \\ \hline \check{D}_2^\top P_{22} \check{C}_2 - \check{B}_2^\top Z & \frac{1}{2} \check{D}_2^\top P_{22} \check{D}_2 \end{array} \right] = \mathrm{He}(\hat{\Psi}_3 \hat{\Psi}_2^{-1}).$$

By finally right-multiplying $\hat{\Psi}_3 \hat{\Psi}_2^{-1}$ with $\hat{\Psi}_2 = (\hat{A}_2, \hat{B}_2, \hat{C}_2, \hat{D}_2)$ we directly obtain $\hat{\Psi}_3$. ∎

It is finally easy to verify that $\hat{\Psi}$ in (G.15) defined with the above constructed matrices satisfies (G.19) with (G.20). Moreover, analogously to Section G.1, the frequency domain condition $\Psi^* P \Psi = \hat{\Psi}^* \hat{P} \hat{\Psi}$ can be seen to be a consequence of (G.14), now with

$$\mathcal{Z} := \begin{pmatrix} 0 & U^\top \begin{pmatrix} J_1^\top & V \end{pmatrix} \\ \star & Z \end{pmatrix} \quad \text{and} \quad J_1 := \begin{pmatrix} I & 0 \end{pmatrix}.$$

References

[1] J. Abedor, K. Nagpal, and K. Poolla. A linear matrix inequality approach to peak to peak gain minimization. *International Journal of Robust and Nonlinear Control*, 6(9-10):895–899, 1996.

[2] B. Anderson. A system theory criterion for positive real matrices. *Siam Journal of Control*, 5(2):171–182, 1967.

[3] P. Apkarian and D. Noll. IQC analysis and synthesis via nonsmooth optimization. *System & Control Letters*, 55(12):971–981, 2006.

[4] P. Apkarian and D. Noll. Nonsmooth $\mathcal{H}_\infty$ synthesis. *IEEE Transactions on Automatic Control*, 51(1):71–86, 2006.

[5] V. Balakrishnan. Linear matrix inequalities in robustness analysis with multipliers. *System & Control Letters*, 25(4):265–272, 1995.

[6] V. Balakrishnan. Lyapunov functionals in complex μ-analysis. *IEEE Transactions on Automatic Control*, 47(9):1466–1479, 2002.

[7] V. Balakrishnan and R. Kashyap. Robust stability and performance analysis of uncertain systems using linear matrix inequalities. *Journal of Optimization Theory and Applications*, 100(3):457–478, 1999.

[8] V. Balakrishnan and L. Vandenberghe. Semidefinite programming duality and linear time-invariant systems. *IEEE Transactions on Automatic Control*, 48(1):30–41, 2003.

[9] V. Balakrishnan and F. Wang. Efficient computation of a guaranteed lower bound on the robust stability margin for a class of uncertain systems. *IEEE Transactions on Automatic Control*, 44(11):2185–2190, 1999.

[10] G. Balas, R. Chiang, A. Packard, and M. Safonov. Robust control toolbox. Technical report, The MathWorks Inc., 2005-2008.

[11] G. Balas, J. Doyle, K. Glover, A. Packard, and R. Smith. *μ-analysis and synthesis toolbox (μ-tools)*. Mathworks, 1991.

[12] H. Balini. *Advanced system theory applied to AMB systems*. PhD thesis, Delft University of Technology, The Netherlands, 2011.

[13] B. Bamieh, E. Paganini, and A. Dahleh. Distributed control of spatially invariant systems. *IEEE Transactions on Automatic Control*, 47(7):1091–1107, 2002.

[14] K. Barbosa, C. d. Souza, and A. Trofino. Robust $\mathscr{H}_2$ filtering for uncertain linear systems: LMI based methods with parametric lyapunov function. *System & Control Letters*, 54(3):251–262, 2005.

[15] K. A. Barbosa, A. Trofino, and C. d. Souza. Robust $\mathscr{H}_2$ filtering for linear system with uncertain time-varying parameters. *in the Proceedings of the 41th IEEE Conference on Decision and Control*, pages 3883–3888, 2002.

[16] H. Bart, I. Gohberg, M. Kaashoek, and A. Ran. *A state space approach to canonical factorization with applications*. Operator theory: advances and applications. Birkhäuser Verlag, Basel, 2010.

[17] B. Bayon, G. Scorletti, and E. Blanco. Robust $\mathscr{L}_2$-gain observation for structured uncertainties: an LMI approach. *in the Proceedings of the 50th IEEE Conference on Decision and Control*, pages 4949–4954, 2011.

[18] B. Bayon, G. Scorletti, and E. Blanco. An LMI solution for a class of robust open-loop problems. *in the Proceedings of the American Control Conference*, pages 5234–5239, 2012.

[19] J. Biannic and S. Tarbouriech. Optimization and implementation of dynamic anti-windup compensators with multiple saturations in flight control systems. *Control Engineering Practice*, 17(6):703–713, 2009.

[20] R. Borges and P. Peres. $\mathscr{H}_\infty$ LPV filtering for linear systems with arbitrarily time-varying parameters in polytopic domains. *in the Proceedings of the 45th IEEE Conference on Decision and Control*, pages 1692–1697, 2006.

[21] G. Boukarim and J. Chow. Modeling of nonlinear system uncertainties using a linear fractional transformation approach. *in the Proceedings of the American Control Conference*, pages 2973–2979, 1998.

[22] S. Boyd, L. El Ghaoui, E. Feron, and V. Balakrishnan. *Linear matrix inequalities in system and control*. Studies in Applied Mathematics, SIAM, 1994.

[23] R. Braatz, P. Young, J. Doyle, and M. Morari. Computational complexity of μ calculation. *IEEE Transactions on Automatic Control*, 39(5):1000–1002, 1994.

[24] M. Cantoni, U. Jönsson, and C. Kao. IQC-robustness analysis for feedback interconnections of unstable distributed parameter systems. *in the Proceedings of the 48th IEEE Conference on Decision and Control*, pages 1124–1130, 2009.

[25] M. Cantoni, U. Jönsson, and C. Kao. Robustness analysis for feedback interconnections of distributed systems via integral quadratic constraints. *IEEE Transactions on Automatic Control*, 57(2):302–317, 2012.

[26] X. Chen and J. Wen. Robustness analysis of LTI systems with structured incrementally sector bounded nonlinearities. *in the Proceedings of the American Control Conference*, pages 3883–3887, 1995.

[27] G. Chesi, A. Garulli, A. Tesi, and A. Vicino. An LMI-based approach for characterizing the solution set of polynomial systems. *in the Proceedings of the* 39th *IEEE Conference on Decision and Control*, pages 1501–1506, 2000.

[28] R. Chiang and M. Safonov. *Robust control toolbox, ver. 2.0*. Mathworks, 1992.

[29] Y. Chou, A. L. Tits, and V. Balakrishnan. Stability multipliers and μ upper bound: connections and implications for numerical verfication of frequency domain conditions. *IEEE Transactions on Automatic Control*, 44(5):906–913, 1999.

[30] J. Cockburn and B. Morton. Linear fractional representation of uncertain systems. *Automatica*, 33(7):1263–1271, 1997.

[31] F. D'Amato, M. Rotea, A. Megretski, and U. Jönsson. New results for analysis of systems with repeated nonlinearities. *Automatica*, 37(5):739–747, 2001.

[32] R. D'Andrea. Generalized l_2-synthesis. *IEEE Transactions on Automatic Control*, 44(6):1145–1156, 1999.

[33] R. D'Andrea. Convex and finite-dimensional conditions for controller synthesis with dynamic integral constraints. *IEEE Transactions on Automatic Control*, 46(2):222–234, 2001.

[34] R. D'Andrea and G. Dullerud. Distributed control design for spatially interconnected systems. *IEEE Transactions on Automatic Control*, 48(9):1478–1495, 2003.

[35] J. de Loera and F. Santos. An effective version of pólya's theorem on positive definite forms. *Journal of Pure and Applied Algebra*, 108:231–240, 1996.

[36] C. de Souza, K. Barosa, and A. Trofino. Robust filtering for linear systems with convex-bounded uncertain time-varying parameters. *IEEE Transactions on Automatic Control*, 52(6):1132–1138, 2007.

[37] C. Desoer and M. Vidyasagar. *Feedback systems: input-output properities*. Academic Press, New York, 1975.

[38] S. Dietz. *Analysis and control of uncertain systems by using robust semidefinite programming*. PhD thesis, Delft University of Technology, The Netherlands, 2008.

[39] J. Doyle. Guaranteed margins for LQG regulators. *IEEE Transactions on Automatic Control*, 23(4):756–757, 1978.

[40] J. Doyle. Synthesis of robust controllers and filters with structured plant uncertainty. *in the Proceedings of the* 22nd *IEEE Conference on Decision and Control*, pages 109–114, 1983.

[41] J. Doyle, K. Glover, P. Khargonekar, and B. Francis. State-space solutions to standard $\mathcal{H}_\infty$- and $\mathcal{H}_2$-control problems. *IEEE Transactions on Automatic Control*, 34(8):831–847, 1989.

[42] G. Dullerud and R. D'Andrea. Distributed control of heterogeneous systems. *IEEE Transactions on Automatic Control*, 49(12):2113–2028, 2004.

[43] G. Dullerud and E. Paganini. *A Course in robust control theory*. Texts in applied mathematics. Springer-Verlag, New York, 1999.

[44] M. K. H. Fan, A. L. Tits, and J. C. Doyle. Robustness in the presence of mixed parametric uncertainties and unmodeled dynamics. *IEEE Transactions on Automatic Control*, 36(1):25–38, 1991.

[45] H. Fang and P. Antsaklis. Distributed control with integral quadratic constraints. *in the Proceedings of the 17th IFAC Wolrd Congress*, pages 574–580, 2008.

[46] M. Fu, S. Dasgupta, and Y. C. Soh. Integral quardatic constraint approach vs. multiplier approach. *Automatica*, 41(2):281–285, 2005.

[47] H. Fujioka and K. Morimura. Computational aspects of IQC-based stability analysis for sampled-data feedback systems. *in the Proceedings of the 38th IEEE Conference on Decision and Control*, pages 3452–3457, 1999.

[48] P. Gahinet and P. Apkarian. A linear matrix inequality approach to $\mathcal{H}_\infty$-control. *International Journal of Robust and Nonlinear Control*, 4(4):421–448, 1994.

[49] P. Gahinet, P. Apkarian, and M. Chilali. Affine parameter-dependent Lyapunov functions and real parametric uncertainty. *IEEE Transactions on Automatic Control*, 41(3):436–442, 1996.

[50] P. Gahinet, A. Nemirovski, A. Laub, and M. Chilali. LMI control toolbox: for use with Matlab. Technical report, The MathWorks Inc., 1995.

[51] J. Geromel. Optimal filtering under parameter uncertainty. *IEEE Transactions on Signal Processing*, 47(1):168–175, 1999.

[52] J. Geromel and M. d. Oliveira. $\mathcal{H}_2$ and $\mathcal{H}_\infty$ robust filtering for convex bounded uncertain systems. *IEEE Transactions on Automatic Control*, 46(1):100–107, 2001.

[53] J. Geromel, M. d. Oliveira, and J. Bernussou. Robust filtering of discrete-time linear systems with parameter-dependent Lyapunov functions. *in the Proceedings of the 38th IEEE Conference on Decision and Control*, pages 570–575, 1999.

[54] A. Giusto and F. Paganini. Robust synthesis of feedforward compensators. *IEEE Transactions on Automatic Control*, 44(8):1578–1582, 1999.

[55] K. Goh. Canonical factorization for generalized positive real transfer functions. *in the Proceedings of the 35th IEEE Conference on Decision and Control*, pages 2848–2853, 1996.

[56] K. Goh. Structure and factorization of quadratic constraints for robustness analysis. *in the Proceedings of the 35th IEEE Conference on Decision and Control*, pages 4649–4654, 1996.

[57] I. Gohberg and M. Kaashoek. *Constructive methods of Wiener-Hopf factorizations*, volume 21 of *Operator Theory: Advances and Applications*. Birkhäuser Verlag, Basel, 1986.

[58] M. Grant and S. Boyd. Graph implementations for nonsmooth convex programs. In V. Blondel, S. Boyd, and H. Kimura, editors, *Recent Advances in Learning and Control*, Lecture Notes in Control

and Information Sciences, pages 95–110. Springer-Verlag Limited, 2008. http://stanford.edu/~boyd/graph_dcp.html.

[59] M. Grant and S. Boyd. *CVX: Matlab Software for Disciplined Convex Programming, version 2.1*, 2014. http://cvxr.com/cvx.

[60] A. Hansson and L. Vandenberghe. Efficient solution of linear matrix inequalities for integral quadratic constraints. *in the Proceedings of the 39th IEEE Conference on Decision and Control*, pages 5033–5034, 2000.

[61] S. Hecker and A. Varga. Generalized LFT-based representation of parametric uncertain models. *European Journal of Control*, 19(4):326–337, 2004.

[62] S. Hecker, A. Varga, and J. Magni. Enhanced LFR-toolbox for Matlab and LFT-based gain-scheduling. Technical report, German Aerospace Center (DLR), Institute of Robotics and Mechatronics, 2004.

[63] A. Helmersson. μ-synthesis and LFT gain-scheduling with mixed uncertainties. *in the Proceedings of the 3rd European Control Conference*, pages 153–158, 1995.

[64] A. Helmersson. An IQC-based stability criterion for systems with slowly varying parameters. *in the Proceedings of the 14th IFAC Wolrd Congress*, 1999.

[65] A. Helmersson. IQC-synthesis based on inertia constraints. *in the Proceedings of the 14th IFAC Wolrd Congress*, 1999.

[66] D. Henrion and A. Garulli. *Positive polynomials in control*, volume 312 of *Lecture Notes in Control and Imformation Science*. Springer, Berlin, Heidelberg, New York, 2005.

[67] D. Henrion and J. Lasserre. Solving nonconvex optimization problems. *IEEE Control Systems Magazine*, 24(3):72–83, 2004.

[68] C. Hoffmann and H. Werner. A survey of linear parameter varying control applications validated by experiments or high-fidelity simulations. *IEEE Transactions on Control Systems Technology (to appear)*, 2014.

[69] C. Huang, H. Meng, and M. Safonov. Positiver real problem with unstable weighting matrices. *in the Proceedings of the AIAA Guidance, Navigation and Control Conference*, 1996.

[70] T. Iwasaki and S. Hara. Generalized KYP lemma: unified frequency domain inequalities with design applications. *IEEE Transactions on Automatic Control*, 50(1):41–59, 2005.

[71] T. Iwasaki, S. Hara, and L. Fradkov. Time domain interpretations of frequency domain inequalities on (semi)finite ranges. *System & Control Letters*, 54(7):681–691, 2005.

[72] T. Iwasaki and R. Skelton. All controllers for the general $\mathscr{H}_\infty$-control problem: LMI existence conditions and state space formulas. *Automatica*, 30(8):1307–1317, 1994.

[73] U. Jönsson. *Robustness analysis of uncertain and nonlinear systems*. PhD thesis, Lund Institute of Technology, 1996.

[74] U. Jönsson. Stability analysis with Popov multipliers and integral quadratic constraints. *System & Control Letters*, 31(2):85–92, 1997.

[75] U. Jönsson. Stability of uncertain systems with hysteresis nonlinearities. *International Journal of Robust and Nonlinear Control*, 8(3):279–293, 1998.

[76] U. Jönsson. A Popov criterion for systems with slowly time-varying parameters. *IEEE Transactions on Automatic Control*, 44(4):844–846, 1999.

[77] U. Jönsson. Robustness of trajectories with finite time extent. *Automatica*, 38(9):1485–1497, 2002.

[78] U. Jönsson, C. Kao, and H. Fujioka. A Popov criterion for networked systems. *System & Control Letters*, 56(9-10):603–610, 2007.

[79] U. Jönsson, C. Kao, and A. Megretski. Robustness of periodic trajectories. *IEEE Transactions on Automatic Control*, 47(11):1842–1852, 2002.

[80] U. Jönsson, C. Kao, and A. Megretski. Analysis of periperiodic fforce uncertain feedback systems. *IEEE Transactions on Automatic Control*, 50(2):244–258, 2003.

[81] U. Jönsson and A. Megretski. The zames-Falb IQC for critically stable systems. *in the Proceedings of the American Control Conference*, pages 3612–3616, 1998.

[82] U. Jönsson and A. Megretski. IQC characterizations of signal classes. *in the Proceedings of the 7th European Control Conference*, 1999.

[83] U. Jönsson and A. Megretski. The zames-Falb IQC for systems with integrators. *IEEE Transactions on Automatic Control*, 45(3):560–565, 2000.

[84] U. Jönsson and A. Rantzer. Systems with uncertain parameters - time-variations with bounded derivatives. *International Journal of Robust and Nonlinear Control*, 6(9-10):969–982, 1996.

[85] U. Jönsson and A. Rantzer. Optimization of integral quadratic constraints. In L. E. Ghaoui and S. Niculescu, editors, *Advances in linear matrix inequality methods in control*, chapter 6, pages 109–127. SIAM, Philadelphia, 2000.

[86] M. Jun and M. Safonov. IQC robustness analysis for time-delay systems. *International Journal of Robust and Nonlinear Control*, 11(15):1455–1468, 2001.

[87] X. Jun and X. Lihua. An improved approach to robust $\mathscr{H}_2$ and $\mathscr{H}_\infty$ filter design for uncertain linear system with time-varying parameters. *in the Proceedings of the 26th Chinees Control Conference*, pages 668–672, 2007.

[88] Y. Kakutani, T. Hagiwara, and M. Araki. LMI representation of the shifted Popov criterion. *Automatica*, 36(5):765–770, 2000.

[89] R. Kalman. A new approach to linear filtering and prediction problems. *Transactions of the ASME - Journal of Basic Engineering*, 82(D):35–45, 1960.

[90] L. Kang-Zhi, Z. Hui, and M. Tsutomu. Solution to nonsingular $\mathscr{H}_2$ optimal control problem with unstable weights. *System & Control Letters*, 32(1):4641–4646, 1997.

[91] C. Kao. *Efficient computational methods for robustness analysis*. PhD thesis, Massachusetts Institute of Technology, 2002.

[92] C. Kao and M. Cantoni. Stability analysis for structured feedback interconnections of distributed-parameter systems and time-varying uncertainties. *in the Proceedings of the 13th European Control Conference*, pages 554–559, 2013.

[93] C. Kao and M. Chen. Robust estimation with dynamic integral quadratic constraints: the discrete-time case. *IET Control Theory and Applications*, 7(12):1599–1608, 2013.

[94] C. Kao, U. Jönsson, and H. Fujioka. Characterization of robust stability of a class of interconnected systems. *Automatica*, 45(1):217–224, 2009.

[95] C. Kao and B. Lincoln. Simple stability criteria for systems with time-varying time-delays. *Automatica*, 40(8):1429–1434, 2004.

[96] C. Kao, A. Megretski, and U. Jönsson. Specialized fast algorithms for iqc feasibility and optimization problems. *Automatica*, 40(2):239–252, 2004.

[97] C. Kao and A. Rantzer. Stability analysis of systems with uncertain time-varying delays. *Automatica*, 43(6):959–970, 2007.

[98] C. Kao, M. Ravuri, and A. Megretski. Control synthesis with dynamic integral quadratic constraints - LMI approach. *in the Proceedings of the 39th IEEE Conference on Decision and Control*, pages 1477–1482, 2000.

[99] H. Köroğlu and C. Scherer. Robust stability analysis against perturbations of smoothly time-varying parameters. *in the Proceedings of the 45th IEEE Conference on Decision and Control*, pages 2895–2900, 2006.

[100] H. Köroğlu and C. Scherer. Robust stability analysis tests for linear time-varying perturbations with bounded rates-of-variation. *in the Proceedings of the 17th IFAC Wolrd Congress*, pages 429–434, 2006.

[101] H. Köroğlu and C. Scherer. Robust performance analysis for structured linear time-varying perturbations with bounded rates-of-variation. *IEEE Transactions on Automatic Control*, 52(2):197–211, 2007.

[102] I. Köse and C. Scherer. Robust feedforward control of uncertain systems using dynamic IQCs. *in the Proceedings of the 46th IEEE Conference on Decision and Control*, pages 2181–2186, 2007.

[103] I. Köse and C. Scherer. Robust $\mathscr{L}_2$-gain feedforward control of uncertain systems using dynamic IQCs. *International Journal of Robust and Nonlinear Control*, 19(11):1224–1247, 2009.

[104] M. V. Kothare and M. Morari. Multiplier theory for stability analysis of anti-windup control systems. *Automatica*, 35(5):917–928, 1999.

[105] V. Kulkarni and M. Safonov. All multipliers for repeated monotone nonlinearities. *IEEE Transactions on Automatic Control*, 47(7):1209–12012, 2002.

[106] C. Lanbort, R. Chandra, and R. D'Andrea. Distributed control design for systems interconnected over an arbitrary graph. *IEEE Transactions on Automatic Control*, 49(9):1502–1519, 2004.

[107] C. Lanbort and R. D'Andrea. Distributed control of spatially reversible interconnected systems with boundary conditions. *Siam Journal of Control and Optimization*, 44(1):1–28, 2005.

[108] K. Liu, M. Hirata, and T. Sato. All solutions to the $\mathscr{H}_\infty$-control synthesis problem with unstable weights. *in the Proceedings of the 36th IEEE Conference on Decision and Control*, pages 4641–4646, 1997.

[109] J. Löfberg. Yalmip: a toolbox for modeling and optimization in matlab. *in the Proceedings of the IEEE International Symposium on Computer Aided Control Systems Design*, pages 284–289, 2004.

[110] D. Luenberger. *Optimization by vector space methods*. John Wiley & Sons, Ltd, New York, 1969.

[111] D. Luenberger. An introduction to observers. *IEEE Transactions on Automatic Control*, 16(6):596–602, 1971.

[112] R. Mancera and M. Safonov. All stability multipliers for repeated MIMO nonlinearities. *System & Control Letters*, 54(4):389–397, 2005.

[113] D. Materassi and M. Salapaka. Less conservative absolute stability criteria using integral quadratic constraints. *in the Proceedings of the American Control Conference*, pages 113–118, 2009.

[114] A. Megretski. New IQC for quasi-concave nonlinearities. *International Journal of Robust and Nonlinear Control*, 11(7):603–620, 2001.

[115] A. Megretski, C. Kao, U. Jönsson, and A. Rantzer. *A guide to IQC-Beta: software for robustness analysis*. Laboratory for Information and Decision Systems, Masschusetts Institute of Technology, Cambrigde, MA, 1997.

[116] A. Megretski and A. Rantzer. System analysis via integral quadratic constraints. *IEEE Transactions on Automatic Control*, 42(6):819–830, 1997.

[117] G. Meinsma. Unstable and nonproper weights in $\mathscr{H}_\infty$-control. *Automatica*, 31(11):1655–1658, 1995.

[118] H. Meng. *Stability analysis and robust control synthesis with generalized multipliers*. PhD thesis, University of California, USA, 2002.

[119] T. Mita, X. Xin, and B. Anderson. Extended $\mathscr{H}_\infty$ control - $\mathscr{H}_\infty$ control with unstable weights. *Automatica*, 36(5):4318–4322, 2000.

[120] T. Nyugen and F. Jabbari. $\mathscr{H}_\infty$-design for systems with input saturation: an LMI-approach. *in the Proceedings of the American Control Conference*, pages 3798–3802, 1997.

[121] A. Packard. Gain-scheduling via linear fractional transformations. *System & Control Letters*, 22(2):79–92, 1994.

[122] A. Packard and J. Doyle. The complex structured singular value. *Automatica*, 29(1):1251–1256, 1993.

[123] E. Paganini. A set based approach for white noise modeling. *IEEE Transactions on Automatic Control*, 41(10):1453–1465, 1996.

[124] E. Paganini. Convex methods for robust $\mathscr{H}_2$ analysis of continuous-time systems. *IEEE Transactions on Automatic Control*, 44(2):239–252, 1999.

[125] E. Paganini. Frequency domain conditions for robust $\mathscr{H}_2$ performance. *IEEE Transactions on Automatic Control*, 44(1):38–49, 1999.

[126] E. Paganini and E. Feron. Linear matrix inequality methods for robust $\mathscr{H}_2$ analysis: a survey with comparisons. In L. El Gahoui and L. Niculesco, editors, *Advances in linear matrix inequality methods in control*, pages 129–151. SIAM, Philadelphia, 2000.

[127] F. Paganini and E. Feron. *LMI methods for robust $\mathscr{H}_2$ analysis: a survey with comparisons*, chapter 7. In advances in linear matrix inequality methods in control SIAM, 1999.

[128] P. Parrilo. *Structured semidefinite programs and semialgebraic geometry methods in robustness and optimization.* PhD thesis, California Institute of Technology, 2000.

[129] J. Partington. Personal communication, 2009.

[130] H. Pfifer and P. Seiler. Integral quadratic constraints for delayed nonlinear and parameter-varying systems. *Automatica (preprint)*, 2014.

[131] H. Pfifer and P. Seiler. Robustness analysis of linear parameter varying systems using integral quadratic constraints. *International Journal of Robust and Nonlinear Control (preprint)*, 2014.

[132] H. Pfifer and P. Seiler. Robustness analysis with parameter-varying integral quadratic constraints. *in the Proceedings of the 53rd IEEE Conference on Decision and Control*, 2014.

[133] H. Pillai and J. Willems. problems and dissipative distributed systems. *Siam Journal of Control and Optimization*, 40(5):1406–1430, 2002.

[134] A. Pinkus. *n-Widths in approximation theory.* Ergebnisse der Mathematik und ihren Grenzgebiete. Springer-Verlag, New York, 1985.

[135] I. Polat and C. Scherer. Stability analysis for bilateral teleoperation: an IQC formulation. *IEEE Transactions on Robotics*, 28(6):1294–1308, 2012.

[136] J. Polderman and J. Willems. *Introduction to mathematical systems theory*, volume 26 of *Texts in Applied Mathematics*. Springer, Berlin, Heidelberg, New York, 1997.

[137] G. Pólya. Über positive darstellung von polynomen. *Vierteljahrschrift der Naturforschenden Gesellschaft in Zürich*, 73:141–145, 1928. reprinted in: Collected Papers, Volume 2, 309-313, Cambridge: MIT Press, 1974.

[138] V. Powers and B. Reznick. A new bound for pólya's theorem with applications to polynomials positive on polyhedra. *Journal of Pure and Applied Algebra*, 164:221–229, 2001.

[139] A. Rantzer. On the Kalman-Yakubovich-Popov lemma. *System & Control Letters*, 28(1):7–10, 1996.

[140] A. Rantzer. Friction analysis based on integral quadratic constraints. *International Journal of Robust and Nonlinear Control*, 11(7):645–652, 2001.

[141] W. Rugh and J. Shamma. Research on gain-scheduling. *Automatica*, 36(10):1401–1425, 2000.

[142] A. Saberi, A. Stoorvogel, P. Sannuti, and G. Shi. On optimal output regulation for linear systems. *International Journal of Robust and Nonlinear Control*, 76(4):319–333, 2003.

[143] M. Safonov. $\mathscr{L}_\infty$-optimal sensitivity vs. stability margin. *in the Proceedings of the 22nd IEEE Conference on Decision and Control*, pages 115–118, 1983.

[144] M. Safonov and R. Chiang. *Real/Complex K_m-synthesis without curve-fitting*, volume 56 (Part 2) of *Control and dynamic systems*. Academic Press, New York, 1993.

[145] M. Safonov and V. Kulkarni. Zames-Falb multipliers for MIMO nonlinearities. *International Journal of Robust and Nonlinear Control*, 10(11-12):1025–1038, 2000.

[146] M. Safonov, J. Ly, and R. Chiang. μ-synthesis robust control: What's wrong and how to fix it? *in the Proceedings of the IEEE Conference on Aerospace Control Systems*, pages 563–568, 1993.

[147] F. Saupe. *Linear parameter varying control design for industrial manipulators*. PhD thesis, Institte for System Dynamics and Control of the Robotics and Mechatronics Center of the German Aerospace Center (DLR), Oberpfaffenhofen, Germany, 2013.

[148] C. Scherer. Multiobjective $\mathscr{H}_2/\mathscr{H}_\infty$-control. *IEEE Transactions on Automatic Control*, 40(6):1054–1062, 1995.

[149] C. Scherer. Design of structured controllers with applications. *in the Proceedings of the 39th IEEE Conference on Decision and Control*, pages 5204–5209, 2000.

[150] C. Scherer. An efficient solution to multi-objective control problems with LMI objectives. *System & Control Letters*, 40(1):43–57, 2000.

[151] C. Scherer. Robust mixed control and linear parameter-varying control with full-block scalings. In L. El Gahoui and L. Niculesco, editors, *Advances in linear matrix inequality methods in control*, pages 187–207. SIAM, Philadelphia, 2000.

[152] C. Scherer. LPV control and full-block multipliers. *Automatica*, 37(3):361–375, 2001.

[153] C. Scherer. LMI relaxations in robust control. special issue on "linear matrix inequalities in control". *European Journal of Control*, 12(1):3–29, 2005.

[154] C. Scherer. Relaxations for robust linear matrix inequality problems with verification for exactness. *SIAM Journal on Matrix Analysis and Applications*, 27(2):365–395, 2005.

[155] C. Scherer. Robust controller synthesis is convex for systems without control channel uncertainties. In P. V. den Hof, C. Scherer, and P. Heuberger, editors, *Model based control, bridging rigorous theory and advanced technology*. Springer, New York, 2009.

[156] C. Scherer. Distributed control with dynamic dissipation constraints. *in the proceedings of the 50th Annual Allerton Conference on Communication, Control and Computing*, pages 55–62, 2012.

[157] C. Scherer. Gain-scheduled synthesis with dynamic positive real multipliers. *in the Proceedings of the 51st IEEE Conference on Decision and Control*, pages 6641–6646, 2012.

[158] C. Scherer. Gain-scheduled synthesis with dynamic generalized strictly positive real multipliers: a complete solution. *in the Proceedings of the 52nd IEEE Conference on Decision and Control*, pages 3901–3906, 2013.

[159] C. Scherer. Gain-scheduled synthesis with dynamic stable strictly positive real multipliers: a complete solution. *in the Proceedings of the 13th European Control Conference*, pages 3901–3906, 2013.

[160] C. Scherer. Structured $\mathscr{H}_\infty$-optimal control for nested interconnections: a state-space solution. *System & Control Letters*, 62(12):1105–1113, 2013.

[161] C. Scherer. Gain-scheduling control with dynamic multipliers by convex optimization. *SIAM Journal on Matrix Analysis and Applications (preprint)*, 2014.

[162] C. Scherer, P. Gahinet, and M. Chilali. Multi-objective output-feedback control via LMI optimization. *IEEE Transactions on Automatic Control*, 42(7):896–911, 1997.

[163] C. Scherer and C. Hol. Asymptotically exact relaxations for robust LMI problems based on matrix valued sum-of-squares. *in the Proceedings of the Mathematical Theory of Networks and Systems Symposium*, 2004.

[164] C. Scherer and C. Hol. Matrix sum-of-squares relaxations for robust semi-definite programs. *Mathematical Programming Series B*, 107:189–211, 2006.

[165] C. Scherer and I. Köse. Robustness with dynamic integral quadratic constraints: An exact state-space characterization of nominal stability with applications to robust estimation. *Automatica*, 44(7):1666–1675, 2008.

[166] C. Scherer and I. Köse. On convergence of transfer matrices and their realizations. *in the Proceedings of the 18th IFAC World Congress*, pages 3348–3353, 2011.

[167] C. Scherer and I. Köse. Gain-scheduled control synthesis using dynamic D-scales. *IEEE Transactions on Automatic Control*, 57(9):2219–2234, 2012.

[168] C. Scherer and I. Köse. From transfer matrices to realizations: convergence properties and parameterization of robustness analysis conditions. *System & Control Letters*, 62(8):632–642, 2013.

[169] C. Scherer, R. Njio, and S. Bennani. Parametrically varying flight control system design with full-block scalings. *in the Proceedings of the 36th IEEE Conference on Decision and Control*, pages 1510–1515, 1997.

[170] C. Scherer and J. Veenman. Robust controller synthesis is convex for systems without control channel uncertainties. *in the Proceedings of the 7th IFAC Symposium on Robust Control Design*, pages 325–330, 2012.

[171] C. Scherer and S. Weiland. *Linear matrix inequalities in control - Lecture Notes*. Delft University of Technology, The Netherlands, 1999 edition, 1999.

[172] C. Scherer and S. Weiland. *Linear matrix inequalities in control - Lecture Notes.* Delft University of Technology, The Netherlands, 2005 edition, 2005.

[173] G. Scorletti. Robustness analysis with time-delays. *in the Proceedings of the 36th IEEE Conference on Decision and Control,* pages 3824–3829, 1997.

[174] G. Scorletti and L. El Ghaoui. Improved LMI conditions for gain-scheduling and related control problems. *International Journal of Robust and Nonlinear Control,* 8(10):845–877, 1998.

[175] G. Scorletti and V. Fromion. Further results on the design of robust $\mathcal{H}_\infty$ feedforward controllers and filters. *in the Proceedings of the 45th IEEE Conference on Decision and Control,* pages 3560–3565, 2006.

[176] P. Seiler. Nonlinear stability analysis with dissipation inequalities and integral quadratic constraints. *IEEE Transactions on Automatic Control (preprint),* 2013.

[177] P. Seiler, A. Packard, and G. Balas. A dissipation inequality formulation for stability analysis with integral quadratic constraints. *in the Proceedings of the 49th IEEE Conference on Decision and Control,* pages 2304–2309, 2010.

[178] S. Skogestad and I. Postlethwaite. *Multivariable Feedback Control.* John Wiley & Sons, Ltd, 2005.

[179] J. F. Sturm. Using SeDuMi 1.02, a matlab toolbox for optimization over symmetric cones. *Optimization Methods and Software,* 11-12:625–653, 1999.

[180] K. Sun and A. Packard. Robust $\mathcal{H}_2$ and $\mathcal{H}_\infty$ filters for uncertain LFT systems. *IEEE Transactions on Automatic Control,* 50(5):715–720, 2005.

[181] G. Szegö. *Orthogonal polynomials.* American Mathematical Society, 1975.

[182] K. Takaba and J. Willems. Stability of dissipative interconnections. *International Symposium on Nonlinear Theory and its Applications,* pages 695–698, 2005.

[183] J. Teng. Robust stability and performance analysis with time-varying perturbations. Master's thesis, University of California Berkeley, 1991. Advisor: Packard, A.

[184] K. Toh, M. Todd, and R. Tütüncü. SDPT3 – a matlab software package for semidefinite programming (1999). *Optimization Methods and Software,* 11:545–581, 1999.

[185] R. Tóth, S. Heuberger, and P. Van den Hof. *Model and identification of linear parameter-varying systems,* volume 403 of *Lecture Notes in Control and Information Sciences.* Springer, Berlin, Heidelberg, New York, 2010.

[186] H. Tuan, P. Apkarian, and T. Nguyen. Robust filtering for uncertain nonlinearly parametrized plants. *IEEE Transactions on Signal Processing,* 51(7):1806–1815, 2003.

[187] R. Tütüncü, K. Toh, and M. Todd. Solving semidefinite-quadratic-linear programs using SDPT3. *Mathematical Programming Series B,* 95:189–217, 2003.

[188] Various. *Control of linear parameter varying systems with applications.* Springer, Berlin, Heidelberg, New York, 2012.

[189] Various. *Robust control and linear parameter varying approaches*, volume 437 of *Lecture Notes in Control and Information Sciences*. Springer, Berlin, Heidelberg, New York, 2013.

[190] J. Veenman, H. Köroğlu, and C. Scherer. An IQC approach to robust estimation against perturbations of smoothly time-varying parameters. *in the Proceedings of the 47th IEEE Conference on Decision and Control*, pages 2533–2538, 2008.

[191] J. Veenman, H. Köroğlu, and C. Scherer. Analysis of the controlled NASA HL20 atmospheric re-entry vehicle based on dynamic IQCs. *in the Proceedings of the AIAA Guidance, Navigation and Control Conference*, 2009.

[192] J. Veenman and C. Scherer. On robust LPV controller synthesis, a dynamic integral quadratic constraint based approach. *in the Proceedings of the 49th IEEE Conference on Decision and Control*, pages 591–596, 2010.

[193] J. Veenman and C. Scherer. IQC-synthesis with general dynamic multipliers. *in the Proceedings of the 18th IFAC World Congress*, pages 4600–4605, 2011.

[194] J. Veenman and C. Scherer. Robust gain-scheduled estimation: a convex solution. *in the Proceedings of the 50th IEEE Conference on Decision and Control*, pages 1347–1352, 2011.

[195] J. Veenman and C. Scherer. Robust gain-scheduled controller synthesis is convex for systems without control channel uncertainties. *in the Proceedings of the 51st IEEE Conference on Decision and Control*, pages 1524–1529, 2012.

[196] J. Veenman and C. Scherer. Stability analysis with integral quadratic constraints: a dissipativity based proof. *in the Proceedings of the 52nd IEEE Conference on Decision and Control*, pages 3770–3775, 2013.

[197] J. Veenman and C. Scherer. A synthesis framework for robust gain-scheduling controllers. *Automatica (to appear)*, 2013.

[198] J. Veenman and C. Scherer. IQC-synthesis with general dynamic multipliers. *International Journal of Robust and Nonlinear Control (to appear)*, 24(17):3027–3056, 2014.

[199] J. Veenman, C. Scherer, and H. Köroğlu. IQC-based LPV controller synthesis for the NASA HL20 atmospheric re-entry vehicle. *in the Proceedings of the AIAA Journal of Guidance, Control and Dynamics*, 2009.

[200] J. Veenman, C. Scherer, and I. Köse. Robust estimation with patial gain-scheduling through convex optimization. In C. Scherer and J. Mohammadpour, editors, *Control of linear parameter varying systems with applications*. Springer, New York, 2011.

[201] P. Viccione, C. Scherer, and M. Innocenti. LPV synthesis with integral quadratic constraints for distributed control of interconnected systems. *in the Proceedings of the 6th IFAC Symposium on Robust Control Design*, pages 13–18, 2009.

[202] F. Wang, H. Pfifer, and P. Seiler. Robust synthesis for linear parameter varying systems using integral quadratic constraints. *in the Proceedings of the 53rd IEEE Conference on Decision and Control*, 2014.

[203] M. Wassink, M. van de Wal, C. Scherer, and O. Bosgra. LPV control for a wafer stage: beyond the theoretical solution. *Control Engineering Practice*, 13(2):231–245, 2005.

[204] J. Willems. Least squares stationary optimal control and the Algebraic Riccati Equation. *IEEE Transactions on Automatic Control*, 16(6):621–634, 1971.

[205] J. Willems. Dissipative dynamical systems, part I: general theory. *Archive for Rational Mechanics and Analysis*, 45(5):321–350, 1972.

[206] J. Willems. Dissipative dynamical systems, part II: linear systems with quadratic supply rates. *Archive for Rational Mechanics and Analysis*, 45(5):352–392, 1972.

[207] J. Willems and K. Takaba. Dissipativity and stability of interconnections. *International Journal of Robust and Nonlinear Control*, 17(5-6):563–586, 2007.

[208] P. Young. Controller design with mixed uncertainties. *in the Proceedings of the American Control Conference*, pages 2333–2337, 1994.

[209] G. Zames and P. Falb. Stability conditions for systems with monotone and slope-restricted nonlinearities. *Siam Journal of Control*, 6(1):89–109, 1968.

[210] K. Zhou, J. Doyle, and K. Glover. *Robust and optimal control*. Prentice Hall, Upper Saddle River, New Jersey, 1996.

Summary

The stability and performance analysis of dynamical systems is highly important for numerous commercial and industrial applications. Rather often, engineers assume that these systems are described by linear time-invariant (LTI) models, which simplifies the analysis drastically due to nice, elementary and well-established theory. Unfortunately, dynamical systems are almost never linear or time-invariant, which calls for more appropriate techniques like e.g. nonlinear or robust analysis. The main idea in robust analysis is to consider a family of models that comprise the true but unknown dynamical behavior. Such a family of models consists of an uncertainty model whose parameters and higher-order, nonlinear and/or time-varying dynamics are assumed to be uncertain and confined to given sets. The key benefit of this approach is that robust stability and performance of the uncertain system also implies stability and performance of the real dynamical process.

A general framework for the analysis of uncertain systems is called the integral quadratic constraint (IQC) approach. Within this framework it is possible to systematically and efficiently verify the stability and performance properties of a rich class of uncertain systems through the use of linear matrix inequalities (LMIs) and convex optimization techniques.

Apart from analysis, also control is essential for numerous engineering applications. Controllers allow us to influence the dynamics of a system by measuring its state and dynamically process it in order to generate suitable commands for the actuators such that the controller achieves some desired behavior. Unfortunately, and in contrast to IQC-analysis, robust controller synthesis based on IQCs is much harder. The essential difficulty is that the robust controller synthesis problem cannot be easily solved with LMIs and convex optimization techniques, which prevents us from efficiently finding globally optimal solutions. On the other hand, it is sometimes possible to consider specialized design scenarios and efficiently solve these with LMI-techniques.

The aim of this thesis was to further develop the IQC-framework on both the analysis as well as on the synthesis side. This has been achieved by first providing a concise tutorial on the main existing analysis tools, followed by an extensive survey on the formulation and parameterization of IQC-multipliers. We presented new insights in this matter, as well as a novel proof of the well-known IQC-theorem with possibly new and interesting applications.

The analysis tools formed a solid basis for our results on the synthesis side. Although the existing design questions within the IQC-framework that can be reformulated as a convex optimization problem are scarce, recent years have witnessed the development of various useful LMI-techniques that can be exploited in order to translate a rich class of robust synthesis problems into convex optimization schemes. Among others, this led to LMI-solutions of the nominal $\mathcal{H}_\infty$-synthesis problem, the gain-scheduling controller synthesis problem, the robust estimation problem and the robust feedforward controller synthesis problem. It was one of the main goals of this thesis to show in which cases these results can be generalized and to reveal for which scenarios robust synthesis is convex.

As the first synthesis result of this thesis, it has been shown how the well-known existing solutions on robust estimation and nominal gain-scheduling controller synthesis can be unified into one general realistic design configuration. In particular, we presented how the synthesis of such robust gain-scheduled estimators can be handled with a convex optimization scheme.

To continue, we formulated a more generic solution for the systematic synthesis of robust gain-scheduling controllers. It has been shown that the underlying problem encompasses a rich class design questions, some of which have already been addressed in the literature, while others have not. For example, it has been illustrated that the robust gain-scheduling controller synthesis problem is convex, under the rough hypothesis that the control channel is not affected by uncertainties. Other useful design scenarios that are covered are e.g. generalized l_2-synthesis, a particular class of multiobjective and structured controller synthesis, open-loop controller synthesis, robust gain-scheduled observer design, gain-scheduling control with uncertain performance weights, robust controller synthesis with unstable weight, among others.

Lastly, analogously to the existing μ-tools, it has been shown how the general robust controller synthesis problem can be heuristically (but still systematically) handled within the IQC-framework with LMI-techniques. The suggested algorithm enables us to perform robust controller synthesis for a significantly larger class of uncertainties if compared to the existing methods. In particular we showed how warm-start techniques can speed-up the synthesis process.

Altogether, we presented new theoretical results, illustrated through various numerical examples, which reveal that the IQC-framework is not only useful for the analysis purposes, but also has great potential for a rather diverse class of synthesis questions.

Zusammenfassung

Die Untersuchung dynamischer Systeme auf Stabilitäts- und Güteeigenschaften spielt eine zentrale Rolle in vielen kommerziellen und industriellen Anwendungen. Häufig treffen Ingenieure die Annahme, dass diese Systeme sich als linear und zeitinvariant beschreiben lassen. Dadurch ermöglicht sich eine drastische Vereinfachung der Analyse, da auf eine grundlegende und etablierte Theorie zurückgegriffen werden kann. Leider verhalten sich dynamische Systeme beinahe nie linear oder zeitinvariant, was eine Behandlung mit anspruchsvolleren Techniken wie z.b. nichtlinearer oder robuster Analyse erfordert. Die grundlegende Idee der robusten Analyse besteht darin, eine Familie von Modellen zu betrachten, welche das tatsächliche aber unbekannte dynamische Verhalten umfasst. Eine solche Familie von Modellen besteht aus einem Unsicherheitsmodell, dessen Parameter und nichtlineare und/oder zeitvariante Dynamik bzw. Dynamik höherer Ordnung als unbekannt aber in einer gegebenen Menge variierend angenommen werden. Dieser Ansatz hat den wesentlichen Vorteil, dass robuste Stabilität und Güte des unsicheren Systems die Stabilität und Güte des realen dynamischen Prozesses implizieren.

Einen allgemeinen Rahmen zur Analyse unsicherer dynamischer Systeme bietet die Methode der quadratintegrablen Nebenbedingungen (IQCs)[1]. Dieser Ansatz ermöglicht die systematische und effiziente Güte- und Stabilitätsanalyse einer großen Klasse unsicherer Systeme mit Hilfe von linearen Matrixungleichungen (LMIs)[2] und Methoden der konvexen Optimierung.

Abgesehen von der Analyse kommt auch der Regelung in zahlreichen Ingenieuranwendungen eine herausragende Bedeutung zu. Durch Messung von Zuständen und deren dynamischer Verarbeitung ermöglichen uns Regler die Einflussnahme auf dynamische Systeme, indem passende Befehle an die Stellglieder generiert werden und

---

[1] Diese Abkürzung kommt vom englischen Begriff 'Integral Quadratic Constraints'

[2] Diese Abkürzung kommt vom englischen Begriff 'Linear matrix inequalities'

so ein gewünschtes Verhalten des Systems erzielt wird. Unglücklicherweise ist die robuste Reglersynthese im Rahmen der IQC-Methode jedoch wesentlich komplizierter als die Analyse. Das Hauptproblem besteht dabei in der Tatsache, dass die robuste Reglersynthese nicht einfach auf die Lösung einer LMI und damit eines konvexen Problems zurückgeführt werden kann, was die effiziente Suche nach einer global optimalen Lösung wesentlich erschwert. Jedoch erlaubt die Betrachtung von Spezialfällen manchmal die effiziente Lösung mit LMI-Techniken.

Das Ziel dieser Arbeit war, die bisherigen Möglichkeiten im Rahmen der IQC-Methode sowohl im Bereich der Analyse als auch der Synthese weiterzuentwickeln. Hierfür wurde zuerst eine prägnante Anleitung zur Benutzung der wichtigsten Analysewerkzeuge erstellt, um dann einen ausführlichen Überblick über die Erstellung und Parametrisierung von IQC-Multiplikatoren zu geben. Hier haben wir neue Einsichten geliefert und darüber hinaus einen neuen Beweis des bekannten IQC-Theorems präsentiert, der neue und interessante Anwendungen ermöglichen kann.

Die vorgestellten Analysewerkzeuge bildeten eine solide Grundlage für unsere Resultate im Bereich der Synthese. Obwohl nur wenige Designprobleme innerhalb der IQC-Methode als konvexes Optimierungsproblem geschrieben werden können, sind in den letzten Jahren einige hilfreiche LMI-Techniken entwickelt worden, um eine vielfältige Klasse robuster Syntheseprobleme in konvexe Optimierungsaufgaben zu überführen. Dies führte unter anderem zu LMI-Lösungen des nominalen $\mathscr{H}_\infty$-Syntheseproblems, der 'gain-scheduling'-Reglersynthese[3], des robusten Schätzproblems und der robusten Führungsreglersynthese. Eines der Hauptanliegen dieser Arbeit lag darin aufzuzeigen, unter welchen Umständen diese Resultate verallgemeinert werden können und wann die robuste Synthese in ein konvexes Problem überführt werden kann.

Im ersten Syntheseresultat dieser Arbeit haben wir beschrieben, wie sich die bekannten Lösungen zur robusten Schätzung und zur nominalen gain-scheduling Reglersynthese zu einer allgemeinen und realistischen Designkonfiguration vereinigen lassen. Insbesondere haben wir dabei auch aufgezeigt, wie die Synthese solcher robuster gain-scheduling Schätzer im Rahmen der konvexen Optimierung behandelt werden kann.

Danach haben wir eine verallgemeinerte Lösung zur systematischen Synthese von robusten gain-scheduling Reglern formuliert. Dabei konnten wir aufzeigen, dass das zugrundeliegende Problem eine große Klasse von Designproblemen abdeckt, von denen einige in der Literatur bereits behandelt wurden, andere aber auch vollkommen neu waren. Beispielsweise haben wir gezeigt, dass unter der etwas groben Annahme, dass der Regelkanal nicht von Unsicherheiten belastet ist, die robuste gain-scheduling Reglersynthese zu einem konvexen Problem führt. Andere hilfreiche Designvarianten die behandelt wurden umfassen zum Beispiel die verallgemeinerte l_2-Synthese, eine spezielle Klasse von Mehrgrößen- und strukturierter Reglersynthese, Steuerungssyn-

---

[3] Dieser Begriff ist nicht übersetzbar.

these, robustes gain-scheduling Beobachterdesign, gain-scheduling Regelung mit unsicheren Gütegewichten und robuste Reglersynthese, um nur einige zu nennen.

Abschließend haben wir, analog zu den bereits existierenden μ-Werkzeugen, gezeigt, wie das allgemeine robuste Reglersyntheseproblem im Rahmen der IQC-Theorie mit LMIs heuristisch (aber immer noch systematisch) behandelt werden kann. Der vorgeschlagene Algorithmus ermöglicht nun die robuste Reglersynthese bei einer weit größeren Klasse von Unsicherheiten als bisher. Insbesonder haben wir gezeigt, wie Warmstarttechniken den Syntheseprozess beschleunigen können.

Insgesamt haben wir neue theoretische Resultate präsentiert, die mit einer Vielzahl an numerischen Beispielen illustriert wurden und verdeutlichten, dass die IQC-Methode nicht auf die reine Analyse beschränkt ist, sondern auch großes Potential für vielfältige Synthesefragen birgt.

About the author

Joost Veenman was born on January 12th, 1979 in Berkel en Rodenrijs, the Netherlands. After receiving his B.Eng. degree in mechanical engineering from the Hogeschool Rotterdam in 2003, he worked for a year in industry. Then he went to the university and received his M.Sc. Degree in mechanical and control engineering Cum Laude from the Delft University of Technology in 2008. The title of his thesis was 'Robust estimation with dynamic integral quadratic constraints'. From September 2008 to May 2010 he stayed in Delft and worked as a researcher within the Delft Center for Systems and Control (DCSC). During this period he was part of the LPVMAD project; a joint project of the European Space Agency (ESA), Deimos Space, the Delft University of Technology and the University of Leicester. The project had as main goal to develop and validate Modeling, Analysis and Design (MAD) tools for Linear Parameter Varying (LPV) systems (hence the name LPVMAD) with a re-entry vehicle as benchmark study-case. After a successful completion of the project, he moved to Germany and started his Ph.D. at the University of Stuttgart in May 2010. His main research interests are robustness analysis, robust (gain-scheduling) control and linear matrix inequality (LMI) techniques. He is currently finalizing his Ph.D. and planning the course of his future.